Polysaccharides for Drug Delivery and Pharmaceutical Applications

ACS SYMPOSIUM SERIES **934**

Polysaccharides for Drug Delivery and Pharmaceutical Applications

Robert H. Marchessault, Editor
McGill University

François Ravenelle, Editor
Labopharm. Inc.

Xiao Xia Zhu, Editor
Université de Montréal

Sponsored by the
ACS Division of Cellulose and Renewable Materials

American Chemical Society, Washington, DC

Chemistry Library

Library of Congress Cataloging-in-Publication Data

Polysaccharides for drug delivery and pharmaceutical applications / Robert H.
Marchessault, editor, François Ravenelle, editor, Xiao Xia Zhu, editor ; sponsored by
the ACS Division of Cellulose and Renewable Materials.

p. cm.—(ACS symposium series ; 934)

"Developed from a symposium sponsored by the ACS Division of Cellulose
and Renewable Materials., at the 228th ACS National Meeting, Philadelphia,
Pennsylvania, August 22–26, 2004"—T.p. verso.

Includes bibliographical references and index.

ISBN 13: 978–0–8412–3960–9 (alk. paper)

ISBN 10: 0–8412–3960–6 (alk. paper)

1. Polysaccharides—Congresses. 2. Excipients—Congresses.

I. Marchessault, R. H. II. Ravenelle, François. III. Zhu, Xiao Xia. IV. American
Chemical Society. Meeting (228th : 2004 : Philadelphia, Pa.) V. American Chemical
Society. Division of Cellulose and Renewable Materials. VI. Series.

[DNLM: 1. Excipients—Congresses. 2. Polysaccharides—Congresses. 3. Biopolymers
—Congresses. 4. Drug delivery systems—Congresses. QU 83 P7832 2006]

QP702.P6P643   2006
572′.566—dc22

                                                                    2005058193

The paper used in this publication meets the minimum requirements of American National Standard
for Information Sciences—Permanence of Paper for Printed Library Materials, ANSI Z39.48–1984.

Copyright © 2006 American Chemical Society

Distributed by Oxford University Press

PRINTED IN THE UNITED STATES OF AMERICA

Foreword

The ACS Symposium Series was first published in 1974 to provide a mechanism for publishing symposia quickly in book form. The purpose of the series is to publish timely, comprehensive books developed from ACS sponsored symposia based on current scientific research. Occasionally, books are developed from symposia sponsored by other organizations when the topic is of keen interest to the chemistry audience.

Before agreeing to publish a book, the proposed table of contents is reviewed for appropriate and comprehensive coverage and for interest to the audience. Some papers may be excluded to better focus the book; others may be added to provide comprehensiveness. When appropriate, overview or introductory chapters are added. Drafts of chapters are peer-reviewed prior to final acceptance or rejection, and manuscripts are prepared in camera-ready format.

As a rule, only original research papers and original review papers are included in the volumes. Verbatim reproductions of previously published papers are not accepted.

ACS Books Department

Contents

Hyaluronan

Chitin and Chitosan

Polysaccharides

Indexes

ix

Preface

American Viscose Corporation (AVC) was a subsidiary of Courtaulds, when it was established as the first North American producer of rayon fiber (regenerated cellulose). The first factory was constructed at Marcus Hook, Pennsylvania, in 1907. "Avisco" was the AVC trademark and it was the origin of the trademark Avicel® known today as the most widely used excipient in the pharmaceutical industry. Avicel® is made from hydrolyzed native cellulose fibers, usually chemical grade wood pulps. The HCl hydrolysis conditions used to convert pulp fibers to Avicel® powder removes about 5–10% by weight of non-crystalline pulp and after neutralization and washing the resultant fine powder, hydrocellulose, is about 80% crystalline. The treatment greatly diminishes the molecular weight, down to a degree of polymerization (DP) of 230 (i.e., about 1/25 of the native cellulose value. Nevertheless Avicel® is completely water insoluble and is often referred to as microcrystalline cellulose (MCC).

Originally intended as a low-calorie food (e.g. as an additive to be used by the baking industries), its use as an excipient was unexpected. Hydrocellulose itself was known as far back as 1854, having been reported by a French scientist, Girard (*1–3*). The discovery was not far behind the seminal paper by Anselme Payen (*4,5*) in which he showed the polyglucan structure of cellulose without providing the famous name: "cellulose" which was coined by the reviewers. A second version of the hydrolyzed cellulose is a colloidal grade used in wet applications for foods.

AVC like many of the manufacturers of cellulose-based fibers and plastics (e.g., Dupont, Celanese, Hercules, and Eastman-Kodak) had extensive analytical facilities for development of tests related to the dissolving grade pulps. These were used in the production of cellulose derivatives such as cellulose acetate, cellulose nitrate, carboxymethyl cellulose, and cellulose xanthate that are used in the film fiber and plastics industries based on chemical grade cellulose. One such test was referred to as "level-off DP", which probably inspired the HCl hydrolysis conditions used for making Avicel®.

xi

Orlando Battista was head of analytical testing at AVC and patented the process (6) for making Avicel® with its numerous examples of food applications. Studies involving puddings, whipped cream, bread, and ice cream for example, all with low-cal properties etc. did not convert to successful marketing results. The Avicel® powder could be pressed into disks to which nails and screws could be fixed. The compressed disks were also resistant to the flame of an oxyacetylene torch. Clearly, these observations were the intellectual seeds that led to excipient applications especially when the tablet-size disks were found to rapidly redisperse when placed in water.

This brief historical review of the origin of Avicel® is to set the stage for the written account of numerous polysaccharide-based excipients, many of which derive from the Avicel® success. The latter is constantly studied and improved as an excipient, for example a typical pharmaceutical formulation today is MCC + colloidal silicone dioxide + magnesium stearate + iron oxides + titanium dioxide + hydroxypropyl methylcellulose. The compositional adjustments take into account properties relating to tablet stability and powder flow into the high speed tableting machines.

However, in keeping with the extensive know-how in cellulose derivatization and the prototypical role of cellulose as a crystalline polysaccharide it is not surprising that cellulose derivatives are used as adjuvents with Avicel® (e.g. cross-linked carboxymethyl cellulose), which controls the tablet dispersion speed. Many different polysaccharide excipients and adjuvents described in the following chapters of this book were chosen for a variety of reasons but most often to complement Avicel®.

In 1963 AVC was acquired by FMC Corporation and the patent portfolio of Avicel® products relating to excipients and food additives was explored by O. A. Battista and his team. Today, many other polysaccharides are used and considered as excipients (e.g., alginate, beta (1–3) glucans, schitosan, hyaluronic acid, xanthan gum, starch). Even a polysaccharide drug such as heparin can be part of the excipient package. The collection of chapters in this book is a celebration of the diversity of polysaccharide platforms, which enable a variety of drug delivery systems and application platforms for pharmaceutical uses.

This is also an occasion to salute the late Orlando Battista whose creative perseverance started it all. He received his graduate training at the Pulp and Paper Research Institute, McGill University and promptly joined the Avisco R&D group. In 1960 he was Chair of the Cellulose Division of the American Chemical Society. In recognition of his work on Avicel® he was awarded the Division's prestigious Anselme Payen Award. His innovative work on MCC is summarized in his well-known book *Microcrystalline Polymer Science* published in 1975 by McGraw-Hill Inc.

The polysaccharides featured in this book, include MCC and starch that are commodity biopolymers. The former is the widely used hydrolyzed cellulose, which derived from Battista's discovery. Large pharmaceutical organizations dedicate considerable effort to exploring the many ways that MCC can be formulated to improve the tableting properties. Chapter 2 by Li and Mei of Boehringer Engelheim provides a survey of the physicochemical processing, called wet granulation, which is at the base of formulation for such pharmaceutical firms where approximately 80% of the products are in solid dosage form. Accordingly, drying, fine powder transport, mixing, and blending, coating, and large-scale manufacture of tablets and capsules are ongoing operations in pharmaceutical companies.

The catalog of available and well-understood polysaccharides was the reason why this symposium was organized. A recent review in

Orlando A. Battista
1917–1995

Chemistry and Engineering News states that "Excipients are the cinderellas of the drug industry... they do all the work but get none of the glory... they are the inactive ingredients that deliver the active ingredient" (*C&E News* Nov. 22, 1999). Examples of new polysaccharides being considered for pharmaceuticals are Hyaluronic acid, a well known physiological material, chitin/chitosan, "a Jekyl and Hyde" material from seafood leftovers which is used for dermatology, arthritics, coacervate gels, etc.

The writer is grateful for the patience of the contributors, the scholarship of the referees, and the support of my colleagues. The ongoing help of Dr. Grégory Chauve was essential to the completion of the book.

References

1. Girard, A. *Comptes rendus* **1875**, *81*, 1105.
2. Girard, A. *Comptes rendus* **1879**, *88*, 1322.
3. Girard, A. *Ann. Chim. Phys.* **1881**, *24*, 333.
4. Payen, A. *Comptes rendus* **1838**, *7*, 1052.
5. Payen, A. *Comptes rendus*, **1838**. 7: p. 1125.
6. Battista, O. A. *Microcrystalline Polymer Science;* McGraw-Hill Inc.: Orlando, FL, 1975.

Robert H. Marchessault
Emeritus E. B. Eddy Professor
Chemistry Department
McGill University
Montréal, Québec H3A 2A7
Canada

Overview

During the past decades, we have seen the pharmaceutical industry grow at a tremendous rate. Along with the discoveries of new drugs for various preventive and therapeutic treatments, the need for novel materials and methods to deliver them energized the field of drug delivery creating new opportunities for material scientists. Drug delivery companies have redefined the picture of the pharmaceutical industry as not only providing tools for the delivery of novel drugs but also as applying their technology to existing formulations available on the market. By doing so, already marketed drugs are given better bioavailability, easier administration, and fewer side effects while giving the patient an optimum treatment.

Among those materials, the pharmaceutical industry has since the beginning relied on starch and cellulose, the most abundant natural polymers on earth. Because of the wide variety and range of physicochemical properties created by Mother Nature, and the possibility to perform a large number of specific modifications, polysaccharides are choice starting materials. Nowadays, many polysaccharides are used, modified and transformed to serve the ever evolving pharmaceutical needs either as excipients, surfactants, adhesives, carriers, protectors, or active substances themselves. This book covers several polysaccharides that are utilized in drug delivery, medical and pharmaceutical applications, together with their sources, properties, preparation, and mechanism of action. By bringing together key researchers in these diverse fields, new dialogues have been initiated for the next generation of drug delivery systems.

François Ravenelle
Manager, New Technologies, Labopharm, Inc.
480 Boulevard Armand-Frappier
Laval, Québec H7V 4B4
Canada

Polysaccharides for Drug Delivery and Pharmaceutical Applications

Cellulose

Chapter 1

Fishing for Proteins with Magnetic Cellulosic Nanocrystals

Robert H. Marchessault, Glen Bremner, and Grégory Chauve

Pulp and Paper Research Institute and Department of Chemistry, McGill University, 3420 University Street, Montréal, Québec H3A 2A7, Canada

A key to proteomics exploitation is magnetic separation of target proteins in a sea of others. The process is similar to fishing, where the target protein is like a specific type of fish. An antigen, covalently linked on a magnetic nanocrystal, is "the bait and hook"; the fishing line is simulated by an applied magnetic field. By selective binding the target protein can be caught, isolated and released from the nanocrystal support once the interaction of antigen-antibody is neutralized. Microcrystalline cellulose (MCC) is responsive to functionalization, as required to tether the desirable bioactive ligand. Chitin nanocrystals with some free surface amino groups are also as suitable as nanocrystal support. To render MCC magnetic, ferrous iron is adsorbed and converted to ferrous hydroxide, which is then oxidized with KNO_3 to form superparamagnetic magnetite. From the magnetization curves, the particle size distribution and magnetite content of ferrites in the polysaccharide matrix can be derived. A trial fishing experiment to isolate a target protein used a magnetized MCC platform to which Protein A, a specific ligand of Immunoglobulin G, IgG, was attached. The resulting magnetic platform, captured the target IgG which was magnetically separated from the supernatant. IgG was isolated by screening the non-bonded interactions between IgG and its specific ligand.

Over the past decades, magnetic particles have come into use when protein separation and purification is needed (*1, 2*). Large carrier beads were part of the toner delivery system in early versions of xerographic engines. However, the genomics proteomics revolution requires magnetic carriers attached to colloidal platforms that capture and classify at the molecular level. In this chapter we will deal with magnetic polysaccharide particles that target specific proteins in a sea of other ones (*3, 4*). The magnetic separations subject the target molecule to very little mechanical stress and compared to other methods, they are rapid, scaleable, low cost, and they avoid hazardous or toxic reagents. By use of selective binding, the technique has become popular for monoclonal antibody and enzyme separations.

The separation process involves two-steps: first is covalent bonding of a ligand protein to the magnetic polysaccharide support, which is then followed by the binding of the ligand to its target antibody or enzyme. This process, as illustrated by the cartoon in Figure 1, is analogous to fishing: the target molecule is like a specific fish in a lake; the ligand protein is the bait which attracts only the target fish; the polysaccharide is the hook to which the bait is attached; and the magnetic property is the fishing line that will pull the target fish out of the protein lake. Finally, a simple pH change liberates the target protein from the affinity ligand.

Polysaccharide particles can achieve magnetic character by complexation of ferrous iron (Fe^{++}) in the insoluble polysaccharide matrix. Ferrites are created by an oxidation reaction (*5*). The small porous openings in the polysaccharide matrix limit the formation of ferrites to small magnetic particles which because of their size exhibit superparamagnetic character. Superparamagnetic materials exhibit magnetic remanence only when subjected to a magnetic field.

Magnetic Polysaccharide-Ligand Platform

This part presents the complete chemical modification applied to microcrystalline cellulose (MCC) to confer magnetic properties and then create the separation platform by attaching the ligand. Obviously, various chemical treatment as well as ligand attachment can be done on MCC and on other polysaccharides. This is treated in Shingel's chapter where other polysaccharide particles, such as chitin and starch, are also used as platforms.

Microcrystalline cellulose

Microcrystalline cellulose is obtained after acid hydrolysis of cellulose microfibrils (*6*). This chemical treatment tends to isolate the crystalline domains

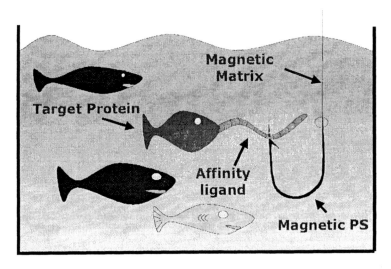

Figure 1: Cartoon of the Fishing for Proteins approach based on game fishing. The response of the magnetic polysaccharide (Magnetic PS) to an applied magnetic field provides the reeling action for removing the "hooked" target

6

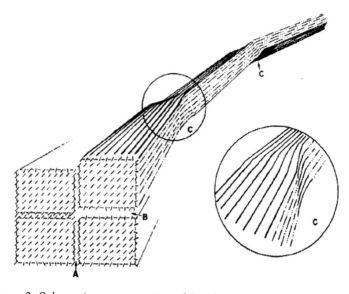

Figure 2: Schematic representation of the elementary fibril of the native cellulose fiber. (A) coalesced surfaces of high order, (B) readily accessible slightly disordered surfaces , (C) readily accessible surfaces of strain-distorted tilt and twist regions (7)

hinged together by less crystalline regions along the microfibril (Figure 2). As the hinging regions are more easily hydrolyzable than the crystalline regions, acid acts as a depolymerization chemical (Figure 2, C).

MCC, commercially sold as Avicel PH 103 (FMC Corp.), was chosen as the polysaccharide matrix. The Avicel PH grade is a pressure sensitive adhesive and therefore an excellent dry binder. Various Avicel PH powder grades are prepared to respond to specific properties when used as an excipient. In particular, Avicel PH 103 exhibits an average particle size of 50 μm (8), although the intrinsic hydrocellulose fragments are colloidal when dispersed in water.

TEMPO-Mediated Oxidation of MCC

TEMPO performs a selective oxidation of primary alcohol groups (9, 10). Several studies were carried out on water-insoluble native cellulose substrates (10-13). The oxidation converts the primary C-6 hydroxyl group to a carboxylic acid. Because MCC is water-insoluble, the oxidation only takes place on the surface cellulose chains.

The role of the oxidation is to create a carboxyl group susceptible to react with an amino group to form an amide linkage. The covalent bond anchors the bait (ligand protein) to the hook (here, MCC).

In order to follow the oxidation, the reaction was repeated with aliquots of the reaction mixture taken at 10, 20, 40, 60, 120, 240 and 330 minutes (Figure 3). For each aliquot, volumetric titration and ^{13}C CP/MAS NMR measurements were used to determine the oxidation yield. The percent carboxylation as determined by ^{13}C NMR was calculated using the procedure described by Kumar and Yang (14). This method is based on a quantitative ratio of the number of C-6 in carboxylic form over the number of C-6 under their hydroxyl form and expressed as a percentage. ^{13}C CP/MAS NMR spectra were acquired using a 300 MHz Chemagnetics CMX-300 instrument operating at 75.4 MHz for the ^{13}C nucleus.

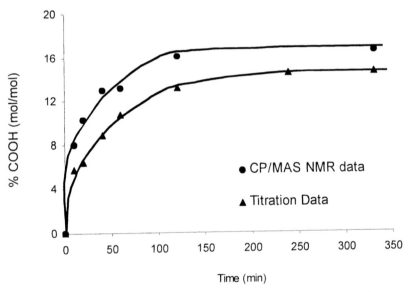

Figure 3: Formation of carboxylic groups followed by ^{13}C CP/MAS NMR and alkaline titration as a function of the oxidation reaction time

The reaction progress curves as determined using ^{13}C CP/MAS NMR and titration possess a similar shape (Figure 3), with the formation of carboxylic groups initially being rapid and subsequently forming a plateau after approximately 150 minutes. The plateau can be explained by the exhaustion of surface hydroxyl groups. Only about 20-25% of the total number of glucose repeat units is located on the crystallite surface of MCC. The remaining hydroxyl groups are located inside the crystallites and are unavailable for oxidation.

In Situ Synthesis of Ferrites

To MCC (5 g) previously suspended in deionized water (100 mL), a 10% (w/w) ammonium iron sulfate hexahydrate solution was added dropwise and the mixture was stirred under low vacuum for 1 h in order to form Fe^{++}-cellulose complex. The Fe^{++}-cellulose particles were centrifuged and washed several times with deionized water. They were re-immersed in deionized water (100 mL) and 0.5 N NH_4OH solution (200 mL) was added. Immediately, the dark-green mixture was immersed in an oil bath set at 70 °C and 10% (w/w) KNO_3 solution was added (100 mL). The reaction mixture was incubated with constant stirring for 1h. Afterwards, the dark grey particles that were produced were centrifuged and successively washed with deionized water, dilute acetic acid, deionized water and finally with acetone. The particles were dried under room conditions.

Two different approaches have been envisaged to prepare the magnetic MCC platform. The first one consists in the magnetization of oxidized MCC (Oxy-MCC) while the other way is the oxidation of the magnetic MCC (Mag-MCC). The former way is labelled Oxy-Mag-MCC and the latter Mag-Oxy-MCC.

X-ray diffractometry was performed in order to record the magnetic derivatizations done on MCC substrates (Figure 4). The instrument used is a Nicolet XRD X-ray Powder Diffractometer, which produced CuKα radiation (λ = 0.1542 nm). Voltage and current were set at 45 kV and 25 mA, respectively. The range of 2θ measured was from 5° to 60°, with an interval of 10 s per 0.1°.

MCC diffraction pattern is typical of cellulose I. Whatever the sequence chosen to prepare the magnetic MCC, the different treatments do not affect the allomorph type of MCC. Nevertheless, the magnetization seems to affect the degree of crystallinity of the cellulosic platform. This is particularly observed when the magnetization step is performed after the oxidation; the global crystallinity is decreased. This can be attributed to the alkaline conditions for the in situ synthesis of ferrites, which is known to decrease crystallinity in the native cellulose. In the magnetized samples, peaks occurring at 2theta of 30.4°, 35.7°, 43.3° and 57.3° correspond to those of Fe_3O_4 (magnetite). These peaks evidence the presence of ferrites in close association with MCC substrates (cf. TEM images in Figure 6).

To complement the X-ray analysis, Fourier Transform Infrared spectra of the MCC samples were recorded using a Bruker IFS-48. Typically, KBr pellets were prepared from 1 mg of sample and 100 mg of KBr, with 100 scans of the samples being taken (Figure 5).

The peaks located at 1616 and 1733 cm^{-1} in the Oxy-MCC spectrum correspond to carbonyl vibrations in the salt and free form of the carboxylic acid, respectively. The Oxy-Mag-MCC spectrum showed only the peak associated with the salt form of the carboxylic acid. This can be attributed to the alkaline

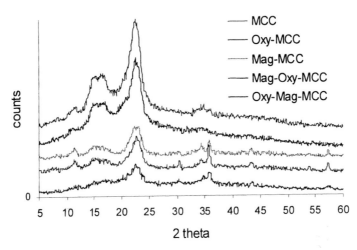

Figure 4: X-ray diffraction traces of the various MCC-based oxidized and magnetized samples. The labels progress from top to bottom for the traces

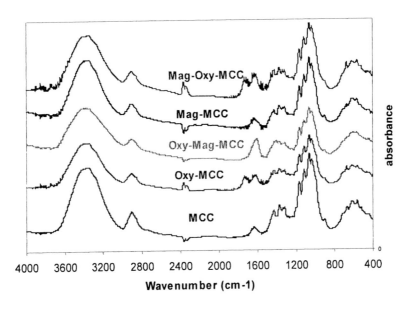

Figure 5: FTIR spectra of the various MCC-based oxidized and magnetized samples

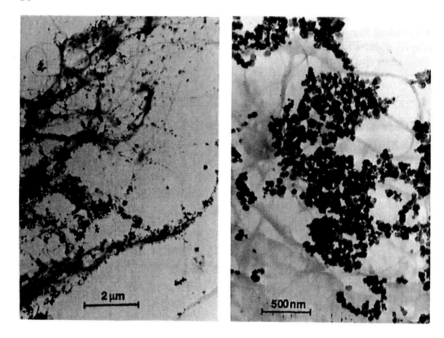

Figure 6: TEM Images of microfibrillar cellulose dispersions equivalent to Mag-Oxy-MCC

conditions involved in the *in situ* synthesis of ferrites, which was performed after the oxidation. A single peak in that area can be found in the MCC and Mag-MCC spectra; however, this peak is located at a slightly longer wavelength (1633 cm^{-1}) and is associated with absorbed water. Thus there is no doubt that oxidation performed after the magnetization is effective.

As a final analysis to check whether the ferrites are selectively attached to MCC substrate, Transmission Electron Microscopy (TEM) micrographs of Mag-MCC, Mag-Oxy-MCC and Oxy-Mag-MCC were recorded using a JEM-2000 FX instrument operating at 80 kV (Figure 6). Prior to the imaging, samples were dispersed using an ultrasound treatment during one minute and then were directly dried onto copper grids.

From the TEM observations, the synthesized ferrite particles can be identified as magnetite (Fe_3O_4) as opposed to maghemite (γ-Fe_2O_3) particles because of their shapes. Magnetite particles have a disc shape whereas maghemite particles are needle-like. The dark spheres are homogeneously localized onto the MCC which is easily visualized with microfibrillar cellulose, as seen in Figure 6 (*15, 16*).

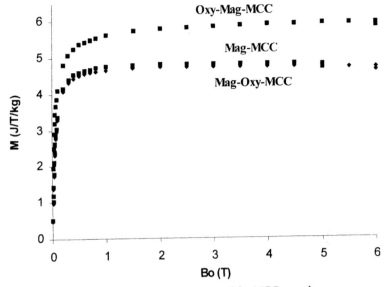

Figure 7: Magnetization curves of the MCC samples

Magnetic Moments of Superparamagnetic MCC Derivatives

The magnetic properties of the samples after *in situ* ferrite synthesis were analyzed using a Quantum Design Physical Properties Materials Extraction Magnetometer.

The superparamagnetic properties of the magnetized products were confirmed by the magnetization curve passing through the origin (Figure 7). Superparamagnetic substances show magnetic properties only in the presence of a magnetic field. Once the field is removed, the substance should show no magnetic moment, as is the case when the curves pass through the origin.

The steep initial slopes of the magnetization curves indicate that the ferrite particles are large and homogenous. Large homogenous magnetic particles behave as a single magnetic moment, and so are very sensitive to an increase in the surrounding magnetic field. Smaller, inhomogeneous particles would respond gradually to an increase in the magnetic field.

The weight percent of ferrites embedded in the polysaccharide matrix can be calculated using the observed saturation magnetization for the samples and by assuming that all ferrites are present in the form of magnetite and have a magnetic moment of 124 $JT^{-1}kg^{-1}$. The ferrite weight content is obtained dividing the sample magnetization by pure magnetite magnetization and is expressed as % Fe_3O_4 (w/w). The ferrite content of the three magnetic MCC platforms were calculated to be 3.89 for the Mag-MCC, 4.79 for Oxy-Mag-MCC and 3.82 % Fe_3O_4 (w/w) for Mag-Oxy-MCC.

Protein Binding

The protein binding capacities of the magnetic MCC were determined using Bovine Serum Albumin (BSA) as a model protein ligand. Four samples (Mag-MCC; Oxy-Mag-MCC; Mag-Oxy-MCC and a control containing no polysaccharide) were placed in a coupling buffer, and stirred for 30 minutes. The polysaccharide particles were then magnetically separated and the supernatant removed.

The capacity of MCC to bind proteins was determined using a Protein Assay. A mixture of bicinchoninic acid solution and copper (II) sulfate pentahydrate 4% solution (50:1 v/v) was added to the supernatants removed from each of the samples after the addition and fixation of the protein solution. Thus, any unbound protein would be located in the supernatant. Proteins are known to reduce the copper complex from Cu (II) to Cu (I) in a concentration dependent manner. The bicinchoninic acid selectively forms a purple complex with Cu (I). By measuring the absorbance at 562 nm and forming an external standards calibration curve from standard protein solutions, the concentration of proteins in the supernatants can be determined. This concentration can then be used to determine the amount of polysaccharide-bound proteins.

Table I presents the protein binding capacities obtained by comparison with the control sample, which did not contain any polysaccharides. Despite not being oxidized, the Mag-MCC showed some binding capacity. This can occur through binding of the hydroxyl group of the polysaccharide to the protein. The Oxy-Mag-MCC had a lower binding capacity than the Mag-Oxy-MCC. Some of the carboxylic groups might be involved in the binding of ferrite particles and thus would not be available for protein binding.

Table I: Protein Binding Capacities of Magnetized Avicel Samples

Sample	% Bound Proteins (w/w)
Mag-MCC	8.02
Mag-Oxy-MCC	21.53
Oxy-Mag-MCC	17.41

Fishing for Proteins

The second part of this chapter is devoted to the operational challenge that the "Fishing for Proteins" process requires. The process is aimed at replacing the large scale chromatographic columns presently used for recovery of valuable proteins. Proper choice of colloidal particles such as described in this chapter and Shingel's chapter could offer a less costly and more rapid approach. This process is divided into three steps:

- preparation of the magnetic platform with bound ligand
- recognition and anchoring the target molecule to the magnetic platform
- magnetic fishing and target molecule separation from the platform

Recognition, Anchoring and Recovery of Human IgG

The purpose of the second step consists in reacting the magnetic platform to allow the specific binding of the ligand with the targeted molecule in a sea of other proteins. We can illustrate this approach using a magnetized chitin substrate to which ProteinA, a specific ligand of Immunoglobulins G, IgG, was covalently attached. IgG are the most abundant immunoglobulins in human blood and represent 75-80% of the circulating antibodies. The resulting MagChi-ProteinA magnetic platform can be dispersed under controlled pH conditions in order to allow the specific binding with the target molecule, IgG. Typically, 10 mg of the MagChi-ProteinA are dispersed in a sodium acetate buffer (50 mM) at pH=7.0. The IgG was previously added to the buffer prior to the introduction of the magnetic platform. Platform and target molecule are then allowed to react for 10 minutes, time during which the specific interaction of IgG with the "leashed" Protein A is allowed to happen. This interaction results in the formation of a stable complex (MagChi-ProteinA/IgG) which in the presence of a magnetic field separates the magnetic platform together with the target molecule from the supernatant. After several washings of the complexed molecules with the same buffer used to allow the reaction (following the same pH conditions). To separate the target molecule from the magnetic platform, MagChi-ProteinA/IgG is placed in a sodium acetate buffer adjusted to pH=3.5. Decreasing the pH screens the non-bonded interaction between the target molecule (IgG) and its specific ligand (Protein A).

Figure 8 represents the amount of IgG bound to the MagChi-ProteinA platform and then eluted, as monitored, using reverse phase chromatography. The mobile phase used was water/acetonitrile/0.1% trifluoroacetic acid. The column temperature was maintained at 30 °C.

Under the conditions used, the MagChi-ProteinA platform showed saturation binding at a level of 2.5 mg of IgG protein per mg of matrix. The plateau in Figure 8 is due to the saturation of binding sites between IgG and Protein A, i.e. a limit in the ligand fixation onto the magnetized platform and consequently in the amount of bound target molecules.

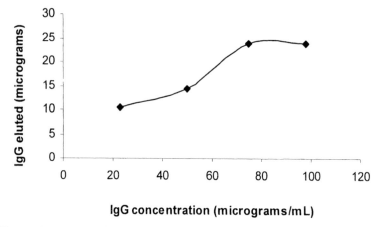

Figure 8: Amount of IgG bound to the the MagChi-ProteinA platform as a function of IgG concentration (for 10 mg of MagChi-ProteinA)

Anchored Ribonuclease

Magnetic fishing of a target molecule is a purification and separation process without time-consuming chromatographic steps. The magnetic polysaccharide-ligand platform can be reused after releasing the target molecule. Its non-invasive feature allows the biological medium to remain uncontaminated and then be subjected to successive separations involving diverse magnetic polysaccharide-ligand platforms, each designed to interact with a specific molecule.

Examples of how magnetic polysaccharides particles can be used to bind a protein and manipulate it for reaction or separation or stepwise synthesis etc. is in the binding of enzymes e.g. ribonuclease (RNase). The chosen magnetic platform was a gelatinized crosslinked high amylose starch excipient called Contramid (Labopharm, Inc., Laval, Qc, Canada). The TEMPO process was used to create a 30 molar substitution in carboxyl groups so the magnetic platform is called: MagCon-30-COOH. The oxidized magnetized starch platform is blended with a solution of polynucleotide. After proper equilibration with sodium acetate, pH 4.0-6.0, or tris-HCl buffer, pH 7.0-8.0, it could be shown by using reverse phase chromatography that all the RNase was completely bound and non-specific binding was not observed. The cartoon shown in Figure 9 implies that Rnase (catalyst) bound to a magnetic platform reacts with soluble ribonucleic acid macromolecule for a controlled time, temperature and pH. The fishing for proteins separation mechanism would find its utility in allowing fast enzyme recovery leaving a pure oligomeric product in solution.

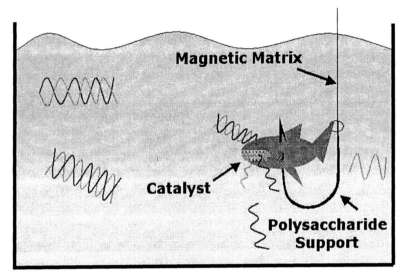

Figure 9: Cartoon showing the "anchored" ribonuclease acting on a polynucleotide solution to create oligomeric entities

Conclusion

If the magnetic platform launch is like casting a baited fishing line to which the target fish "bites", the magnetic field effect on the stable complex MagChi-ProteinA/IgG is equivalent to the "strike" action that hooks the fish and removes it from the supernatant. Fishing for Proteins has been successfully demonstrated *(17, 18)*. Repetitive use of the magnetic platform is feasible. Versatility in this platform design and ease of functionalization suggests that the process is adaptable to a wide variety of biological systems. Figure 9 is a simple example involving a well known enzyme. The RNase is covalently attached to the magnetic platform and assuming that covalent attachment does not harm the integrity of the enzyme its recovery and reuse is feasible. Reverse phase chromatography demonstrates the amount of RNase bound and eluted. Saturation was observed and nonspecific binding was absent. Recovery of enzyme and its separation from the supernatant of reacted molecules is straightforward.

MCC and chitin are solid crystalline substrates whose difference in surface functionalization potential offers multiple choices in experimental strategy. Surprisingly, chitin is seldom used as an excipient but it offers great versatility for functionalization especially if different levels of deacetylation are used.

In Shingel's chapter the variety of polysaccharide particles that can be used as a magnetic platforms are described. Given the status of avalaible reagents and

16

known functionalization reactions for carbohydrates, the potential of magnetic separation using polysaccharides is vast.

An important aspect of the various steps in Fishing for Proteins is the ferrite synthesis which yields magnetite predominantly with superparamagnetic properties. By definition, the superparamagnetic state yields ferrites that remain well dispersed and insure good surface interactions between proteins and ligands. This part is probably a *sine qua non* for the success of this novel separation method.

Acknowledgements

Financial support from NSERC, Labopharm Inc., Xerox Corp. and H3Pharma Inc. is acknowledged. Professor Dominic Ryan of the McGill Physics Dept. recorded and advised concerning the magnetization data.

References

1. Safarik, I.; Safarikova, M. *Biomag. Res. Tech.* **2004**, 2, 7-24.
2. Bucak, S.; Jones, D. A.; Laibinis, P. E.; Hatton, T. A. *Biotech. Prog.* **2003**, 19(2), 477-484.
3. Safarikova, M.; Roy, I.; Gupta, M. N.; Safarik, I. *Int. J. Biotech.* **2003**, 105(3), 255-260.
4. Safarik, I.; Safarikova, M. *J. Biochem. Biophys. Methods* **1993**, 27, 327-330.
5. Raymond, L.; Revol, J. F.; Marchessault, R. H.; Ryan, D. H. *Polymer* **1995**, 36(26), 5035-5043.
6. Battista, O. A., *Microcrystal Polymer Science.* 1975, McGraw-Hill: New York. pp 208.
7. Rowland, S. P.; Roberts, E. J. *J. Polym. Sci., Part A: Polym. Chem.* **1972**, 10(8), 2447-2461.
8. FMC-Corporation, *Avicel Microcrystalline Cellulose: Problem Solver and Reference Manual.* 1984.
9. Chang, P. S.; Robyt, J. F. *J. Carbohydr. Chem.* **1996**, 15(7), 819-830.
10. Isogai, A.; Kato, Y. *Cellulose (London)* **1998**, 5(3), 153-164.
11. Tahiri, C.; Vignon, M. R. *Cellulose (Dordrecht, Netherlands)* **2000**, 7(2), 177-188.
12. Araki, J.; Wada, M.; Kuga, S. *Langmuir* **2001**, 17(1), 21-27.
13. Montanari, S.; Roumani, M.; Heux, L.; Vignon, M. R. *Macromolecules* **2005**, 38, 1665-1671.
14. Kumar, V.; Yang, T. *Int. J. Pharm.* **1999**, 184(2), 219-226.

15. Sourty, E.; Ryan, D. H.; Marchessault, R. H. *Chemistry of Materials* **1998**, 10(7), 1755-1757.
16. Sourty, E.; Ryan, D. H.; Marchessault, R. H. *Cellulose (London)* **1998**, 5(1), 5-17.
17. Marchessault, R. H.; Shingel, K.; Vinson, R. K.; Coquoz, D. G. US Patent Application 2004/146855, 2004.
18. Marchessault, R. H.; Shingel, K.; Ryan, D.; Llanes, F.; Coquoz, D. G.; Vinson, R. K. US Cont-in-part of US Ser No 2004/146855 2005/019755, 2005.

Chapter 2

Applications of Cellulose and Cellulose Derivatives in Immediate Release Solid Dosage

Jinjiang Li and Xiaohui Mei

Research and Development, Boehringer Ingelheim Pharmaceuticals Inc., 175 Briar Ridge Road, Ridgefield, CT 06877

In this paper, the use of cellulose crystallites and cellulose derivatives in manufacturing immediate release dosage forms is reviewed from three aspects: tablet binder, tablet disintegrant and wet granulation binder. Cellulose crystallites such as microcrystalline cellulose (MCC) are typically derived from pure cellulose via a hydrolysis process. As a tablet binder, cellulose crystallites exhibit plastic deformation under compression owing to their crystalline and hydrogen bonded structure. The compressibility of cellulose crystallites (MCC) and the effect of their physical properties such as density and particle size on the compressibility and tablet hardness have been summarized. Some new developments in tablet binder using cellulosic materials are also reviewed. Secondly, cellulose crystallites and cellulose derivatives such as croscarmellose as tablet disintegrants have been compared with other disintegrants from a polymer chemistry point of view. Croscarmellose has been shown to have certain advantages when used in a small quantity. The disintegration mechanism and other affecting factors are discussed. Thirdly, cellulose derivatives such as hydroxypropylmethylcellulose (HPMC), hydroxypropylcellulose (HPC), methylcellulose (MC), sodium carboxymethylcellulose (Na-CMC) and ethylcellulose (EC) have been extensively used as wet granulation binders. The theory of wet granulation and in particular the effect of binder characteristics on the physical properties of dry granules has been reviewed. Under certain circumstances cellulose derivatives are better binders than other binders such as povidone. For instance, HPC, which has low surface tension, serves as a good binder for wet granulating poorly water-soluble compounds.

Introduction

Cellulose is the most abundant natural polymer on the earth with an annual production of 50 billion tons (1). Cellulose consists of linear chains of $\beta(1\text{-}4)$-linked –D-glucopyranosyl units (2) as shown in Schematic 1. The degree of polymerization for cellulose isolated from wood, which is the source for manufacturing pharmaceutical exicipients, varies and it can be as high as about 9000 (3). Consequently, this affects the processing properties of cellulose derivatives such as viscosity used in the pharmaceutical industry. In its native state, cellulose's macromolecular chains are packed in layers and held together mainly by strong intra-molecular and inter-molecular hydrogen bonds (see Schematic 2 for hydrogen bonding structure) (4). The inter-molecular hydrogen bonds between polymer chains are responsible for the mechanical properties of cellulose crystallites. They are also responsible for some physical behaviors, such as swelling, of cellulose crystallites and cellulosic materials used in pharmaceutical applications. Natural cellulose has a fibrous structure and is typically semi-crystalline (see Schematic 3) (5). Both crystallinity and fibrous structure significantly impact the compaction and flow behavior of cellulosic materials including microcrystalline cellulose, powder cellulose, and others. These properties are essential for their applications in pharmaceutical dosage form preparation. There are three functional groups available for derivatization in each repeating unit of cellulose: the primary hydroxyl group at the C-6 position, the secondary hydroxyl group at the C-2 position and the secondary hydroxyl group at the C-3 position (6). To derivatize cellulose, it is typically first reacted with concentrated NaOH to form alkali-cellulose followed by chemical reactions such as etherification, etc. Since cellulose is so abundant, both crystalline cellulose and its derivatives have been extensively used in the biomedical and pharmaceutical fields (7). Microcrystalline cellulose (MCC), produced by acid hydrolysis from high purity cellulose, has been extensively used in the pharmaceutical industry as a tabletting excipient since its introduction to the market in the early 1960's (8). Cellulose derivatives such as hydroxypropylmethylcellulose (HPMC), hydroxypropylcellulose (HPC), hydroxyethylcellulose (HEC), carboxylmethycellulose, etc. are widely used as binders in wet granulation (9,10) where granulating liquid is added to the powder starting material to produce granules (11). Wet granulation is the most versatile granulation process even though it is labor intensive and relatively expensive. After wet granulation, the compaction and flow property of pharmaceutical powders can be improved and cellulose derivative are advantageous as granulation binders compared with other synthetic binders. In this chapter, the applications of cellulose crystallites and cellulose derivatives in the traditional tablet technology such as wet granulation, tabletting, etc. will be discussed from both fundamental and application aspects.

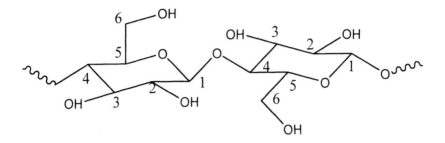

Scheme 1. The chemical structure of cellulose.

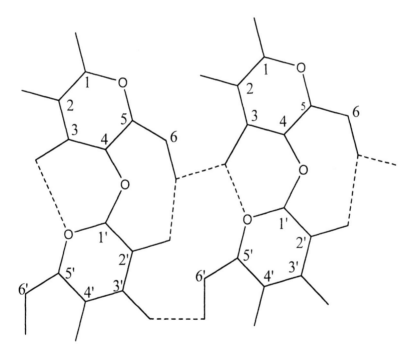

Scheme 2. The inter-molecular and intra-molecular hydrogen bonds between parallel chains in native cellulose.

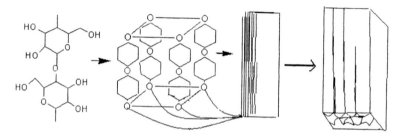

Scheme 3. The crystalline and fibrous structure of cellulose.

Cellulosic materials as Tablet Binders

Binding Mechanism

In a tablet, solid particles including drug substance are bound together by excipients in order to preserve the tablet integrity for transport and patient handling. There are three types of bonds (12) involved in tablet binding (a) surface forces (van der Waals forces, electrostatic force and hydrogen bonding) (b) inter-mechanical locking and (c) solid bridge formation. Van der Waals forces, which exist among all solid particles and are generally attractive, are relatively weak compared to hydrogen bonding and inter-mechanical-locking. They have a long interaction range (greater than 10 nm) and their magnitude decreases with distance (13). On the other hand, electrostatic interaction is a strong force with a long interaction range. Electrostatic force is associated with charged surfaces and its interaction range can extend to more than 100 nm (14). The force magnitude of electrostatic interaction strongly depends on the surface charge density. Relative to electrostatic and van der Waals forces, hydrogen bonding is a short range force whose interaction range is typically within a few Angstroms (15). During tabletting, inter-mechanical locking force becomes significant when particles are compressed into each other. At a high compression force, sintering occurs for materials with low melting point that can form a solid bridge at the drug/excipient interface. For tablets containing cellulosic materials, hydrogen bonding and inter-mechanical locking interactions are the predominant forces involved in tablet binding.

Macroscopically, the tensile strength of agglomerates in a tablet can be expressed in eq. 1 (16):

$$\sigma_t = \frac{1-\varepsilon}{\pi} k \frac{A}{x^2} \tag{1}$$

where ε is the specific void volume of the agglomerates, also referred to as porosity. k is the average coordination number, x is the size of the particles forming the agglomerate, and A is the adhesion force between particles. As shown in eq.1, for given solid particles, both contact area and adhesion force need to be increased in order to increase tablet strength. In addition, porosity reduction can also increase tablet strength, which can be achieved by increasing compression force. One of the common methods for increasing the contact area of solid particles in a tablet is to add deformable materials (plastic materials) to formulations in addition to increasing the compression force. Plastic excipients are easily deformed under compression so that close contact between drug particles and other ingredients can be achieved during the tabletting process. This consequently results in increasing the adhesion force between particles and ultimately the tablet strength. Cellulose crystallites, such as MCC, powder cellulose, and modified cellulose crystallites are widely used as tablet binders in the pharmaceutical industry. Both cellulose and modified cellulose crystallites are readily deformable owing to their inter-hydrogen bonding and fibrous structure.

MCC was designed to be used mainly in direct compression processes although using MCC as a tablet binder for preparing tablets containing granules made from wet granulation is now a common practice. For the purpose of balancing the compressibility and flowability of MCC, a variety of grades of MCC, including Avicel PH-101, PH-102, PH-301 and PH-302, were introduced by Food Machinery Corp. (FMC). In addition, other modified cellulosic products were invented for the improvement of MCC (such as problems with compaction). In the following text, the manufacturing methods, physical properties, and compression properties of these materials will be discussed.

Physical Properties

Microcrystalline cellulose is purified and depolymerized cellulose with a crystallinity generally between 60 and 65%, and a true density of about $1.5g/cm^3$. The typical physical properties of MCC from FMC (17) and other cellulosic materials (18) are shown in Table I.

Table I. The Physical Properties of Various Products of MCC and SMCC

Samples	Surface Area (m^2/g)	Moisture (%)	Norminal Particle size (μm)	$\rho_b(g/cm^3)$	ρ_t (g/cm^3)
Avicel PH-101	1.182	3.8	50	0.31	0.45
Avicel PH-102	1.042	3.6	100	0.31	0.42
Avicel PH-301	0.339	2.8	50	0.41	0.58
Avicel PH-302	0.837	2.9	100	0.40	0.56
Emcocel 90 M	No data	No data	91	0.32	0.35
SMCC90	No data	No data	90	0.31	0.39
PH-M06	No data	No data	7	0.56	0.76
L-HPC	No data	No data	50-150	No data	No data

ρ_b and ρ_t are bulk and tap density

As seen from Table I, *Avicel PH-101* and *PH-102* differ from *PH-301* and *PH-302* mainly in terms of their surface area and densities. The difference between *PH-101* and *PH-102*, and *PH-301* and *PH-302* is their particle size, which affects not only their flow but also their binding properties. SMCC, depending on the grade of MCC being silicified in which MCC and colloidal silicon dioxide suspension are co-spray dried (5), should have similar surface area and particle size as the parent product. After silification the compaction property is generally improved. PH-M 06 has very small particle size, which was designed to be used in fast-disintegrated tablets. As Table I indicates, the moisture level of MCC is around 3.0-4.0%.

Generally MCC is hygroscopic. The interaction of MCC with water not only affects MCC as a disintegrant but also its flow property. The general interaction mechanism of MCC with water is that water adsorbs on the surface of cellulose and binds to the free hydroxyl groups (5). Then the heat released from the adsorption breaks the inter-hydrogen bonds and water further interacts with the hydroxyl groups. A recent reference (19) suggests that at the molecular level, water is associated with cellulose in a triphasic process: tightly bound (structured), loosely bound, and as free water. Firstly, water binds to adjacent anhydroglucose units of MCCs, forming two hydrogen bonds in the process. As the water content increases, some of the initial bonds between water and the cellulose molecules are broken, vacating some of the anhydrous units for binding with the newly added water molecules. When the available binding sites of the cellulose molecules are saturated with water molecules, further addition of water will result in the formation of hydrogen bonds between water molecules. There is also condensation in the pores of MCC. In Table II, the amount of structured water in various grades of MCC and their monolayer capacities are listed.

Table II. Amount of Structured Water per Unit Weight of Sample and Monolayer Capacities of MCC of Various Grades

MCC	Absorption (g)		$g\ H_2O_{(s)}/g$ dry sample	$g\ H_2O_{(s)}/g$ $M_{o(corr)}$
	M_o	$M_{o(corr)}$		
PH-101	0.034	0.112	0.2454	2.1984
	(0.001)	(0.00194)	(0.0374)	(0.365)
PH-102	0.036	0.118	0.206	1.74
	(0.000)	(0.00128)	(0.0417)	(0.368)
PH-301	0.029	0.124	0.2598	2.098
	(0.000)	(0.00127)	(0.0248)	(0.2034)
PH-302	0.033	0.216	0.2112	0.9798
	(0.0005)	(0.00293)	(0.0103)	(0.0415)

Values in parenthesis represent the standard deviation. * The $g\ H_2O_{(s)}/g\ M_{o(corr)}$ values between the 4 MCC grades were found to be statistically significant.

The data from Table II indicate that *PH-302* is different from the others in terms of its monolayer capacity (Mo), and the amount of structured water. As shown in Table I, the particle size of L-HPC, which is manufactured by reacting alkaline-cellulose with propylene oxide under a nitrogen atmosphere (18), ranges from 50 to 150 μm. This material was developed as a tablet binder and a disintegrant for direct compression formulations. For it to function, the degree of substitution needs to be low (between 0.1 and 0.5). In this case, coiled helical structure of the fiber is partially maintained. Material with a lower substitution than 0.1 does not give enough binding property. The L-HPC also exhibits high fluidity, which is an advantage over MCC. This material also has both fibrous and granular structures. Its disintegration property is provided by the granular portion of the material, which can provide sufficient capillary effect.

Compression Properties

What are the mechanical properties required for making good tablets? To answer this question, one has to look at the compression profile of tablets experienced during tabletting. In tabletting, pharmaceutical powders typically undergo consolidation, deformation, and relaxation as shown in Figure 1. Thus, the ideal mechanical property needed should be plastic-elastic. A newly discovered compound, having a certain crystal lattice and crystal habit, may exhibit either elastic or plastic, or plastic-elastic behavior. Therefore, excipients such as cellulose crystallites (MCC) or other materials (lactose) are added to formulations to compensate the mechanical properties of the active species. This is to ensure that the formulation has a balanced mechanical property to be successfully manufactured. To evaluate the compaction property of an excipient, a porosity-compression pressure plot (Heckle plot) can be constructed according to the following equation (20):

$$\ln \frac{1}{1-\rho} = KP + A \qquad (2)$$

where ρ is the relative density of the compact and P is the compression pressure. Both K and A are constants that are determined from the slope and intercept, respectively, of the extrapolated linear region of the plot. In fact, $1/K$ is a yield pressure that appears to be a material constant. A high value of K is indicative of a material that is ductile or plastic in nature. On the other hand, a low value of K indicates a material that has a brittle nature. As shown in Figure 2, MCC is a plastic or ductile material relative to dicalcium phosphate dihydrate (DCP) and crystalline lactose that undergo extensive fragmentation under compression (20). Furthermore, the various mixtures of MCC and DCP have plastic-elastic properties between MCC and DCP. This is well illustrated in Figure 3 where MCC is mixed with DCP in various proportions. Similarly, the crystallinity of MCC also affects the compaction behavior (see Table III) (21). In Table III, the B values mainly express the magnitude of fracture of the particles caused after densification by movement and reorientation of particles. The yield pressure ($1/K$ in the Heckel equation) is mainly related to the plastic deformation of particles during compression. It was found that the yielded pressure decreased as the degree of crystallinity of MCC became smaller. In other words, the yielded pressure decreased as the amorphous region increased. This is because the amorphous region of MCC is easier to deform than its crystalline region even though both can be deformed plastically. This is further confirmed with the crushing strength results when tablets made of these materials were tested. As seen from Table III, the crushing strength decreased when the amorphous region increased. The data are consistent with the results obtained from the Heckle equation.

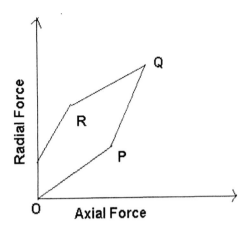

Figure 1. Compression force profile during tabletting: (1) consolidation (OP), (2) deformation (PQ) and (3) relaxation (QR).

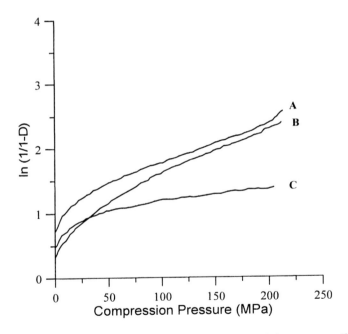

Figure 2. Heckle plot for plain materials (A) lactose, (B)microcrystalline cellulose and (C) dicalcium phosphate dihydrate.

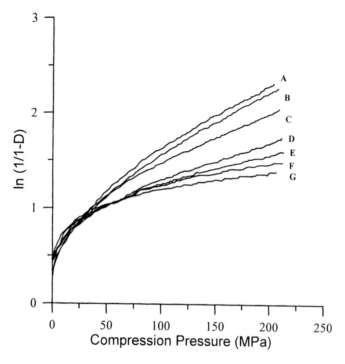

Figure 3. Heckle plots for the binary mixtures of microcrystalline cellulose (MCC) and dicalcium phosphate dihydrate (DCP). (A) MCC 100%, (B)MCC:DCP 80:20, (C)MCC:DCP65:35(D)MCC:DCP 50:50, (E) MCC:DCP 35:65, (F)MCC:DCP 20:80, and (G) DCP 100%.

Table III. Results of Heckle Plot Analysis of MCC with Various Degrees of Crystallinity

Degree of Crystallinity (MCC)	B*	Yield pressure** (MPa)	Crushing strength of tablet compressed at 155.6 MPa (kg/mm^2)
65.5% (I-MCC)	0.1415	232.0	0.44
54.6%	0.0899	194.1	0.34
48.2%	0.0811	191.8	0.32
37.6%	0.0751	178.4	0.28
25.8%	0.0697	174.0	0.23
12.1%	0.0663	148.5	0.22

*Differences between the relative density at the point where measurable force was applied and the relative density calculated from the intercept of the Heckle plot linear portion

**The reciprocal of the slope in the Heckle plot's linear portion, which would reflect plasticity of the sample.

As noted from Figures 2 and 3, there are two regions in a Heckel plot: a curved region at a low-pressure region, and a linear region at a high compression pressure. The Heckle plot works fairly well at a high compression pressure region. However, in the low compression pressure region, the Heckel equation is not valid. This is because in the low compression pressure range, a transition from a powder to a compact occurs, and the pressure susceptibility of a material exhibits a curvature, which is treated as a constant in the Heckle equation. A new function, the pressure susceptibility, is introduced as $\chi_p = C/(\rho - \rho_c)$ and a modified Heckle equation can be expressed as follows (22):

$$P = \frac{1}{C}\left[\rho_c - \rho - (1 - \rho_c)\ln\left(\frac{1 - \rho}{1 - \rho_c}\right)\right] \qquad (3)$$

where C is a constant, ρ and ρ_c are the relative density and the critical relative density, respectively around which the critical transition occurs. In Table IV, the parameters derived from both the Heckle and modified Heckel equations are compared for cellulose crystallites. As seen from Table IV, The modified Heckle equation shows a much better fit to experimental data. Table IV shows that cellulose crystallites including Emcocel 50 M, *Avicel PH-101* and *PH-102* all exhibit similar plastic deformation behavior since they all have similar values of

30

Table IV. Comparison between the Heckle Equation and the Modified Heckle Equation

Sample name	Heckel equation			Modified Heckel equation		
	K (MPa^{-1})	A	r^{2*}	$C\ (MPa^{-1})$	$\rho_c(g/cm^3)$	R^{2*}
Emcocel 50M	0.015	0.449	0.960	0.006	0.217	0.998
Avicel PH-101	0.016	0.453	0.978	0.007	0.180	1.000
Avicel PH-102	0.016	0.442	0.969	0.007	0.180	0.998
Klucel	0.044	0.682	0.983	0.014	0.379	0.986

*Corrected squared correlation coefficient

K and C constants. However, Klucel is more plastic as compared to other cellulose crystallites listed in the table because of higher K and C obtained for this material. In addition, the critical density value (ρ_c) indicates that the compression pressure for forming a rigid structure for this material is higher, which is consistent with higher K and C values. To further compare the Heckle equation and the modified Heckle equation, a new cellulose material made from alkali hydrolysis of cellulose (referred to as UICEL-A) was investigated (23). Its Heckle constants and modified Heckles constants are listed in Table V. The parameters derived from both the Heckle equation and modified Heckle equation indicates that Avicel PH-102 is more of a plastic material than UICEL-A. This could be attributed the fact that Avicel PH-102 and UICEL-A have different lattice structures and therefore, different intermolecular forces are involved. Again, the modified Heckle equation shows a better fit with experimental data. The compression behavior of enzyme β-galactosidase and its mixture with MCC were evaluated from both the Heckle and modified Heckle equations (See Table VI) (24). As seen in Table VI, β-galactosidase is closer to DCP than MCC based on both K and C values even though fitting with experimental data for both the Heckle and the modified Heckle equations is poor. Mixing β-galactosidase with MCC can increase the plastic behavior of this material.

In practice, tablet hardness is typically evaluated instead of constructing the Heckle plot since tablet hardness is directly related to friability and disintegration of the tablet. When various grades of MCC (i.e. PH-101, PH-102, PH-301 and PH-302) are used to make a tablet, the tablet hardness generally increases with compression force as shown in Figures 4 and 5 (17). The results

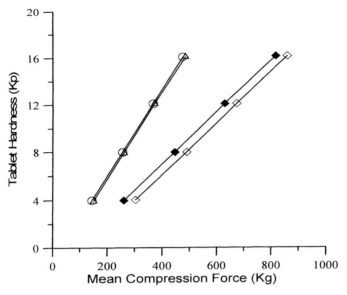

Figure 4. Tablet hardness vs. compression force for various grades of MCC in the absence of lubricants: (●)PH-102, (o)PH-101, (◆)PH-302 and (◇)PH-301.

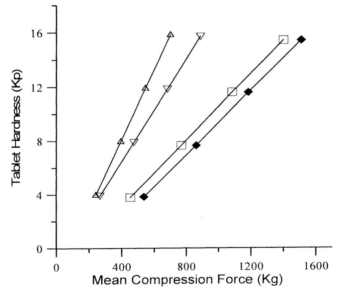

Figure 5. Tablet hardness vs. compression force for various grades of MCC in the presence of 0.5% magnesium stearate: (△)PH-101, (▽)PH-102, (☐)PH-301 and (◆)PH-302.

Table V. Values of the Heckle Equation and Modified Heckle Equation Parameters

	UICEL-A		Avicel PH-102	
Compression pressure range (MPa)	1.06-111.00	37-108	1.06-111.00	37-111
$K(10^{-3}, MPa^{-1})$	7.96 (0.32)	6.62	14.12 (0.78)	10.64
A	0.463 (0.016)	0.564	0.413 (0.038)	0.681
r^2	0.984	0.999	0.972	0.997
$C (10^{-3}, MPa^{-1})$	2.57 (0.09)	—	6.86 (0.20)	—
$\rho_c(g/cm^3)$	0.235 (0.008)	—	0.142 (0.011)	—
r^2	0.998	—	0.999	—

*Data presented as mean (s.e.m)

Table VI. The Parameters from Heckle and Modified Heckle Equations

Parameters	DCP	Mixture of MCC with β-galactosidase	β-galactosidase	MCC
$K (MPa^{-1})$	0.00453	0.00832	0.00465	0.02004
r^2	0.811	0.920	0.947	0.944
$C (MPa^{-1})$	0.00058	0.00068	0.00040	0.00114
r^2	0.924	0.847	0.897	0.701

from Figure 4 show that, in the absence of lubricant, the hardness for tablets made with *PH-101* or *PH-102* increases faster with compression force compared to those made with *PH-301* or *PH-302* as evidenced by their difference in slope. In addition, the hardness-pressure curve for both *PH-101* and *PH-102* superimpose on each other. In the presence of 0.5% magnesium stearate in the tablet blend, different grades of MCC respond to compression force differently. As shown in Figure 5, tablets made of *PH-101* and *PH-102* no longer have superimposed compressibility. Magnesium stearate affects *PH-102* to a greater degree than *PH-101*, as evidenced by the obviously lower slope value for the *PH-102* line. Figure 5 also shows that tablets made with *PH-301* or *PH-302* have parallel lines in the hardness-compression pressure curve, indicating that those tablets are less sensitive to lubricant magnesium stearate even though the surface area for *PH-302* is greater than *PH-301*.

When SMCC is used instead of MCC, the compaction property of the tablet is changed (25). In Table VII, the mechanical properties of SMCC are compared with MCC. The results in Table VII indicate that under comparable.

Table VII. The Mechanical Properties of MCC, SMCC and Blends of MCC with CSD

Sample	Test Rate (mm/min)	Deflection (mm)	Tensile strength (MPa)	E (J)	Density (g/cm^3)
MCC90	5	0.88	10.3	1.7	1.45
MCC90	0.05	0.73	8.9	1.2	1.44
SMCC90	5	1.13	12.7	2.6	1.44
SMCC90	0.05	0.96	11.2	2.0	1.45
Blend	5	0.70	9.1	1.2	1.44

test conditions, compacts of silicified MCC (SMCC) exhibit greater strength and stiffness than those of MCC, and require considerably more energy for tensile failure to occur. The functional properties of SMCC are not simply due to a combination of the properties of two individual components as seen in the comparison of dry blend made with MCC and colloidal silicon dioxide in the same proportion as the two components in SMCC.

Disintegrants

Disintegration Mechanism

In vivo disintegration of a tablet is an integral part of dissolution strongly influencing bioavailability (26). The disintegration mechanism includes three steps: (1) uptake of liquid from gastrointestinal fluid, (2) the swelling of a disintegrant in the tablet and (3) ultimately the breaking-down of the tablet. The liquid uptake by a porous bed from an aqueous medium is controlled by capillary action. This phenomenon is well described by the Washburn equation (27):

$$\frac{dh}{dt} = \frac{\gamma a \cos\alpha}{4\eta h} \qquad (4)$$

where h is the capillary length and t is the time for liquid to fill the capillary. a, γ, α and η are the capillary radius, interfacial tension, contact angle and the medium viscosity, respectively. A tablet is a porous bed. The equation can be used for modeling the liquid uptake by a tablet from gastro-intestinal fluid. For tablet disintegration, the liquid uptake from its dissolution medium is an essential step for the subsequent swelling of the disintegrant, thus controlling the capillary action can effectively influence a tablet disintegration, dissolution and bioavailability. For instance, a tablet can be made very hard to reduce its porosity and thereby slow down its disintegration and dissolution. For the purpose of controlling tablet disintegration, co-disintegrants such as zeolites, cellulose crystallites and other porous materials are frequently added to the tablet formulation to facilitate the liquid uptake. Cellulose crystallites such as MCC can be used as co-disintegrants primarily due to their porous, fibrous and hydrogen bonding structure. In addition, a class of super-disintegrants, including crosslinked PVP (crospovidone), cross-linked sodium carboxymethylcellulose (croscarmellose) and sodium starch glycolate (SSG), were introduced and frequently used in the pharmaceutical industry. These materials have a polymeric network structure. Since they expand quickly upon contacting an aqueous medium, only a small percentage is needed in a formulation to achieve tablet disintegration. The action mechanism of these super-disintegrants is based on the swelling of a polymer network. Thermodynamically, the swelling of a polymer network was modeled by Flory and Rhener (see eq. 5) (28):

$$\ln(\frac{Q-1}{Q}) + \frac{1}{Q} + \chi\frac{1}{Q^2} = -\frac{V_d G_0}{RT}\left[\left(\frac{1}{Q}\right)^{1/3} - \frac{2}{\langle f \rangle}\frac{1}{Q}\right] \qquad (5)$$

where Q, χ, V_d are the swelling ratio, Flory interaction parameter and the molar volume of a diluent, respectively. For in vivo tablet disintegration the diluent is gastrointestinal fluid and for in vitro dissolution it is the dissolution medium. G_0, $\langle f \rangle$, T and R are the shear modulus of the disintegrant used, junction functionality, temperature and gas constant, respectively. As indicated in eq. 5, the swelling of a disintegrant, which is proportional to its disintegration capability, depends on the interaction of the disintegrant with the medium, the degree of cross-linking, and the material property of the disintegrant. In fact, the swelling pressure, which is responsible for the tablet disintegration, should be the same as the stress generated from the disintegrant. The relation of this stress to the swelling ratio and the crosslink functionality can be expressed in eq. 6 (28):

$$\pi = kTN_v\left[(\frac{V}{V_0})^{-1/3} - \alpha\frac{V_0}{2V}\right] \tag{6}$$

where V_0 and V are the volumes of a disintegrant before and after swelling, respectively, and α and N_v are the material extension ratio and the number of cross-links, respectively. As shown in eq. 6, the stress of a disintegrant exerted on neighboring particles in a tablet is proportional to the swelling ratio, the density of cross-links and the physical property of the disintegrant itself. Since disintegration is a dynamic process, the kinetics of tablet swelling also plays a very important role in addition to the final equilibrium swelling volume. In fact, the kinetics of a tablet swelling directly affects its disintegration time, which is the time for a tablet to reach the critical tablet breaking force. The disintegration force is proportional to the rate of swelling as shown in eq. 7 (29):

$$\frac{dF}{dt} = K\frac{dV}{dt} \tag{7}$$

where dV/dt is the rate of swelling and K is a constant related to material properties such as porosity. As shown from eq.7, a faster swelling tablet would cause quicker tablet disintegration due to a faster buildup of the tablet breaking stress. For selecting a tablet disintegrant for formulation based on the above principles, the practical factors to be considered are: adsorption, absorption of water, porosity, swelling, effect of surfactants, properties of the active species, other fillers, etc. These will be discussed in the following section:

Cellulosic Disintegrants

As indicated by the physical properties of cellulose, cellulose crystallites interact strongly with water. For MCC such as Avicel PH-101, it can absorb as much as twice its weight in water before turning into slurry. However, its volume expansion in primary particles is limited due to its hydrogen bonded structure and therefore, cellulose crystallites are relatively inefficient disintegrants as compared to the super-disintegrants (i.e. croscarmellose, etc.). In addition to the final swelling volume, the rate of swelling for MCC is also slow compared to super-disintegrants. As shown in eq. 6, the stress exerted by a disintegrant on neighboring particles in a tablet is proportional to the swelling ratio of the disintegrant. Thus, the disintegration stress resulting from the swelling of cellulose crystallites is relatively small in comparison with that from super-disintegrants. Furthermore, disintegration time is related to the rate of stress buildup, which is linearly proportional to the swelling rate as shown in eq. 7. In Table VIII, the swelling properties of MCC (Avicel PH-101 and Elcema G 250) and SSG are compared (30).

36

Table VIII. The Effect of Disintegrant on the Penetration of Water into Ascorbic Acid Tablets. Vm/Vp= the ratio of volumetric maximum water uptake to calculated pore volume. K= rate coefficient obtained from the modified Washburn equation (30).

Disintegrant	Concentration (%)	Porosity (%)	V_m/V_P	$K_{(ml/g/s)}$
Avicel PH-101	1.0	14.7	109	2.20×10^{-4}
	2.5	14.2	153	1.22×10^{-3}
	5.0	13.6	191	4.91×10^{-3}
Elcema G 250	1.0	14.3	163	1.20×10^{-4}
	2.5	15.3	235	1.45×10^{-3}
	5.0	14.4	342	4.26×10^{-3}
SSG	1.0	14.1	303	2.36×10^{-3}
	2.5	14.7	381	3.34×10^{-3}
	5.0	13.8	626	4.66×10^{-3}

Table VIII shows that at the same disintegrant concentration the ratio of volumetric maximum water uptake to pore volume (Vm/Vp) for tablets made with SSG is much higher than that for the tablets made with MCC (Avicel PH-101 and Elcema G 250). In addition, V_m/V_p is higher for tablets made with Elcema G 250 than those made with Avicel PH-101 because Elcema G 250 has a larger particle size and is composed mainly of mechanically aggregated cellulose fibers. The swelling due to an aggregated structure is more than that of the Avicel PH-101. However, the swelling of aggregated cellulose particles is apparently not a true swelling but is rather the expansion of particle structure owing to the hydrogen and mechanical bonds broken by the water penetration. It was observed that the primary particles of Avicel PH-101 swell very little even though there is a significant amount of water uptake. Besides the swelling ratio, the rate of water uptake is also essential for tablet disintegration (see eq.7). The faster a tablet can reach its critical breaking force, the quicker a tablet disintegrates as shown in eq.7. As seen in Table VIII, the rate coefficient obtained from the modified Washburn equation (30) for the SSG tablet is almost ten times that of the Avicel PH-101 tablet at a low disintegrant concentration. Based on the Washburn equation, the pore size, and therefore the porosity of tablets, can strongly affect the rate of water uptake, which subsequently influences the swelling and disintegration of the tablet. This has been demonstrated for nitrazepam tablets made with both cellulose crystallites (Avicel PH-101 and Elcema G 250) and SSG (see Table IX). Table IX shows that for the same disintegrant at the same concentration, the rate coefficient for water uptake increases with porosity. Again, at a similar porosity, the rate coefficient for a tablet made with SSG is much higher than those made with Elcema G 250, followed by Avicel PH-101.

The effect of the rate of water uptake on disintegration time was demonstrated with other tablets. As shown in Table X (31), Avicel PH-101 and PH-102 could have the

Table IX. Effect of Disintegrant and Porosity of Tablet on the Penetration of Water into Nitrazepam Tablets

Disintegrant	Concentration (%)	Porosity (%)	V_m/V_P	$K_{(ml/g/s)}$
Avicel PH-101	2.5	17.8	104	6.58×10^{-4}
		14.9	122	9.01×10^{-5}
		11.7	110	8.07×10^{-5}
Elcema G 250	2.5	19.3	160	3.37×10^{-3}
		15.0	163	1.38×10^{-3}
		13.3	158	9.64×10^{-4}
Primojel STD	2.5	18.0	826	3.05×10^{-2}
		15.5	975	2.12×10^{-2}
		12.8	1203	1.93×10^{-2}

Table X. The Swelling Capacity and the Initial Water Uptake Rate for Avicel PH-101 and Avicel PH-102

Excipient	Swelling capacity $V_{water}/V_{paraffin}$	Initial uptake rate of water$\times 10^{-3}$ (mL/s)
Avicel PH-101	0.95	5.9
Avicel PH-102	0.95	1.8

exact same swelling capacity, but significantly different water uptake rates. This is probably due to the difference in particle size between two grades as shown in Table I. Thus, lactose and DCP tablets made with those disintegrants show significantly different disintegration time (see Table XI).

Table XI. Disintegration Time of DCP and Lactose Tablets Made with Avicel and SSG Disintegrants

Filler	Disintegrant (SSG)	*Avicel PH-101*	*Avicel PH-102*
Emcompress (DCP)	15 (second)	266 (second)	>1200 (second)
Lactose 100-mesh	15.6 (second)	16 (second)	39.7 (second)

Also indicated in Table XI, tablets made with SSG (super-disintegrant) have the fastest disintegration of all. As mentioned before, croscarmellose is a very popular super-disintegrant used in the pharmaceutical industry. In the following text, cellulosic polymers as disintegrants are compared with other super-disintegrants: SSG and crospovidone. The maximum intrinsic and bulk swelling for these materials (carboxymethylcellulose, anionic cellulose polymer, soy cellulose derivative, SSG and crospovidone) are shown and compared in Table XII. Table XII shows that SSG has the highest maximum swelling, both intrinsic and bulk, of five disintegrants followed by carboxymethylcellulose, anionic cellulose polymer, crospovidone and soybean cellulose derivative. The same trend was observed for the bulk swelling rate of the disintegrants (see Table XIII). In Table XIII, both intrinsic and bulk swelling rates of five disintegrants are listed. Table XIII shows sodium starch glycolate has the highest bulk swelling rate, of all five disintegrants followed by carboxymethylcellulose, anionic cellulose polymer, crospovidone and soybean cellulose derivative while carboxymethylcellulose has the highest intrinsic swelling rate of all.

Table XII. The Maximum Intrinsic and Bulk Swelling of Five Disintegrants

Disintegrant	Intrinsic[1]	Bulk[2]
Carboxymethylcellulose	541.4	165
Anionic cellulose polymer	208.7	176
Soybean cellulose derivative	—	45
Crospovidone, U. S. P	39.2	112
SSG	726.7	1260

1. Area was measured as mm^2 on film.
2. Volume was measured as mL.
Both indices are presented as a percent of the original area or volume.

Table XIII. The Maximum Intrinsic and Bulk Swelling Rates of Five Disintegrants

Disintegrant	Intrinsic	Bulk
Carboxymethylcellulose	271.8	32.5
Anionic cellulose polymer	216.4	25.0
Soybean cellulose derivatice	—	5.0
Crospovidone U. S. P	57.9	15.0
SSG	98.5	140.0

Both indices are presented as a percent of the original area or volume per second

Table XIV. Disintegration Time of DCP/Lactose Tablets Made with 0.5 and 1.0 Percent Disintegrants

Disintegrants	Carboylmethyl cellulose		Sodium starch glycolate		Crospovidone	
Concentration (%)	0.5	1.0	0.5	1.0	0.5	1.0
Tablet weight (mg)	404.1	401.9	394.2	405.4	398.1	397.3
Hardness (kg)	9.45	8.80	9.35	9.65	9.25	7.90
Disintegration time (min)	2.59	1.38	4.14	2.36	3.81	2.56

Disintegrant	Anionic cellulose polymer		Soybean cellulose derivative	
Concentration (%)	0.5	1.0	0.5	1.0
Tablet weight	401.0	400.4	401.6	400.1
Hardness (kg)	9.05	9.40	9.32	9.34
Disintegration time (min)	7.17	1.97	9.21	2.96

In addition, DCP (75%) and lactose tablets (25%) tablets were made with these materials as disintegrants. Their disintegration time is shown in Table XIV. Table XIV shows that for all disintegrants, disintegration time decreases with increasing concentration. At the same concentration, carboxymethyl cellulose is a more effective disintegrant than others. This seems different from the swelling volume and rate as shown in Table XII and XIII, where SSG has the highest swelling volume and rate. However, as indicated by eq. 6, the stress generated by the swelling of a disintegrant is not only proportional to swelling ratio but also to the density of crosslinks. In practice, for preparing a tablet with desirable disintegration properties, there are many other factors to be considered.

Particle size and size distribution of disintegrants are always important factors to consider. For preparing fast dissolving tablets, small particle size of MCC such as PH-M series was introduced. The effect of particle size on disintegration time of acetaminophen and ascorbic acid tablets prepared using MCC (PH-M series) via direct compression method was examined (32). It was found that tablets made with PH-M 06 (particle size, 7 μm) had faster disintegration compared to the tablets made with *Avicel PH-102* . Furthermore, wetting is a very important first step for tablet disintegration. As shown in the Washburn equation, wetting is related to the interfacial tension, which ultimately affects the water uptake. Surfactant solutions are often used as wet granulation liquid to improve wetting. Sodium lauryl sulfate (SLS) and polysorbate 80 are frequently incorporated into tablet formulation to improve the wetting property of MCC, for which SLS is more effective than polysorbate 80 (33). High molecular weight nonionic surfactants could have binding effect, which can slow down the tablet wetting. When using surfactants for improving wetting, the surface properties of the active species need to be considered. As shown before, higher porosity appears to promote water uptake according to the Washburn equation (34). If a tablet has too much porosity, tablet disintegration could be prolonged simply because the efficiency of the stress buildup is slowed at a high porosity. Also when too soluble filler is incorporated into a tablet formulation, due to a fast dissolution of filler, tablet porosity goes up quickly and the effectiveness of a disintegrant could be reduced (34). Generally, disintegration time decreases with increasing concentration. For different materials, their working concentration range is different. MCC is an inefficient disintegrant and therefore, there is a need for a high concentration of MCC in the formulation to be effective. Compared with MCC, super-disintegrants are more effective. They typically only need a few percent to be used in a tablet formulation. It was found that for ethenzamide tablet, croscarmellose and carboxymethyl cellulose were effective disintegrants even at a concentration less than 1% while crospovidone needs more than 1% to be effective (35). SSG needs to work together with MCC to be more effective. The effectiveness of super-disintegrants are also affected by aging. It was found that for naproxen and ethenzamide tablets, croscarmellose was affected to a greater extent than crospovidone and SSG upon storage (36).

Cellulose Derivatives as Wet Granulation Binders

Aqueous solutions of cellulose derivatives are often used as wet granulation liquids. If the drug substance is water sensitive, organic solvents such as ethanol will be considered instead of water. Typically a wet granulation process includes the following basic steps: (1) dry mixing in which the drug substance and one or more other ingredients are mixed together; (2) addition of binding liquid where the granulating fluid is added to the dry materials with the help of mechanical forces; (3) drying and (4) milling.

Mechanism of Wet Granulation

The increase in particle size of a powder mix is a result of the formation of liquid bridges between the powder particles. Therefore, wet granulation processes are determined by the nature and extent of these bridges. The process of agglomeration can be divided into four different stages (37,38) based on the amount of liquid added, and on the bonding mechanism as shown in Schematic 4. At the pendular state, the granulating fluid forms discrete liquid bridges at the points of contact of the particles. The particles are held together by surface tension at the solid-liquid-air interface and the hydrostatic suction pressure of the liquid bridge. Assuming a spherical shape for the particles, the tensile strength of the individual pendular bond can be calculated:

$$\sigma = \frac{\pi d \gamma}{\left[1 + \tan\left(\frac{\phi}{2}\right)\right]} \tag{8}$$

where d is the diameter of the spherical particle, γ is the surface tension of the liquid, and ϕ is the angle formed between the line joining the center of the spheres and the edge of the pendular bond. According to this equation, the bond strength is proportional to the surface tension of the granulating fluid. As the liquid content is increased to a certain level so that the liquid bridges extend beyond the contact points between particles, the funicular state is reached. This state is characterized by a continuous network of liquid interspersed with air. With a further increase in water content, the capillary state is reached where granulate becomes paste-like. At this stage, all the pore spaces in the granule are completely filled with liquid. Thus, the granules are held together by capillary suction developed on the granule surface. Finally, the droplet state is reached when the granules are completely surrounded by liquid. In such a slurry system, the strength of the droplet is determined by the surface tension of the liquid phase. The theoretical tensile strength of agglomerates in the funicular and capillary states has been derived by Rumpf et al. (39) with the assumption that the agglomerates are formed from monosized spheres:

$$\sigma = SC\left[\frac{(1-\varepsilon)}{\varepsilon}\right]\left(\frac{\gamma}{d}\right)\cos\theta \tag{9}$$

where d is the diameter of spheres which are packed with porosity of ε. The degree of liquid saturation is S, and the liquid has surface tension γ, and a contact angle θ with the solid. C is a material constant. According to this equation, the agglomerate strength increases with reduced particle size and increased material density.

Overall, wet granulation can be divided into three phases: nucleation, transition, and ball growth. The first stage is the start of granulation when a powder gets wet, and particles contact each other through liquid bridges. With increased liquid saturation, the agglomerates become less brittle and more plastic. Eventually the mean particle size of the granules grows quickly and large. Spherical granules are formed at this point. The amount of binder liquid just needed to reach the upper limit for the funicular state is considered to be optimal. In other words, the end point of wet granulation is generally reached when the capillary state is initiated (40).

Practically, the optimum amount of binding liquid can be determined by means of measuring the power consumption. Various instruments, such as mixer-torque rheometers (41), shear cells (42), split bed shear testers (43), and split plate tensile testers (44), have been used to identify the different stages of agglomerates during wet granulation.

Podczeck and Wood (45) studied the agglomerate growth in wet granulation using an automated split bed shear tester. Lactose monohydrate was chosen as the substrate, and a 5% HPMC solution was used as the binding liquid in this study. It was found that the threshold between the pendular and funicular state was at about 6% (w/w) of liquid binder added to the wet mass. The end point was reached when approximately 15% (w/w) of binding liquid was introduced. Dry granule characteristics such as granule density and compressive Young's modulus were shown to be a function of the amount of binder liquid added. For example, in Figure 6, with addition of 7% of HPMC solution, the Young's modulus increased sharply, and then remained fairly constant between 7 and 13%. It reached the maximum at 15% binding liquid and dropped down afterwards. These results suggested that the transition between pendular state and funicular state happened at 6% binding liquid, and that the funicular state changed gradually into the capillary state between 13 and 15% binding liquid in the wet mass. This conclusion was further confirmed by the changes in other parameters, including angle of internal friction (a measure of interparticulate friction during shear), and cohesion coefficient.

Effect of Binders

Excipients for tablet formulations include fillers, binders, disintegrants, lubricants, glidants etc. Among these excipients, the binders are critical, particularly for the wet granulation process. Binders are the adhesives which provide the cohesiveness essential for the bonding of the solid particles under compaction to form a tablet (46). In a wet granulation process, binders promote size enlargement to provide granules,

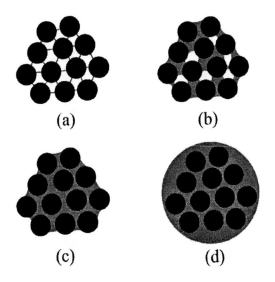

Schematic 4. Stages during wet granulation: (a) pendular state, (b) funicular state, (c) capillary state, (d) droplet state.

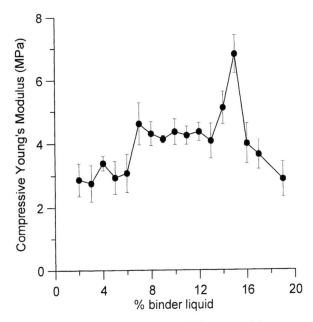

Figure 6. Compressive Young's modulus as a function of the amount of binder liquid added (calculated as % w/w of the wet mass using a 5% HPMC colloidal solution in demineralised water).

and thereby improve flowability of the blend during the manufacturing process. Binders may also improve the hardness of the tablets by enhancing intragranular as well as intergranular forces. Commonly used binders for wet granulation include natural polymers, synthetic polymers, and sugars. Cellulose and its derivatives have been drawing more attention than the others due to their strong binding power, well controlled quality, variety of grades, flexibility in method of incorporation, and compatibility with different solvents. The properties of granules, as well as the compressed tablets, depend on characteristics of binders such as type and content, viscosity, surface tension, molecular weight, etc.

Type and Content

The physical properties of granulations have been shown to be dependent on the types and concentrations of binders used in the process (47). Table XV summarizes commonly used cellulose derivatives for the wet granulation process. Typically, increases in the binder content in the formula will lead to: (1) increased average granule size, (2) decreased granule friability, (3) increased interparticulate porosity, (4) decreased granule flowability. Table XVI summarizes the effect of binder content in a tablet formula on the physical properties of fluidized bed granulations prepared using Hydroxypropylcellulose (HPC) as a binder (dry mixing).

Kokubo et al. (48) used a model system containing lactose-corn starch to evaluate the effect of five cellulosic binders (HPC, 6 cp; HPMC, 3, 6, 15 cp; and MC, 15 cp) on the particle size distribution of granules prepared in an agitating fluidized bed under fixed operating conditions. As shown in Figure 7, the median particle size of granules increased as the binder content was raised, except for MC. The anomalous behavior of MC was attributed to its low thermal gelation temperature. The MC solutions lose their adhesiveness earlier than the other binder solutions. Figure 8 shows the relationships between particle size and granule hardness when using HPC and HPMC in the solution method and the dry mixing method. The two binders behaved very similarly with the solution method. However, in the dry mixing method the granules using HPC were much harder than those prepared with HPMC. This difference was explained by differences in the flexibility of the film, and the adhesiveness of the binder solution. HPC film has been shown to be more flexible than HPMC film. Also, HPC maintained its adhesiveness in the gel state at low moisture levels whereas HPMC lost its adhesiveness.

Viscosity and Surface Tension

Wells and Walker (49) suggested that the wet granulation process is affected by the viscosity and surface tension of the binder due to their influence on the strength of the liquid bridges between particles, and the distribution of the binder during granulation. Typically an increase in viscosity of the binding solution will result in larger (50) and

Table XV. Commonly used cellulose derivatives for wet granulations

Binder	Structural formula	Method of incorporation	Concentration (%)	Solvent
HPMC		Solution Dry mixing	2-5	Water or hydroalcoholic solution
HPC		Solution Dry mixing	2-6	Water or hydroalcoholic solution
MC		Solution Dry mixing	1-5 5-10	Water
Na-CMC		Solution Dry mixing	1-5 5-10	Water
EC		Solution Dry mixing	1-5 5-10	Ethanol

Table XVI. Physical Properties of Fluidized Bed Granules Prepared using HPC (Klucel EF) as a Binder

Physical properties	Formula weight of binder (% w/w)			
	2.00	2.75	3.50	4.25
Average granule size (μm)	257	291	343	406
Friability (%)	12.2	10.0	6.2	3.3
Bulk density (g/mL)	0.44	0.43	0.43	0.46
Granule density (g/mL)	1.502	1.499	1.492	1.483
Porosity (%)	70.71	71.32	71.18	68.98
Flow rate (g/mL)	167.4	151.9	135.8	130.4

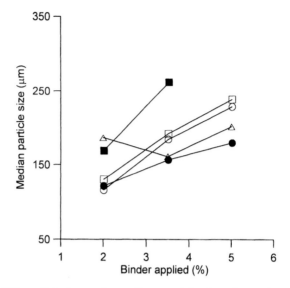

Figure 7. Effect of binder on the median particle size of granules prepared by the solution method in an agitating fluidized bed. ○, HPC; ●, HPMC (3 cP); □, HPMC (6 cP); ■, HPMC (15 cP); △, MC.

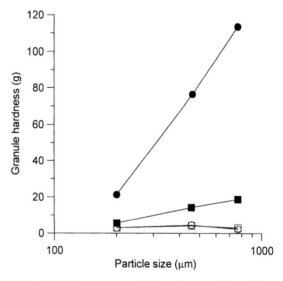

Figure 8. Relationship between granule hardness and particle size. ○, HPC solution (5%); ●, HPC dry mixing (8%); □, HPMC (3 cP) solution (5%); ■, HPMC (3 cP) dry mixing (8%).

stronger (51) granules. This is due to the increased amount of binder per bond, and subsequently increasing bond strength. It has been shown that (52) the higher surface tension of a binder solution will generate moist agglomerates with lower porosity andfurthermore increased granule strength through the effect on liquid saturation. This correlation between surface tension and the strength of moist agglomerates was further demonstrated by Parker et al. (53) The investigators utilized an instrumented mixer torque rheometer to study the interaction between substrate and binders, which have different viscosity and surface tension (Table XVII). Figure 9 illustrates the effect of viscosity and surface tension of binder solutions on the maximum torque which represented the shear force in the capillary state, and thereby was correlated to the strength of the moist agglomerate. The maximum torque increased with increasing viscosity, which indicated that a more viscous binding solution led to stronger granules. On the other hand, at the same viscosity, a higher shearing force was required by the binder with higher surface tension. Therefore, a higher surface tension of the binding solution resulted in granules with higher strength.

Table XVII. Viscosity and surface tension of polymer binders

Binder	Concentration (% w/v)	Viscosity (cP)	Surface tension (mN/m)
Water	-	1.0	72.00
PVP K25	1	1.19	58.48
	3	1.58	56.43
	5	2.25	55.14
	8.24	3.82	-
	12.78	6.87	-
HPMC 603	0.1	-	49.30
	0.5	-	48.50
	1.3	2.26	48.00
	3	5.41	47.57
	5	12.56	47.53

The relationship between surface tension and granule strength can be reversed when the wetting of the substrate becomes a major issue. Poor wetting of the substrate in a wet granulation typically leads to weak and porous granules, which will compromise granule flow as well as tablet mechanical properties (54). Therefore, surface activity of different cellulose derivatives can be employed to improve wetting of a poorly water soluble drug. In this circumstance, lower surface tension of the binder liquid would increase granule strength.

It has been shown that the surface tensions of HPC and HPMC solutions, and the contact angles between poorly water soluble drugs and HPC and HPMC are significantly smaller than PVP (55). Binder solution containing HPC or HPMC can wet Ibuprofen and Naproxan more efficiently than PVP (Table XVIII). As a result, the Naproxan tablets using HPC or HPMC as binder exhibited superior hardness and friability than those prepared with PVP (Table XIX). The poor granule flowability and low fill-weight of tablets using PVP was due to low bulk density caused by poor wetting during the granulation process.

Tablet XVIII. Wetting characteristics on Ibuprofen and Naproxan

Wetting Solution	Surface Tension (mN/m)	Viscosity (cp)	Contact Angle with Ibuprofen (°)	Contact Angle with Naproxan (°)
HPC^a	40.0	2.3	68	0
HPMC^b	48.4	1.9	81	37
PVP^c	53.6	1.5	88	63

a 0.5% Klucel® EF Pharm Hydroxypropylcellulose

b 0.5% Methocel® E5 Premium Hypromellose Type 2910

c 0.5% Plasdone® K29/32 Povidone

Tablet XIX. Physical properties of Naproxan granules and tablets

Binder (4%, w/w)	Mean time to avalanche (sec)	Bulk density (g/mL)	Compression force (kN)	Tablet weight (mg)	Tablet hardness (kP)	Tablet Friability (%)
HPC	9.8	0.38	3	605	5.7	2.0
			5	596	14.1	0.9
			10	596	29.9	0.5
			15	596	37.5	0.8
HPMC	9.3	0.39	3	602	8.3	2.2
			5	601	12.7	1.4
			10	601	24.7	0.7
			15	600	26.4	0.9
PVP	14.5	0.34	3	564	7.3	3.3
			5	569	15.8	2.0
			10	567	21.8	1.5
			15	564	15.1	3.4

Molecular Weight

Typically higher molecular weights of polymers will result in higher viscosity of their solutions. Therefore, a similar relationship between properties of granules and viscosity or molecular weight of binders is expected. However, Park et al. (56) found that with respect to the rheology of wet mass during granulation, different molecular weight grades of the same type of binder behaved differently at equivalent viscosity. This observation could not be completely explained by the traditional theories, which relate binder surface tension to granule strength, since the surface tensions are almost the same among these binders. They studied two molecular weight grades of HPMC (HPMC 603, Mw 45×10^3; HPMC 606, Mw 93×10^3). As shown in Figure 10, the curve for the low molecular weight HPMC passes through a minimum, followed by a positive slope until the curve crosses that of the high molecular weight grade. The different grades of HPMC exhibit different behavior at equivalent viscosity. They attributed this difference to the change of intra-granular viscosity caused by preferred adsorption of high molecular weight binder to the substrate.

Danjo et al. (50) studied the influence of the molecular weight of HPMC (TC5E, MW 12,600; TC-5RW, MW 29,400; TC-5S, MW 64,800), and another binder PVP (K-25, MW 29,000; K-30, MW 45,000; K-90, MW 1,100,000) on physical properties of granules and tablets. Figure 11 shows the granule size D50 gradually increased as the molecular weight of the binding agents increased. Therefore, less liquid is required for wet granulation when higher molecular weight binders are used. Figure 12 shows that the tensile strength of HPMC film increased with increasing molecular weight of the binder which suggested granule strength was also affected by the molecular weight of binders. Furthermore, it was found that the required compression force for granules made with high molecular weight binders is higher than that for granules made with low molecular weight binders. The investigators also observed that the radial tensile strength of a compressed tablet gradually increased with increasing granule tensile strength, which can be correlated with the increasing molecular weight of binders

Conclusions

In this chapter, the mechanisms of tablet binding and disintegration, and wet granulation were discussed from a polymer chemistry and practical point of view. Cellulose crystallites are the most widely used tablet binders, due to their physical properties. Various grades of microcrystalline cellulose have different particle size and surface area. As a consequence, they exhibit different properties under compression. Cellulose crystallites are generally plastic with a high Heckle constant.

50

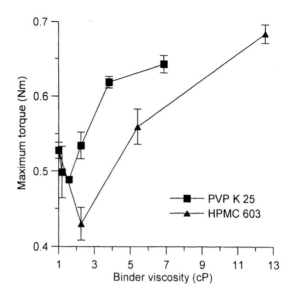

Figure 9. Effect of binder viscosity on maximum torque for experimental materials.

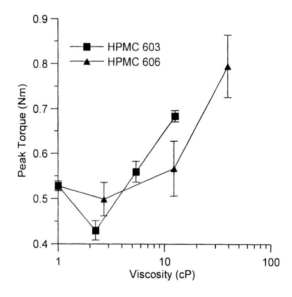

Figure 10. Relationship between binder viscosity and maximum torque for Avicel/HPMC systems.

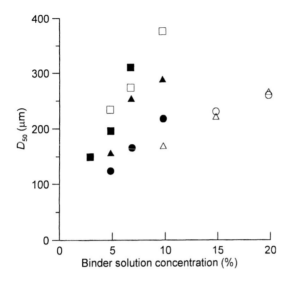

Figure 11. Relationship between binder solution concentration and granule size (D50) as a function of the molecular weight of the binding agents. . ○, K-25; △, K-30; □, K-90; ●, TC-5E; ▲, TC-5RW; ■.

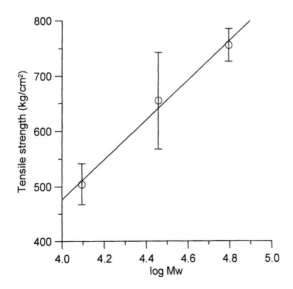

Figure 12. Relationship between the logarithm of the molecular weight (Mw) and the tensile strength of polymer HPCM films.

Mixing cellulose crystallites with other brittle materials can modulate the mechanical property of the final mixture. MCC is an inefficient disintegrant but it can be a very effective co-disintegrant. Cellulose derivatives can work as super-disintegrants. The most effective cellulosic superdisintegrant is croscarmellose. The disintegration mechanism was analyzed based on the Flory swelling theory for polymer network materials. In general, disintegration is related to water uptake, swelling, and swelling rate. Croscarmellose is found to be an effective disintegrant because it swells, and it has a high swelling ratio. Cellulose derivatives with various chemical structures are used as wet granulation binders, in particular for poorly water-soluble compounds. Both chemical structure and molecular weight of binders significantly impact the film flexibility the friability and hardness of tablets made from these granules. Other physical properties such as viscosity, surface tension, etc. also affect their performance as wet granulation binders. In general, cellulose derivatives exhibit a good wetting property and tablets made with cellulose derivative binders show superior friability and hardness to those prepared with PVP.

Acknowledgement

The authors would like to thank Ms. D. Chiappetta for reading the manuscript.

References

1. Carragher, C. E., Jr. *Polymer Chemistry*, Marcel Dekker: New York, NY, 2000; p170.
2. Kondo, T. In *Polysaccharides*; Dumitriu; S. Ed.; Marcel Dekker: New York, NY, 1998; pp131-172.
3. Hon, D. N.-S. In *Polysaccharides in Medicinal Applications*; Dumitriu, S. Ed.; Marcel Dekker: New York, NY, **1996**, pp87-105.
4. Liang, C. Y.; Marchessault, R. H. *J. Polymer. Sci.* **1959**, 37, 385.
5. Luukkonen, P. Ph.D. thesis, Pharmaceutical Technology Division, Department of Pharmacy, University of Helsinki, Helsinki, Finland, 2001.
6. Robyt, J. F. *Essentials of Carbohydrate Chemistry*; Springer: New York, NY, 1998; pp228-244.
7. Dumitriu, S. In *Polymeric Biomaterials*; Dumitriu, S.; Ed. 2nd Edition, Marcel Dekker: New York , NY, 2002; pp1-61.
8. Battista, O. A. US Patent 3,146,170, 1964.

9. Grover, J. A. In *Industrial Gums: Polysaccharides and Their Derivatives*; Whistler, R. L.; BeMiller, J. N.; Eds.; Academic Press: New York, NY, 1993; pp475-504.
10. Pietsch, W. *Agglomeration Processes: Phenomena, Technologies, Equipment*; Wiley-VCH: New York, NY, 2002; pp43-55.
11. Gordon, R. E; Rosanske, T. W; Fonner, D. E.; Anderson, N. R.; Banker, G. S. In *Pharmaceutical Dosage Forms: Tablets*, Lieberman, H. A.; Lachman, L.; Schwartz, J. B. Eds.; Marcel Dekker: New York, NY, 1990; Vol. 2, pp245-316.
12. Carstensen, J. T. *Advanced Pharmaceutical Solids*; Marcel Dekker: NewYork, NY, 2001; pp411-412.
13. Israelachvili, J. *Intermolecular and Surface forces*; 2nd edition; Academic Press: New York, NY, 1992; pp83-108
14. Hunter, R. J. *Foundations of Colloid science*; 2nd edition; Oxford University Press: New York, NY, 2001; pp305-369.
15. Dill, A. K.; Bromberg, S. *Molecular Driving Forces: Statistical Thermodynamics in Chemistry and Biology*; Garland Science: New York, NY, 2003; pp563-568.
16. Pietsch, W. In *Handbook of Powder Science and Technology*; 2nd Edition; Fayed, M. E.; Otten, L. Eds.; Chapman & Hall: New York,NY, 1997; p209.
17. Reier, G. E. ; Wheatley, T. A. *In Chemical Aspects of Drug Delivery Systems*; Karsa, D. R.; Stephenson, R. A. Eds.; The Royal Society of Chemistry: Cambridge, UK, 1996; pp116-126.
18. Maruyama, N.; Umezawa, H.; U.S. Pat. Appl. Publ. **2003**
19. Heng, P. W. S.; Liew. C. V.; Soh, J. L. P. *Chem. Pharm. Bull.* **2004**, 52, 384-390.
20. Ilkka, J.; Paronen, P. *Int. J. Pharm.*, **1993**, 94, 181-187.
21. Suzuki, T.; Nakagami, H. *Euro. J. Pharm. Biopharm.*, **1999**, 47, 225-230.
22. Kuentz, M.; Leuenberger, H. *J. Pharm. Sci.*, **1999**, 88, 174-179.
23. Reus-Medina, M.; Lanz, M.; Kumar, V.; Leuenberger, H. *J. Pharm. Pharmacol.*, **2004**, 56, 951-956..
24. Kuny, T.; Leuenberger, H. *Int. J. Pharm.*, **2003**, 260, 137-147.
25. Edge, S.; Steele D. F.; Chen, A.; Tobyn, M. J.; Staniforth, J. N. *Int. J. Pharm.*, **2000**, 200, 67-72.
26. Niazi, S. *Textbook of Biopharmaceutics and Clinical Pharmacokinetics*; Appleton-Century-Crofts: New York, NY, 1979; pp45-46.
27. Morrison, I. D.; Ross, S. *Colloidal Dispersions : Suspensions, Emulsions, and Foams;* Wiley-interscience, New York, NY, 2002; pp210-212.
28. Graessley, W. W. *Polymeric Liquids & Networks: Structure and Properties*; Garland Sciences: New York, NY, 2004; pp490-500.

54

29. Rudnic, E. M.; Rhodes, C. T. *Drug Dev. Ind. Pharm.* **1982**, 8, 87-109.
30. Paronen, P.; Juslin, M.; Kasnanen, K. *Drug Dev. Ind. Pharm.* **1985**, 11, 405.
31. Chebli, C.; Cartilier, L. *Int. J. Pharm.* **1998**, 171, 101.
32. Ishikawa, T.; Mukai, B.; Shiraishi, S.; Utoguchi, N.; Fuji, M.; Matsumota, M.; Watanabe, Y. *Chem. Pharm. Bull.* **2001**, 49, 134-139.
33. Wan, L. S. C.; Heng, P. W. S. *Pharmaceutica Acta Helvetiae*, **1986**, 61, 157-163.
34. Lowenthal, W. *J. Pharm. Sci.* **1972**, 61, 1695-1711.
35. Tagawa, M.; Chen, R.; Chen, P.; Kobayashi, M.; Okamoto, H.; Danjo, K. *Yakuzaigaku*, **2003**, 63, 238-248.
36. Gordon, M. S.; Rudraraju, V. S.; Rhie, J. K.; Chowhan, Z. T. *Int. J. Pharm.* **1993**, 97, 119-131.
37. Newitt, D. M.; Conway-Jones, J. M. *Tran. Inst. Chem. Eng.* (London), **1958**, 36, 422-442.
38. Barlow, C. G.; Chem. Eng. (London), **1968**, 220, CE296.
39. Rumpf, H. In *Agglomeration*; Knepper, W. A. Ed.; Wiley Science: New York, NY, 1962; pp379-418.
40. Leuenberger, H. In *Powder Technology and Pharmaceutical Processes*; Chulia, D.; Deleuil, M.; Pourcelot, Y. Eds.; Elsevier: Amsterdam, 1994; pp377-389.
41. Rowe, R. C.; Sadeghnejad, G. R. *Int. J. Pharm.* **1987**, 38, 227-229.
42. Pilepel, N. In *Advances in Pharmaceutical Sciences*; Bean, H. S.; Beckett, A. H.; Carless, J. E. Eds.; Academic Press: London, 1971, p173-219.
43. Shah, R. D.; Kabuki, M.; Pope, D. G. Augsburger, L. L. *Pharm. Res.* **1995**, 12, 496-507.
44. Khan, F.; Pilpel, N. L. *Powder Technol.* **1986**, 48, 145-150.
45. Podczeck, F.; Wood, A. *Int. J. Pharm.* **2003**, 257, 57-67.
46. Pariah, D. M. *Handbook of Pharmaceutical Granulation Technology*; Marcel Dekker: New York, NY, 1997; pp59-73.
47. Davies, W. L.; Gloor, W. T. *J. Pharm. Sci.* **1972**, 61 (4), 618-622.
48. Kokubo, H.; Nakashima, C.; Sunada, H. *Chem. Pharm. Bull.* **1998**, 46, 488-493.
49. Well, J. L.; Walker, C. V. *Int. J. Pharm.* **1983**, 15, 97-111.
50. Danjo, K.; Kozaki, K.; Sunada, H.; Otsuka, A. *Chem. Pharm. Bull.* **1994**, 42, 2121-2125.
51. Cutt, T.; Fell, J. T.; Rue, P. J.; Spring, M. *Int. J. Pharm.* **1986**, 33, 81-87.
52. Ritala, M.; Holm, P.; Schaefer, T.; Kristensen, H. G. *Drug Dev. Ind. Pharm.* **1988**, 14, 1041-1060.

53. Parker, M. D.; York, P.; Rowe, R. C. *Int. J. Pharm.* **1990**, 64, 207-216
54. Rowe, R. C. *Int. J. Pharm.* **1990**, 58, 209-213.
55. Lusvardi, K. M.; Durig, T.; Skinner, G. W.; Harcum, W. W. *Pharmaceutical Technology Report* PTR-026; Hercules: 2003.
56. Park, M. D.; York, P.; Rowe, R. C. *Int. J. Pharm.* **1991**, 72, 243-249.

Chapter 3

Hydroxypropylcellulose in Oral Drug Delivery

Mira F. Francis[1], Mariella Piredda[1], and Françoise M. Winnik[1,2,*]

[1]Faculty of Pharmacy and [2]Department of Chemistry, University
of Montreal, C.P. 6128 Succ. Centre-ville, Montréal,
Québec H3C 3J7, Canada

Peroral drug administraion is by far the most common and most convenient route of drug delivery. The need of polymeric carriers permitting controlled release of a desired drug following oral administration has led to the screening of a large variety of synthetic and natural polymers. In oral solid dosage forms, hydroxypropylcellulose (HPC) is widely used as binder due to its excellent physico-chemical characteristics combined with mucoadhesive properties. Graft copolymers of HPC, decorated at random with short amphiphilic chains, form nanoparticles in water, consisting of a hydrophobic core surrounded by a hydrophilic, mucoadhesive shell. Such HPC-based assemblies can act as carrier of highly lipophilic drugs. The molecular design and characterization of HPC-based drug delivery systems are discussed, together with their ability to entrap cyclosporin A (CsA), and to carry it through model intestinal membrane.

58

Introduction

Hydroxypropylcellulose (HPC) is a non-ionic water-soluble polymer, formed by reaction of cellulose, the most abundant naturally occurring polymer, with propylene oxide inder conditions of high temperature and pressure (*1*). It presents amphiphiphilic properties: it is only soluble in cold water, precipitating out of solution above a critical temperature of ~ 42 °C (*2*). It is also soluble in a number of organic solvents. It is a commercial inexpensive product, extensively used in many commercial formulations. Moreover, HPC is essentially non-toxic and non irritant (*3,4*). The World Health Organization (WHO) specifies no acceptable daily intake for HPC, since the levels consumed are not considered to represent any hazard to health (*5*).

Therefore, HPC has been widely used as excipient in many pharmaceutical oral solid dosage forms (such as tablets, capsules, etc) due to its bioadhesive properties (*6,7,8*). In these formulations, HPC acts as a binder in granulation (*9*). Its widespread use stems from its ease of compression, ability to accommodate a large percent of drug and negligible influence of the processing variables and on on drug release rates (*10*). In soild dosage manufacturing, HPC is mainly used as a dry binder in roller compaction processing. It is generally incorporated into the tablet formula at 4-8% *w/w*. Skinner et al. showed that, as the binder concentration of HPC increased, tablet capping and fragility decreased, but only slight differences were noted in tablet hardness (*11*).

During the last two decades, the use of HPC as hydrophilic matrix has become extremely popular in controlling the release of drugs from solid dosage forms (*12,13*). Although many different polymers have been used in swellable controlled release systems, water-soluble cellulose ethers are probably the most frequently encountered in pharmaceutical formulations as matrices for drug delivery system (*14*). In matrix devices, the drug is often released by a diffusion process such that a receding drug boundary will exist within the device. The rate of drug release is influenced by several factors, which include the pH of the dissolution media, drug loading, structural and geometric factors, and the presence of a co-active agents (*15*). For example, Ebube et al. (*16*) reported that the rate of acetaminophen release and the physicochemical attributes of the HPC hydrophilic matrix formulations were considerably altered when pseudoephedrine is present in the formulation as co-active, since the presence of a highly soluble compound, such as pseudoephedrine, in a hydrophilic cellulose matrix can generate an additional osmotic gradient, which results in a faster rate of polymer swelling and increase in gel thickness. Furthermore, the mobility of the polymer chain is enhanced with gradual transformation of a glassy matrix into rubbery swollen gel, thereby creating a viscous barrier between the tablet core and dissolution medium (17). Thus, drug release from the matrix tablet is retarded considerably.

It is important to point out that the mucoadhesive properties of HPC have been widely used in the development of bioadhesive drug delivery systems, such as peroral extended-release bioadhesive tablet formulation of verapamil HCl (18), buccoadhesive tablets for insulin (19) and propranolol (20) delivery, as well as vaginal bioadhesive tablets of acyclovir (21).

In thet study described here, we set out to investigate the suitability of HPC-based carriers in the oral delivery of Cyclosporin A, a poorly water-soluble drug used as immunosupressant agent for the prevention of graft rejection following organ transplantation (22). CsA has an extremely low solubility in water (23 µg/mL at 20 °C), and a very low oral bioavailability. HPC was the polymer of choice, not only because of its known clinical safety, but also due to its bioadhesive property which is expected, in vivo, to increase the intimate contact with the absorbing intestinal membrane, thus prolonging residence time within the small intestine (23,24), and increasing the extent of drug transport into the systemic circulation (24).

Despite the inherent hydrophobicity of HPC due to the presence of isopropoxy substituents, known as "the hydrophobic pockets" within the HPC structure (25), the affinity of CsA for unmodified HPC was found to be rather poor (see below and Figure 2). Therefore, a modification of the polysaccharide structure was required in order to obtain high drug incorporation levels. For this, a small number of polyoxyethylene alkyl chains, from a nonionic surfactant consisting of a hydrophobic alkyl chain attached via an ether linkage to a hydrophilic POE chain (Figure 1.), was grafted on the HPC chains.

The aim of this work is to assess the effectiveness of such amphiphically-modifed HPC copolymer as delivery vehicle for CsA. Furthermore, using model intestinal epithelial cells, Caco-2 cells, we investigated the permeation of HPC-solubilized CsA through Caco-2 cell membranes, in order to gain insight into the absorption of the micelle-entrapped drug from the gastrointestinal tract into the blood stream.

Materials and Methods

Materials

The polysaccharide used in this study was prepared based on design guidelines we had established through a systematic evaluation of key structural parameters controlling the effectiviness of hydrophobically modified HPC (HM-HPC) as CsA carriers (26,27). Briefly, the HM-HPC graft copolymer was synthesized *via* ether formation between a tosylated poly(oxyethylenecethyl ether (POE-C_{16}) and hydroxyl groups of HPC (MW 80 000 Da) (Figure 1.). As the polymer and POE-C_{16} have similar solubility characteristics, the coupling could be carried out in homogeneous solution. Under these conditions, high levels of hydrophobic modification were achieved and the distribution of alkyl chains along the polymer chain tends to be random rather than "blocky". The resulting HPC-g-POE-C_{16} copolymer was purified by soxhlet extraction with *n*-hexane to remove free POE-C_{16} residues. The level of POE-C_{16} grafting was determined by ^{1}H-NMR spectroscopy measurements carried out with polymer solutions in DMSO-d_6.

Synthesis of Fluorescein-labelled HM-polysaccharide

The labelled polymers were prepared by modification of HPC-g-POE-C_{16} (Figure 1) as follows: HPC-g-POE-C_{16} (300.0 mg) was dissolved in a 1/1 v/v of water/acetone mixture (50 ml). The pH of the solution was adjusted to 10 using aqueous NaOH (5 N). A solution of 5-([4,6-dichlorotriazin-2-yl]amino)-fluorescein (DTAF, 8.0 mg, 0.015 mmol) in aqueous NaOH (15 ml, pH=10) was added, in five portions, to the polymer solution at time intervals of 30 min. At the end of the addition, the reaction mixture was kept at room temperature for 17 h. Subsequently, it was dialyzed extensively against distilled water (MW cut-off: 6,000-8,000 dalton, Spectrum Laboratories Inc., Rancho Dominguez, CA) and isolated by freeze-drying; yield: 107.2 mg. The degree of DTAF substitution was determined by quantitative UV/Visible spectrophotometry. The labelled polymer was dissolved in aqueous solution (pH 9). DTAF was used as reference (molar extinction coefficient ε_{492nm}: 70 000 $cm^{-1}.mol^{-1}$ at λ = 492 nm) (28).

Characterization of HM-HPC in Aqueous Solution

When placed in an aqueous environment, HPC-g-POE chains spontaneously assemble in the form of polymeric micelles consisting of a hydrophobic core surrounded by a hydrophilic corona made up by highly hydrated polysaccharide chains exposed to water. Unlike surfactant micelles, which tend to disintegrate upon dilution triggering lysis of cell membranes (29), polymeric micelles are remarkably stable towards dilution (30).

The critical association concentration (CAC) of the polymeric micelles in water was estimated by a steady-state fluorescence spectroscopy assay, using polymer solutions of increasing concentrations in pyrene-saturated water equilibrated overnight. The CAC value was obtained by monitoring the changes

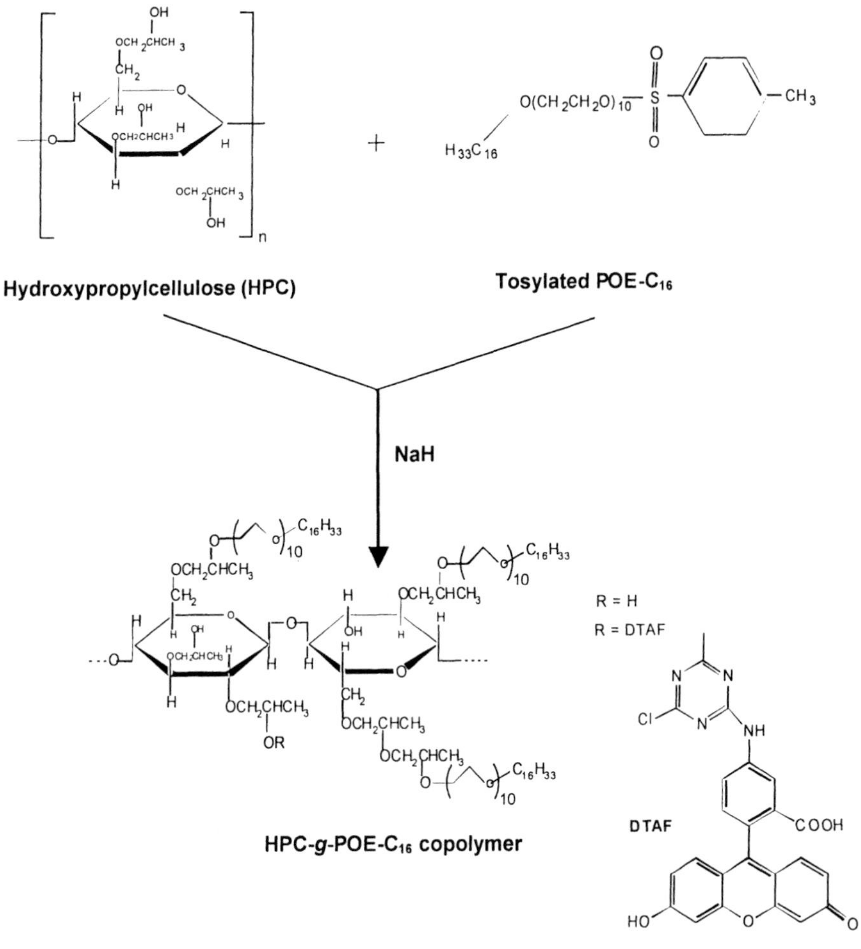

Figure 1. Synthesis of unlabelled and fluorescein-labelled HM-Hydroxypropylcellulose (5.4 mol %) copolymer.

in the ratio of the pyrene excitation spectra intensities *(31)* at λ = 333 nm (I_{333}) for pyrene in water and λ =336 nm (I_{336}) for pyrene in the hydrophobic medium within the micelle core.

The hydrodynamic diameter of drug-free and drug-loaded polymeric micelles in water was evaluated by dynamic laser light scattering (DLS) at 25°C with a scattering angle of 90°.

Physical Loading of CsA in HM HPC Polymeric Micelles

A dialysis method was employed to prepare CsA-loaded polymeric micelles. A polymer solution in deionized water and a CsA solution in ethanol were prepared separately. Subsequently, different mixtures of polymer with varying CsA initial concentrations (2.5 - 40% w/w) were prepared by mixing the two solutions. Following 48 h of dialysis, each solution was filtered through a 0.22-μm pore-size filter and the filtrate was freeze-dried yielding a free flowing powder, readily resolubilized in water or aqueous buffer solutions.

High Performance Liquid Chromatography (HPLC) Analysis

CsA was extracted from freeze-dried micelles using acetonitrile (ACN). The resulting suspensions were sonicated for 10 min then agitated for 8 h. They were then filtered and assayed by HPLC using a symmetry® octadecyl-silane C_{18} column *(32)*. The mobile phase consisted of ACN : water (80:20 v:v) with a flow rate of 1.2 mL/min. The column was thermostated at 70 °C. The CsA peak, monitored at 210 nm, appeared at a retention time of 6.5 min. A CsA calibration curve was prepared using standard solutions of concentrations ranging from 3.12 - 400 mg/L, with a first order correlation coefficient (r^2) greater than 0.99. Drug loading (DL) was calculated using Eq. 1:

$$DL\ (\%) = 100\ (W_c/W_M) \qquad (1)$$

where W_c is the weight of CsA loaded in micelles and W_M is the weight of micelles before extraction.

Stability study

The stability of CsA-loaded polymeric micelles in gastrointestinal fluid was monitored by measuring the release of CsA entrapped within micelles as a function of contact time with simulated gastric or intestinal fluids (prepared according to USP XXIV). Dialysis bags (MW cut-off: 6,000-8,000 dalton) containing a solution of CsA-loaded micelle in simulated gastric or intestinal fluid were placed into flasks containing 180 mL of the corresponding simulated fluid. The flasks were shaken at 100 rpm and the temperature was maintained at 37°C during 8 h. At specific time intervals, 10 mL-aliquots were taken from the release medium (dialysate) and replaced by the corresponding fresh simulated fluid in order to keep the system under sink conditions. At the end of the experiment, the dialysis bags were cut open and their content was allowed to leak into the release medium. An aliquot of this solution was sampled to determine the concentration corresponding to 100 % release. The aliquots were freeze-dried. The CsA content of the residue isolated was assayed by the HPLC method described above. Release of free CsA was also performed as a control. All stability tests were performed in triplicates; the data presented are the mean ± S.D.

Cell Culture

The human colon adenocarcinoma cells, Caco-2, were routinely maintained in Dulbecco's modified Eagle medium, supplemented with 10% (v/v) heat-inactivated fetal bovine serum, 1% (v/v) non-essential amino acids and 1% (v/v) penicillin-streptomycin antibiotics solution. Cells were allowed to grow in a monolayer culture at 37 °C, 5% CO_2 and 90% relative humidity.

Cytotoxicity Assay

The Caco-2 cell viability in the presence of increasing concentrations (0 – 10 g/L) of HPC, POE-C_{16} or HPC-g-POE-C_{16} was evaluated using the MTT colorimetric assay (33,34) at 570 nm. The assay is based on the reduction of MTT by mitochondria in viable cells to water insoluble formazan crystals.

CsA Transport Study

Absorption of orally administered drugs, a major determinant of bioavailability, is mainly controlled by two key factors; drug solubility in the intestinal lumen and its permeation across the intestinal barrier. In order to assess, whether polymeric micelles promote drug bioavailability it is necessary to quantify the transport of the micelles through the intestinal barrier (35). To model the in-vivo situation, we investigated the permeability across Caco-2 cells, since a strong correlation was observed between in vivo human absorption and in vitro permeation across Caco-2 cells for a variety of compounds (36). CsA transport across Caco-2 cells was assayed by liquid scintillation counting. Briefly, cells were grown in Transwell dishes until a tight monolayer was formed as measured by transepithelial electrical resistance (TEER). The integrity of the monolayers following the transport experiments was evaluated similarly. Free or micelle-loaded CsA was added at a final concentration of 1 μM to the apical (AP) compartments and aliquots were withdrawn from the basal (BL) chamber at predetermined time points. After sample withdrawal, an equivalent volume of the transport medium (Hank's buffered salt solution, HBSS) was added to the receiving compartment to keep the receiver fluid volume constant. The intestinal efflux caused by P-gp acts as an absorption barrier to limit the oral bioavailability of hydrophobic drugs from the gastrointestinal tract (37). Since, the efflux transporter protein, P-gp, is expressed at the apical side of the Caco-2 monolayer and is not expressed at the basolateral side (38), Pluronic P85® (P85) was added as a P-glycoprotein inhibitor (PGI) to the apical (AP) compartment. Caco-2 monolayers were then solubilized in 1% Triton X-100, and aliquots were taken for determining protein content using the Pierce BCA method.

HM-HPC Transport Study

The polymer permeation across Caco-2 cells was determined by a fluorescence assay using the intrinsic fluorescence of fluorescein labelled HPCg-PEO. Solutions of CsA-loaded fluorescein-labelled polymeric micelles in HBSS transport medium (24×10^{-3} mg/ml) were placed within the AP or BL (donor) compartments. For permeability studies in the presence of a P85 solution (30 μM) in the transport medium was placed on the AP side of the cell monolayers. The transport of CsA and polymer was allowed to proceed for 4 h under the same conditions as above. The fluorescence intensity of the solutions recovered from the AP and/or BL (receiver) compartments was determined. Samples were excited at λ_{ex} = 493 nm, and the emission intensity was monitored at λ_{em} = 519 nm. The amount of transported polymer was calculated using a predetermined standard curve and calibrated with the protein content of the cells in each well, as described above.

Statistical analysis

All experiments were performed in triplicate; the data presented are the mean ± S.D., standardized on individual well protein concentrations. The differences between the mean values were analyzed for significance using ANOVA test. Results were considered statistically significant from the control when $P<0.05$.

Results and Discussion

Synthesis of the Polymers

A initial relative concentration of HPC and POE-C_{16}-Tos, was selected to prepare a HM-HPC copolymer carrying between 5 and 6 mol % POE-C_{16}

(relative to the number of glucose units). Previous studies told us that this level of amphiphilic graft incorporation was such that HM-HPC is soluble in water in the form of polymeric micelles. Samples oif HM-HPC with higher levels of modification were not soluble in water. The synthesis took place in excellent yield and led to a polymer of high purity. Note that the POE-C_{16} residues are linked to HPC chains *via* ether linkages, functional groups that are resistant to temperature and pH changes, and rather inert towards enzymatic degradation in biological fluids (*39,40,41*). This issue was an important one to consider for the modified polymers in order to ensure their stability in the gastrointestinal environment.

The resulting copolymer was further labeled with 5.7 x 10^{-6} mol DTAF/g polymer to assess the permeability of the polymeric micelles across Caco-2 cell monolayers (Table 1.).

Table 1. Characteristics of the HPC-*g*-POE-C_{16} copolymers

Sample	*Grafted POE-C_{16}*[a] *(mol %)*	*DTAF*[b] *(mol DTAF/g polymer)*	*CAC*[c] *(mg/L)*	*POE-C_{16} at CAC (x10^6 mol/L)*
POE-C_{16}	-	-	4.3 ± 1[d]	6.3 ± 1.4[d]
HPC-*g*-POE-C_{16}	5.4 ± 0.5		17 ± 2	2.6 ± 0.3
DTAF-HPC-*g*-PEO-C_{16}	5.4 ± 0.5	5.7 x 10^{-6}	17 ± 2	2.6 ± 0.3

[a]Determined by ^{1}H-NMR measurement; [b]Determined by UV-Vis spectrophotometry; [c]Determined by changes of I_{336nm}/I_{333nm} ratio of pyrene fluorescence with polymer concentration (25 °C, water); [d]These values refer to the critical micelle concentration (CMC) of free POE-C_{16} surfactant.

Micellar Properties of the Polymer

In aqueous solution, HM-HPC samples form polymeric micelles. The major driving force for the assembly of amphiphilic copolymers in water is the removal of hydrophobic fragments from the aqueous surroundings in the hydrophobic core of micelles carrying a shell of highly hydrated hydrophilic chains exposed into water and acting as steric stabilizers to ensure the colloidal stability of he micelles (*42*).

The onset of micellization takes place in solutions of very low polymer concentration (~ 17 mg/L, as determined by a fluorescence spectroscopy assay). We note that the CAC value for the HPC graft copolymer, reported in units of hexadecyl group concentration, is significantly lower than the critical micelle concentration (CMC) of POE-C_{16} (Table 1). The low HM-HPC concentration corresponding to micelle formation ensures their stability againd the dilution encounterd upon oral ingestion.

The size of the micelles is another key parameter to consider in designing drug delivery systems. For particulate drug formulations, size is one of the key factors determining the extent of drug absorption and much has been debated on the optimal size of micro- and nano-particles in relation to their uptake by the intestine. It is generally assumed that uptake is inversely proportional to particle size, and most published data support this hypothesis (43). However, several studies on the uptake of nanoparticles, such as dendrimers 2 to 5 nm in diameter, point to the possible existence of an optimal colloidal size for the efficient entrapment of particles in the mucous and subsequent transport through intestinal epithelial cells (44). The HPC-g-POE-C_{16} micelles investigated are substantially larger than dendrimers, but their size (Table 2) remained within the range considered to be ideal for mucosal uptake.

Table 2. Characteristics of HPC-g-POE-C_{16} Micelles

Sample	Maximum CsA loading[a] (% w/w)	Mean diameter[b] (nm ± SD)	
		CsA-free micelles	CsA-loaded micelles
POE-C_{16}	17.5 ± 0.5	-	-
HPC	1.3 ± 0.1	-	
HPC-g-POE-C_{16}	5.5 ± 0.6	76 ± 2	55 ± 1

[a]Determined by HPLC analysis with UV detection at 210 nm; [b]Determined by DLS measurements with a scattering angle of 90° (25 °C, water).

Stability of CsA loaded micelles in simulated biological fluids.

The design of an oral drug delivery system should include stability testing in simulated gastric fluid (pH 1.2) and intestinal fluid (pH 6.8). In order to evaluate the stability of the HPC-micelles in simulated GI fluids, the release rates of CsA from polymeric micelles in simulated biological fluids were monitored by an *in vitro* release assay, in which CsA-loaded micelles (5.5 w/w %) captured in dialysis bags were placed in contact with simulated fluids during 8 h. The membrane allowed permeation of free drug present in equilibrium with the CsA loaded in micelles in the releasing cell, but not to the CsA-loaded micelles themselves. An identical amount of free CsA was placed in contact with the fluids, serving as control. The amount of free CsA in the dialysate was monitored as a function of contact time.

The release data recorded for each type of fluid present the same features: a small fraction of CsA (4 %) was released from the micelles after 4 h; but this amount was much lower than that recorded in control experiments (~ 85 %), indicating that the micellar system is quite stable towards drug release during the average residence time period of the macromolecular carriers in the GIT before reaching the systemic circulation. Compared to various nanocarrier systems, such as niosomes and liposomes (45,46,47), HM-polysaccharide based micelles represent a significant improvement by eliminating physical stability problems, and may therefore offer improved bioavailability of poorly soluble drugs. A possible explanation for the high stability of these micelles in simulated biological fluids of acid pH, is that the POE-C$_{16}$ residues are linked to the polysaccharide backbone through ether linkages, which, unlike commonly-used ester linkages, are stable towards pH changes and enzymatic degradations.

Zuccari et al. (48) reported that drug release from the micelles takes place slowly over time, due to the drug partition between the micelles and the aqueous solution, a dynamic process, which promotes drug release from the micelles when the free drug present in solution is continuously removed by the absorption *in vivo*.

Cytotoxicity of HM DEX and HPC Copolymers

From previous reports, we knew that while HPC presents no toxicity (4,49), free POE-C$_{16}$ inhibits cell growth due to its surfactant properties which affect membrane integrity (50). It was important to assess if linking POE-C$_{16}$ chains to a polysaccharide framework would alleviate their toxicity. Thus, we carried out tests of the toxicity of HPC, POE-C$_{16}$ and HPC-g-POE-C$_{16}$ towards human intestinal epithelial cells (Caco-2 cells), which are widely used to investigate the intestinal absorption mechanisms of drugs (51). As expected, HPC shows no toxicity at concentrations as high as 10 g/L, whereas POE-C$_{16}$ inhibits cell growth, when added to cells at concentrations as low as 0.5 g/L. Like HPC, the modified polysaccharide, HPC-g-POE-C$_{16}$, exhibits no significant cytotoxicity at a concentration as high as 10 g/L (data not shown) (26). These results confirm 1) that upon linking to a polymer chain, POE-C$_{16}$ looses its cytotoxicity, and 2) that the polymer purification method efficiently removed any free POE-C$_{16}$ from the polymer.

Incorporation of CsA in Polymeric Micelles

The CsA incorporation within polymeric micelles, expressed in w/w % (CsA/polymer), reached ~5.5 %, as evidenced in Figure 2, where we present the CsA loading in the polymeric micelles, as a function of the initial CsA/copolymer ratio in the case of HPC-g-POE-C$_{16}$ as well as unmodified HPC.

Note the non-negligible affinity of CsA to unmodified HPC This trend is consistent with the inherent hydrophobicity of HPC ascribed to the presence of isopropoxy substituents.

Permeability Studies

Recently, the use of Caco-2 cell monolayers has emerged as a leading method to investigate absorption mechanisms of several classes of potential drugs in the early development stages of formulations for oral delivery vehicles (51,52,53,54,55,56,57). In order to gain insight into the absorption of solubilized CsA from the GI tract into the blood stream, the bi-directional transport of CsA

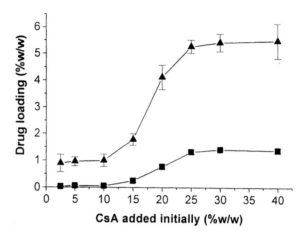

Figure 2. CsA final loading (w/w %) in micelles of (▲)HPC-g-POE-C$_{16}$ as a function of the initial CsA/ copolymer weight ratio w/w %). For comparison, CsA was incorporated in (■) unmodified HPC polymer. Mean ± SD (n = 3).

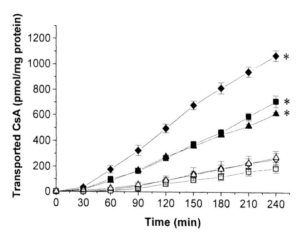

Figure 3. Amount of CsA (pmol/mg protein) transported across Caco-2 cell monolayers after 240 min-incubation with free CsA (open symbols) and CsA-loaded HPC micelles (closed symbols) in the AP-BL direction in absence (triangles) and presence (diamonds) of P85, as well as in the BL-AP direction (squares). Mean ± S.D. (n = 3). (*) Statistically significant compared to free CsA.

entrapped in polymeric micelles across Caco-2 cell monolayers was assayed by liquid scintillation counting. As shown in Figure 3, the transport of CsA across Caco-2 monolayers is significantly enhanced when it is incorporated in polymeric micelles,: The AP to BL permeation of CsA across Caco-2 cells increased by a factor of 3 when incorporated in HPC-g-POE-C$_{16}$ micelles compared to that of free CsA, after an incubation time of 240 min.

It has been demonstrated that, for compounds that are substrates of P-gp, the use of a P–gp inhibitor resulted in a better estimate of absorption in humans (36). Natural or synthetic fatty acid ester based surfactants, such as polysorbates and solutol, have been investigated and were found to inhibit P-gp mediated drug efflux (58,59). Recently, amphiphilic triblock copolymers of poly(ethylene oxide)-b-poly(propylene oxide)-b-poly(ethylene oxide), also known as Pluronic block copolymers, have been shown to enhance cellular accumulation, membrane permeability, and to modulate multidrug resistance of numerous P-gp substrates (60,61,62). In the present study, we chose to use Pluronic P85 as P-gp inhibitor since it has been recently approved by the US Food and Drug Administration (FDA) (63). Pluronic P85 has been observed to block P-gp mediated efflux in Caco-2 and bovine brain microvessel endothelial cells (64) which suggested that this agent may be useful for formulations to enhance oral and brain absorption. In the presence of P85, CsA transport was significantly enhanced in the AP-BL direction (Figure 3.), in agreement with recent reports on the major role of the P-gp efflux mechanism in determining CsA transport across Caco-2 cells (65).

Next, we investigated the bidirectional permeation acrossCaco-2 cell monolayers of the host polysaccharide, using a fluorescein-labelled copolymer in order to assess whether CsA is transported across the Caco-2 cells in free form or entrapped within micelles. Indeed, the polymer is transported across Caco-2 monolayers, as indicated by the detection of fluorescence in the receiver compartment. Under all circumstances (AP-BL, BL-AP, without P85, with P85) the amount of transported HPC-g-PEO-C$_{16}$ was high (Figure 4). The permeability of HM-HPC was higher in the BL-AP direction, compared to the AP-BL permeability. These results are comparable to those reported for the in vitro permeability of polyamidoamine (PAMAM) water soluble dendrimers (66,67,68). It should be pointed out, though, that this assay allowed us to detect and quantify the amount of polymer transported across a Caco-2 cell membrane, but gives no indication on the preserved integrity of the polymeric micelles as they cross the membrane.

Conclusions

In this study, an optimized polysaccharide-based polymeric micelle formulation is proposed as carrier for the oral delivery of poorly soluble therapeutic agents such as CsA. By coupling amphiphilic groups to a water–soluble polysaccharide,

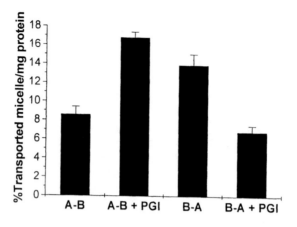

Figure 4. Permeation of DTAF-HPC-g-PEO-C$_{16}$ across Caco-2 monolayers following 4h incubation with Caco-2 cells. Mean ± S.D. A: apical, B: basal; PGI: P-gp inhibitor pluronic P85

the solubilizing power of hydroxypropylcellulose towards CsA was significantly improved. The permeation enhancement upon entrapment of the drug within polysaccharide-based micelles across model intestinal membranes was demonstrated in the case of CsA. Initial results indicate that this effect is more pronounced in the case of HPC, compared to other HM-polysaccharides, such as HM-dextran (69), possibly in view of the inherent bioadhesive properties of HPC. In summary, polysaccharide-based micelles offer unique opportunities for the oral delivery of poorly-water soluble drugs, in vue of intrinsic characteristics: (1) their small size and very low onset of micellization, (2) their high encapsulation capacity, (3) their stability in simulated biological fluids, (4) their lack of toxicity towards intestinal epithelial cells, and (5) their ability to enhance the permeation of entrapped drugs across the intestinal barrier *versus* free drug. The use of this delivery approach can, in principle, modulate both the pharmacokinetic behaviour and bioavailability of drugs, resulting in an overall increase of the drug therapeutic index.

Acknowledgements

This work was financially supported by the Natural Sciences and Engineering Research Council of Canada under its strategic grants program. M.F. Francis acknowledges a scholarship from the Rx&D Health Research Foundation (HRF)/Canadian Institutes of Health Research (CIHR).

References

1. Alvarez-Lorenzo, C., Gomez-Amoza, J. L., Martinez-Pacheco, R., Souto, C. and Concheiro, A. Interactions between hydroxypropylceluloses and vapour/liquid water. *Eur. J. Pharm. Biopharm.* **2000**, *50*, 307-318.
2. Winnik, F. M. Effect of temperature on aqueous solutions of pyrene-labeled (hydroxypropyl)cellulose. *Macromolecules* **1987**, *20*, 2745-2750.
3. Final report on the safety assessment of hydroxyethylcellulose, hydroxypropylcellulose, methylcellulose, hydroxypropyl methylcellulose and cellulose gum. *J. Am. Coll. Toxicol.* **1986**, *5*, 1-60.
4. Obara, S., Muto, H., Kokubo, H., Ichikawa, N., Kawanabe, M. and Tanaka, O. Primary dermal and eye irritability tests of hydrophobically modified hydroxypropyl methylcellulose in rabbits. *J. Toxicol. Sci.* **1992**, *17*, 21-29.
5. FAO/WHO Evaluation of certain food additives and contaminants: thirty-fifth report of the joint FAO/WHO expert committee on food additives. *Tech. Rep. Ser. Wld. Hlth. Org.* **1990**, *No. 789.*
6. Joneja, S. K., Harcum, W. W., Skinner, G. W., Barnum, P. E. and Guo, J. H. Investigating the fundamental effects of binders on pharmaceutical tablet performance. *Drug Dev. Ind. Pharm.* **1999**, *25*, 1129-1135.
7. Lee, J. W., Park, J. H. and Robinson, J. R. Bioadhesive-based dosage forms: the next generation. *J. Pharm. Sci.* **2000**, *89*, 850-866.
8. Repka, M. A. and McGinity, J. W. Bioadhesive properties of hydroxypropylcellulose topical films produced by hot-melt extrusion. *J. Control. Release* **2001**, *70*, 341-351.
9. Machida, Y. and Nagai, T. Directly compressed tablets containing hydroxypropyl cellulose in addition to starch or lactose. *Chem. Pharm. Bull.* **1974**, *22*, 2346-2351.
10. Alderman, D. A. A review of cellulose ethers in hydrophilic matrices for oral controlled-release dosage forms. *Int. J. Pharm. Tech. Prod. Mfr.* **1984**, *5*, 1-9.
11. Skinner, G. W., Harcum, W. W., Barnum, P. E. and Guo, J. H. The evaluation of fine particle hydroxypropylcellulose as a roller compaction binder in pharmaceutical applications. *Drug. Dev. Ind. Pharm.* **1999**, *25*, 1121-1128.
12. Huber, H. E., Dale, L. B. and Christenson, G. L. Utilization of hydrophilic gums for the control of drug release from tablet formulations. I. Disintegration and dissolution behavior. *J. Pharm. Sci.* **1966**, *55*, 974-976.
13. Al-Hmoud, H. L., Efentakis, M. and Choulis, N. H. A controlled release matrix using a mixture of hydrophilic and hydrophobic polymers. *Int. J. Pharm.* **1991**, *68*, R1-R3.
14. Rao, K. V. R., Devi, K. P. and Buri, P. Cellulose matrices for zero-order release of soluble drugs. *Drug Dev. Ind. Pharm.* **1988**, *14*, 2299-2320.

15. Skoug, J. W., Mikelsons, M. V., Vigneron, C. N. and Stemm, N. L. Qualitative evaluation of the mechanism of release of matrix sustained release dosage forms by measurement of polymer release. *J. Control. Release* **1993**, *27*, 227-245.

16. Ebube, N. K. and Jones, A. B. Sustained release of acetaminophen from a heterogeneous mixture of two hydrophilic non-ionic cellulose ether polymers. *Int. J. Pharm.* **2004**, *272*, 19-27.

17. Pham, A. T. and Lee, P. I. Probing the mechanisms of drug release from hydroxypropylmethyl cellulose matrices. *Pharm. Res.* **1994**, *11*, 1379-1384.

18. Elkheshen, S., Yassin, A. E. and Alkhaled, F. Per-oral extended-release bioadhesive tablet formulation of verapamil HCl. *Boll. Chim. Farm.* **2003**, *142*, 226-231.

19. Hosny, E. A., Elkheshen, S. A. and Saleh, S. I. Buccoadhesive tablets for insulin delivery: in-vitro and in-vivo studies. *Boll. Chim. Farm.* **2002**, *141*, 210-217.

20. Celebi, N. and Kislal, O. Development and evaluation of a buccoadhesive propranolol tablet formulation. *Pharmazie* **1995**, *50*, 470-472.

21. Genc, L., Oguzlar, C. and Guler, E. Studies on vaginal bioadhesive tablets of acyclovir. *Pharmazie* **2000**, *55*, 297-299.

22. Laupacis, A., Keown, P., Ulan, R., McKenzie, N. and Stiller, C. Cyclosporin A: A powerfull immunosuppressant. *Can. Med. Assoc. J.* **1982**, *126*, 1041-1046.

23. Ueda, K., Sakagami, H., Ohtaguro, K. and Masui, Y. Studies on the retention of the mucous-membrane-adhesive anticancer agent hydroxypropylcellulose doxorubicin. *Eur. Urol.* **1992**, *21*, 250-252.

24. Eiamtrakarn, S., Itoh, Y., Kishimoto, J., Yoshikawa, Y., Shibata, N., Murakami, M. and Takada, K. Gastrointestinal mucoadhesive patch system (GI-MAPS) for oral administration of G-CSF, a model protein. *Biomaterials* **2002**, *23*, 145-152.

25. Klug, E. D. Some properties of water-soluble hydroxyalkyl celluloses and their derivatives. *J. Polymer Sci.: PART C* **1971**, *36*, 491-508.

26. Francis, M. F., Piredda, M. and Winnik, F. M. Solubilization of poorly water soluble drugs in micelles of hydrophobically modified hydroxypropylcellulose copolymers. *J. Control. Release* **2003**, *93*, 59-68.

27. Francis, M. F., Piredda, M., Cristea, M. and Winnik, F. M. Synthesis and evaluation of hydrophobically-modified polysaccharides as oral delivery vehicles of poorly-water soluble drugs. *Polymeric Materials: Science & Engineering* **2003**, *89*, 55-56.

28. Blakeslee, D. Immunofluorescence using dichlorotriazinylaminofluorescein (DTAF). II. Preparation, purity and stability of the compound. *J. Immunol. Methods* **1977**, *17*, 361-364.

29. Hofland, H. E. J., Bouwstra, J. A., Verhoef, J. C., Buckton, G., Chowdry, B. Z., Ponec, M. and Junginger, H. E. Safety aspects of non-ionic surfactant vesicles-a toxicity study related to the physicochemical characteristics of non-ionic surfactants. *J. Pharm. Pharmacol.* **1992**, *44*, 287-294.

30. Jevprasesphant, R., Penny, J., Attwood, D., McKeown, N. B. and D'Emanuele, A. Engineering of dendrimer surfaces to enhance transepithelial transport and reduce cytotoxicity. *Pharm. Res.* **2003**, *20*, 1543-1550.

31. Zhao, C. L., Winnik, M. A., Riess, G. and Croucher, M. D. Fluorescence probe techniques used to study micelle formation in water-soluble block copolymers. *Langmuir* **1990**, *6*, 514-516.

32. Francis, M. F., Cristea, M., Winnik, F. M. and Leroux, J. C. Dextran-*g*-Polyethyleneglycolcetyl Ether Polymeric Micelles For Oral Delivery of Cyclosporin A. *Proceed. Intern. Symp. Control. Rel. Bioact. Mater.* **2003**, *30*, 68-69.

33. Mosmann, T. Rapid colorimetric assay for cellular growth and survival: Application to proliferation and cytotoxicity assays. *J. Immunol. Methods* **1983**, *65*, 55-63.

34. Hansen, M. B., Nielsen, S. E. and Berg, K. Re-examination and further development of a precise and rapid dye method for measuring cell growth/cell kill. *J. Immunol. Methods* **1989**, *119*, 203-210.

35. Audus, K. L., Bartel, R. L., Hidalgo, I. J. and Borchardt, R. T. The use of cultured epithelial and endothelial cells for drug transport and metabolism studies. *Pharm. Res.* **1990**, *7*, 435-451.

36. Yee, S. In vitro permeability across Caco-2 cells (colonic) can predict in vivo (small intestinal) absorption in man: fact or myth. *Pharm. Res.* **1997**, *14*, 763-766.

37. Hunter, J. and Hirst, B. H. Intestinal secretion of drugs. The role of P-glycoprotein and related drug efflux systems in limiting oral drug absorption. *Adv. Drug Deliv. Rev.* **1997**, *25*, 129-157.

38. Hosoya, K. I., Kim, K. J. and Lee, V. H. Age-dependent expression of P-glycoprotein gp170 in Caco-2 cell monolayers. *Pharm. Res.* **1996**, *13*, 885-890.

39. Sovak, M. and Ranganathan, R. Stability of nonionic water-soluble contrast media: implications for their design. *Invest. Radiol.* **1980**, *15*, S323-S328.

40. Cavallaro, G., Pitarresi, G., Licciardi, M. and Giammona, G. Polymeric prodrug for release of an antitumoral agent by specific enzymes. *Bioconjug. Chem.* **2001**, *12*, 143-151.

41. Oishi, M., Sasaki, S., Nagasaki, Y. and Kataoka, K. pH-responsive oligodeoxynucleotide (ODN)-poly(ethylene glycol) conjugate through acid-labile beta-thiopropionate linkage: preparation and polyion complex micelle formation. *Biomacromolecules* **2003**, *4*, 1426-1432.

42. Gao, Z. and Eisenberg, A. A model of micellization for block copolymers in solution. *Macromolecules* **1993**, *26*, 7353-7360.
43. Carr, K. E., Hazzard, R. A., Reid, S. and Hodges, G. M. The effect of size on uptake of orally administered latex microparticles in the small intestine and transport to mesenteric lymph nodes. *Pharm. Res.* **1996**, *13*, 1205-1209.
44. Florence, A. T. and Hussain, N. Trancytosis of nanoparticle and dendrimer delivery systems: evolving vistas. *Adv. Drug Deliv. Rev.* **2001**, *50*, S69-S89.
45. Hu, C. and Rhodes, D. G. Proniosomes: a novel drug carrier preparation. *Int. J. Pharm.* **1999**, *185*, 23-35.
46. Ozpolat, B., Lopez-Berestein, G., Adamson, P., Fu, C. J. and Williams, A. H. Pharmacokinetics of intravenously administered liposomal all-trans-retinoic acid (ATRA) and orally administered ATRA in healthy volunteers. *J. Pharm. Pharm. Sci.* **2003**, *6*, 292-301.
47. Manjunath, K., Reddy, J. S. and Venkateswarlu, V. Solid lipid nanoparticles as drug delivery systems. *Methods Find. Exp. Clin. Pharmacol.* **2005**, *27*, 127-144.
48. Zuccari, G., Carosio, R., Fini, A., Montaldo, P. G. and Orienti, I. Modified polyvinylalcohol for encapsulation of all-trans-retinoic acid in polymeric micelles. *J. Control. Release* **2005**, *103*, 369-380.
49. Couch, N. P. The clinical status of low molecular weight dextran: a critical review. *Clin. Pharmacol. Ther.* **1965**, *6*, 656-665.
50. Dimitrijevic, D., Shaw, A. J. and Florence, A. T. Effects of some non-ionic surfactants on transepithelial permeability in Caco-2 cells. *J. Pharm. Pharmacol.* **2000**, *52*, 157-162.
51. Krishna, G., Chen, K. J., Lin, C. C. and Nomeir, A. A. Permeability of lipophilic compounds in drug discovery using in-vitro human absorption model, Caco-2. *Int. J. Pharm.* **2001**, *222*, 77-89.
52. Delie, F. and Werner, R. A human colonic cell line sharing similarities with enterocytes as a model to examine oral absorption: advantages and limitations of the Caco-2 model. *Crit. Rev. Ther. Drug Carrier Syst.* **1997**, *14*, 221-286.
53. Hidalgo, I. J., Raub, T. J. and Borchardt, R. T. Characterization of the human colon carcinoma cell line (Caco-2) as a model system for intestinal epithelial permeability. *Gastroenterology* **1989**, *96*, 736-749.
54. Artursson, P., Palm, K. and Luthman, K. Caco-2 monolayers in experimental and theoretical predictions of drug transport. *Adv. Drug Deliv. Rev.* **2001**, *46*, 27-43.
55. Artursson, P. and Borchardt, R. T. Intestinal drug absorption and metabolism in cell cultures: Caco-2 and beyond. *Pharm. Res.* **1997**, *14*, 1655-1658.

56. Kamm, W., Jonczyk, A., Jung, T., Luckenbach, G., Raddatz, P. and Kissel, T. Evaluation of absorption enhancement for a potent cyclopeptidic alpha(nu)beta(3)-antagonist in a human intestinal cell line (Caco-2). *Eur. J. Pharm. Sci.* **2000**, *10*, 205-214.
57. Faassen, F., Kelder, J., Lenders, J., Onderwater, R. and Vromans, H. Physicochemical properties and transport of steroids across Caco-2 cells. *Pharm. Res.* **2003**, *20*, 177-186.
58. Woodcock, D. M., Linsenmeyer, M. E., Chojnowski, G., Kriegler, A. B., Nink, V., Webster, L. K. and Sawyer, W. H. Reversal of multidrug resistance by surfactants. *Br. J. Cancer* **1992**, *66*, 62-68.
59. Nerurkar, M. M., Burton, P. S. and Borchardt, R. T. The use of surfactants to enhance the permeability of peptides through Caco-2 cells by inhibition of an apically polarized efflux system. *Pharm. Res.* **1996**, *13*, 528-534.
60. Seeballuck, F., Ashford, M. B. and O'Driscoll, C. M. The effects of Pluronics block copolymers and Cremophor EL on intestinal lipoprotein processing and the potential link with P-glycoprotein in Caco-2 cells. *Pharm. Res.* **2003**, *20*, 1085-1092.
61. Batrakova, E. V., Han, H. Y., Miller, D. W. and Kabanov, A. V. Effects of pluronic P85 unimers and micelles on drug permeability in polarized BBMEC and Caco-2 cells. *Pharm. Res.* **1998**, *15*, 1525-1532.
62. Batrakova, E., Lee, S., Li, S., Venne, A., Alakhov, V. and Kabanov, A. Fundamental relationships between the composition of pluronic block copolymers and their hypersensitization effect in MDR cancer cells. *Pharm. Res.* **1999**, *16*, 1373-1379.
63. BASF Pluronic P85 block copolymer surfctant. **2002**.
64. Batrakova, E. V., Li, S., Miller, D. W. and Kabanov, A. Pluronic P85 increases permeability of a broad spectrum of drugs in polarized BBMEC and Caco-2 cell monolayers. *Pharm. Res.* **1999**, *16*, 1366-1372.
65. Augustijns, P., Bradshaw, T. P., Gan, L. S., Hendren, R. W. and Thakker, D. R. Evidence for a polarized efflux system in Caco-2 cells capable of modulating cyclosporin A transport. *Biochem. Biophys. Res. Commun.* **1993**, *197*, 360-365.
66. Wiwattanapatapee, R., Carreno-Gomez, B., Malik, N. and Duncan, R. Anionic PAMAM dendrimers rapidly cross adult rat intestine in vitro: a potential oral delivery system? *Pharm. Res.* **2000**, *17*, 991-998.
67. El-Sayed, M., Ginski, M., Rhodes, C. and Ghandehari, H. Transepithelial transport of poly(amidoamine) dendrimers across Caco-2 cell monolayers. *J. Control. Release* **2002**, *81*, 355-365.
68. El-Sayed, M., Rhodes, C. A., Ginski, M. and Ghandehari, H. Transport mechanism(s) of poly (amidoamine) dendrimers across Caco-2 cell monolayers. *Int. J. Pharm.* **2003**, *265*, 151-157.
69. Francis, M. F., Cristea, M., Yang, Y. and Winnik, F. M. Engineering polysaccharide-based polymeric micelles to enhance permeability of cyclosporin A across Caco-2 cells. *Pharm. Res.* **2005**, *22*, 209-219.

Starch

Chapter 4

Contramid®: High-Amylose Starch for Controlled Drug Delivery

François Ravenelle and Miloud Rahmouni

Labopharm Inc., 480 Boulevard Armand-Frappier, Laval, Québec H7V 4B4, Canada

Contramid® is a novel pharmaceutical excipient obtained by chemically and physically modifying a high-amylose corn starch. When a compressed tablet of Contramid® powder is placed in an aqueous medium, the material forms a hard gel displaying sponge-like viscoelastic properties. When observed by x-ray tomography, the gel reveals a membrane at its surface which governs the slow release of medication. Basic starch properties responsible for Contramid®'s slow release behavior are reviewed and used to explain the mechanism of self-assembly behind the controlled drug delivery technology. Characteristics of Contramid® such as diffusion and enzymatic susceptibility are also presented.

Introduction

The first publication on the Contramid® technology was reported by Lenaerts et al in 1991 (1). Starting from the well-understood principle postulated by Flory in 1953 (2) by which one can limit the swelling of a network by increasing its crosslink density, modified starches of different crosslinking degrees were prepared and theophylline release rate were evaluated. In doing so, the researchers attempted to prepare a crosslinked network of amylose to form a hydrogel able to control the release of medication, the rate of which would be a function of the crosslinking degree. Typically, by increasing the degree of crosslinking, one would create a tetrafunctional polymer network with a higher density of crosslinks and thus reduce its swelling capacity. It was soon discovered that crosslinking the natural polymer yielded a nonlinear response for swelling capacity and drug release (3,4,5,6). In fact, for higher crosslinking degrees, higher swelling capacity and faster drug delivery were observed. At moderate crosslinking degrees, the matrix efficiently sustained release of medications for 12-24 hours. This peculiar phenomenon sparked the interest of researchers and Contramid® was born (7).

Contramid® is a hydroxypropyl and crosslinked high-amylose starch pharmaceutical excipient used to prepare oral controlled release formulations. A brief review of starch composition and properties will allow the reader to efficiently capture the essence of this technology that relies essentially in harnessing the natural properties of starch. This paper provides an update and expands on the current level of knowledge reported over the years by the original inventors and researchers (8), using results from fundamental starch research to better explain and describe the Contramid® technology. Contramid® is used to prepare pharmaceutical tablets for controlled delivery of medication. This technology has been mainly applied to oral drug delivery but the reader may find interesting to know that Contramid® may also be used to prepare implants. The latter demonstrated excellent biocompatibility, bioabsorption and are effective in controlling subcutaneous delivery of medication for several weeks (9,10,11,12).

Rationale for Controlled Oral Delivery

Soon after their discovery, drugs are formulated in their fastest, most convenient form for rapid access to market. With time however, their reformulation to the most convenient form for patients, posology and treatment is a trend that is undeniably logic-based. The significant increase in new products based solely on new formulations using existing or new technologies is a trend that has given birth to a new face of the pharmaceutical industry: drug delivery companies. These companies generally use proprietary technologies to prepare value-added formulation of new or existing drugs. For drug delivery companies, oral delivery is a very competitive market as it represents the least invasive formulation, coupled with the highest patient compliance. However, this is not the end of the line, as prescriptions often require several tablets a day. Taking multiple doses of the same medication causes a series of highs and lows of the drug concentration in the blood, hitting the therapeutic window with alternating periods of limited efficacy and overdose (Figure 1). There is also the risk of forgetting to take one of the recommended doses, the result of which may alter treatment duration, cost and potentially the patient's health and safety. As presented in Figure 1, oral controlled release technologies such as Contramid® help to overcome these limitations by allowing slow release of the medication during its transit in the gastrointestinal tract in order to reach an efficient drug concentration in the blood and maintain it at therapeutic level for hours. Overall benefits of oral controlled drug delivery includes reduction in drug blood level variations, reduction in dosing frequency, enhanced patient convenience and compliance, reduction in adverse side effects and reduction in healthcare costs.

Cellulose and starch: poly (D-glucopyranose) in Drug delivery

The pharmaceutical industry has a long history of using cellulose and its derivatives as binders and adjuvents for tablet formulations (13,14,15,16). Nowadays, microcrystalline cellulose (MCC) is widely used because of its high compactibility property, i.e. its ability to form hard tablets at low compression force. This hardness is due in part to the numerous hydrogen bonds formed between particles of cellulose microcrystals. However, in tablet formulation, MCC readily disperses in water and instant drug release is obtained. Other cellulose derivatives (e.g. Hydroxypropylmethyl cellulose or HPMC) have been shown to allow controlled drug delivery due to continuously swelling matrix systems, i.e. drug release is governed by the kinetics of swelling. On the other hand, modified starches also possess great compactibility in the dry state, but adds the possibility of forming limited swelling gels to control the release of drugs. Because polysaccharides such as modified celluloses and starches are readily available "starting material" they hold an advantage over other more expensive synthetic polymer systems. In particular, the food industry has been using modified starches for a long time and has paved the way in terms of large-scale manufacturing. Starch is among the most important industrial biopolymers. It is the main nutrient consumed by humans worldwide, it is our main source of energy representing 70-80% of calories ingested (17). It is found in rice, pasta, bread, corn, potatoes and a wide variety of other grains, vegetables and plants. Because starches from various sources have different properties which can then be modified by a large number of processes, there is a staggering variety of applications for starch based materials. Following is a brief overview of important aspects of starch chemistry that play important roles in the preparation and performance of Contramid®.

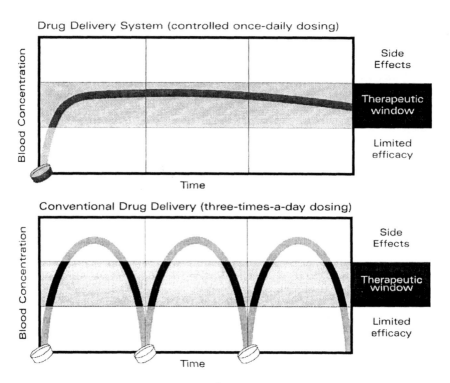

Figure 1. Advantage of Contramid® controlled release technology versus immediate release formulations

Starch Composition

Starch is composed of a mixture of two natural polymers: amylose and amylopectin (Figure 2). Amylose is a linear polymer of D-glucopyrranose units joined by α-1,4 acetal linkages. Depending on its botanical source, molecular weight of amylose strands are generaly found to be less than 0.5 million and have degree of polymerization varying from 1,500 up to 6,000. Amylopectin molecules are much larger with typical molecular weights between 50 to 100 million and degree of polymerization of about 300,000 to 3,000,000 (17,18). Also, as shown in Figure 2, amylopectin has a racemose architecture where α-

1,4-linked linear segments are branched together through α-1,6 linkages. Proportions of amylose and amylopectin vary according to their botanical source. Typically, corn and potato starches will have a 30%(w/w) content of amylose and 70% amylopectin. Contramid® is prepared from a high-amylose corn starch, a hybrid containing 70% amylose and 30% amylopectin.

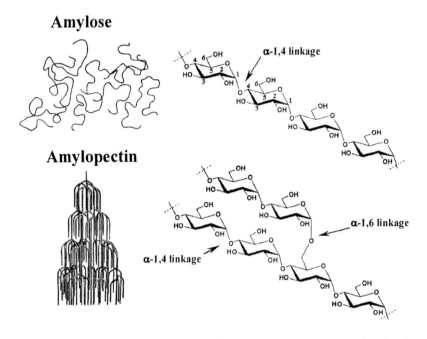

Amylose

Amylopectin

Figure 2. Amylose and amylopectin schematic representations and molecular architecture.

In a native starch granule, amylose and amylopectin are found in intimate contact with each other, randomly interspersed (19). Furthermore, both are organized in a semi-crystalline arrangement where amorphous and crystalline regions are formed by neighboring amylose and amylopectin molecules (20). The hybrid high-amylose starch used to prepare Contramid® is found to exist as three main morphologies: Non-crystalline regions, V-type single helices and B-type double helices. The V-type is a single strand 6_1 helix whose exterior is hydrophilic and interior is a hydrophobic cavity able to complex fatty acids (21) and small organic molecules (22,23). In fact, orientation of the glucose unit in the V-type helix is highly similar to a β-cyclodextrin molecule. This single helix is present in amorphous regions in starch. Crystal arrangements and packing in crystalline regions are made of clusters of double helices arising from the association of

84

neighboring branches of amylopectin molecules, and amylose chains, in such double helices (24). Figure 3 presents molecular models of a V-type single helix and a double helix made of linear segments of starch molecules (all molecular model graphics found in this chapter were graciously provided by Prof. Stefan Immel, Technical University of Darmstadt[1]). More detailed molecular models of both these helical arrangements have previously been published by Immel et al. (25).

Starch Gels

The food industry has for a long time relied on starches for viscosity enhancement and gelling properties. At room temperature, starch granules are insoluble in water and swells relatively little. However, if a starch suspension is heated to temperatures above its gelatinization temperature and/or suspended in alkaline solution, starch undergoes an irreversible transition: gelatinization. Gelatinization is a general term used to describe a series of thermal events that occur in starch granules upon heating or in alkaline aqueous media. Most importantly, during gelatinization, the starch granule's crystalline arrangement is destroyed and the granule swells to its maximum as an amorphous suspension (26). For an in-depth review on the complex subject of gelatinization of starches, please refer to (27). This transition is often referred to as being irreversible. However, if the order found in the native granule may never reform as once nature intended it, the formation of double helices is not entirely compromised. In fact, reformation of these double helices is a common phenomenon in starch science called retrogradation. Retrogradation of starch is a term used to define the changes that occur in gelatinized starch from an initially amorphous state to a more ordered or crystalline state (28). More importantly here, it is the process by which Contramid® self-assembles in water to form a gel.

Swelling of Contramid® tablets in water is presented as a self-assembly process that differs from other swelling polymeric systems: swelling increases with increasing crosslinking degree, which is opposite to typical crosslinked polymer networks (2). This is explained by starch retrogradation. When a tablet of Contramid® swells in water, self-assembly of its polymeric chains into double helices confers to the gel high mechanical strength and limited swelling capacity.

---

[1] *http://caramel.oc.chemie.tu-darmstadt.de/~lemmi*

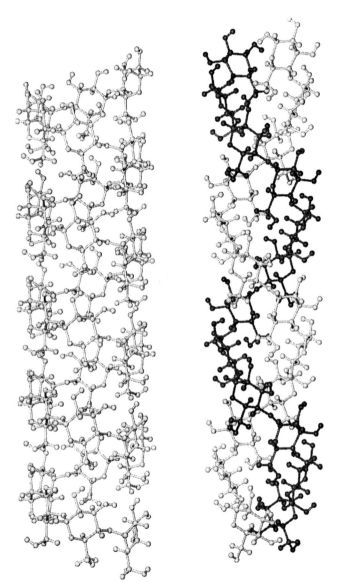

Figure 3. Molecular model representation of the V-type single helix (top) and double helical (bottom) arrangements found in starch.

In other words, the limited swelling and formation of a tridimensional network is caused by the formation of physical or pseudo-crosslinks: double helices and chain entanglement. By increasing the crosslinking degree, one is actually reducing the number of possible double helix formation, and thus oppositely reducing the number of physical crosslinks. Consequently, the tablet is able to swell more and, at an excessive crosslinking degree, no tridimensional network is formed and the tablet completely loses cohesion. This principle was previously observed by Hollinger et al. who measured an increase in water sorption in crosslinked starch (29), and by Moussa and Cartilier who measured an increase in swelling capacity of crosslinked starch tablets (30), with increasing crosslinking degree.

Contramid® Preparation

Contramid® is a crosslinked and hydroxypropylated high-amylose starch based excipient prepared according to a US patent (31). During the first preparation step (Figure 4), high-amylose starch is slurried in a weak sodium hydroxide suspension and heated to ca. 30°C. In these conditions, starch granules are partially swollen by the combined action of heat and sodium hydroxide, but are not gelatinized. At this point, hydroxyl groups on amylose and amylopectin molecules are available for modifications and an alkaline pH assures activation of future chemical reactions. This step is hereby referred to as activation step. Phosphorous oxychloride crosslinking agent is then added to the slurry, followed by propylene oxide. This reaction further functionalizes amylose and amylopectin molecules with hydroxypropyl side chains.

Crosslinking and hydroxypropylation

Crosslinking and hydroxypropylation of starch are both used to improve the stability and hardness of gels obtained from starch materials. Instability of starch-containing products is often associated with retrogradation of starch over time which changes the properties of the intial product. The best example of this is bread staling. Covalent crosslinks and hydroxypropyl side chains allow greater stability in terms of temperature and pH by hindering retrogradation over time (17). Crosslinking also plays the role of "grafting" amylose chains on the giant amylopectin molecules (32). As it will be seen further, both increased stability in the dry state and properties in the swollen state are important factors altered by chemical modifications such as crosslinking and hydroxypropylation.

Spray-Drying

The last step, and not the least important, is the drying of the material. After washing with water, the previous slurry is heated to a very high temperature (ca. 160°C) and gelatinized to an entirely swollen and amorphous state. Drying is then performed by spraying the hot slurry in hot air (>200°C). This results in rapid water evaporation and consequently, Contramid® is trapped in a non-crystalline state. The chemical modifications discussed above and the speed at which water is evaporated effectively prevents retrogradation. Resulting average

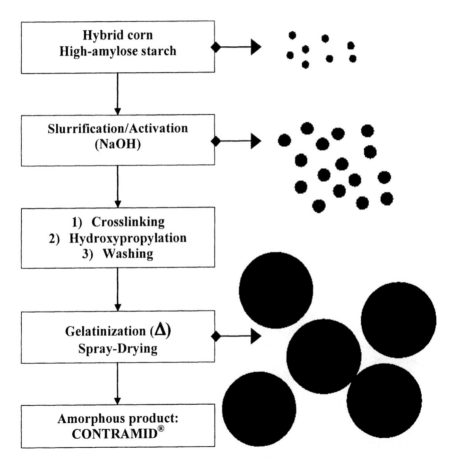

Figure 4. Contramid® preparation flow chart and progression of particle size representation.

particle size is around ten times larger than the starting material's average granule size (Figure 4)

.Safety and regulatory acceptance

Modified starches, such as Contramid®, have a long history of safe use as food additives. Contramid® conforms to the specifications for modified starch under the Food Chemicals Codex and is considered a food additive by the FDA and the equivalent European authorities. Under these regulations, all modified starches are generally regarded as safe for oral administration and generally may be used with orally administered drugs without significant modification to the safety profile of the medication in unlimited quantities. Furthermore, the International Pharmaceutical Excipient Council guidelines indicate that pre-clinical toxicity and safety studies should not be required for the approval of Contramid® as a pharmaceutical excipient.

How Contramid® works

Most of the cited studies in this section were performed on starch suspensions. We believe these events also apply to a directly compressed starch tablet, albeit at a further extent because of the proximity of starch particles that allow more interparticular (intermolecular) interactions. This correlation between systems is also in line with the findings showing that an increase in starch concentration results in an increase of gel viscosity (33). Contramid® is prepared from high-amylose starch that is proven to yield stronger gels. Crosslinking and hydroxypropylation further increase the strength of gels (17). Rahmouni et al. (34) demonstrated that the mechanical strength of Contramid® gels, as measured by the force needed for a probe to penetrate the swollen tablets, was eight-fold greater than that of a widely used cellulose derivative. Ravenelle et al. (35) described swollen Contramid® tablets as sponge-like. In the latter study, a uniaxial unconfined compression of a tablet demonstrated the quasi reversible deformation of the gel. During a step compression, water slowly flows out of the tablet, and upon decompression, the water is sucked back in.

Visually, when a Contramid® tablet is swollen to equilibrium, its shape increases by 60-80% in thickness and 15-35% in width (8,35) depending on tablet preparation (compression force, tablet size). This anisotropy of swelling is simply explained by the flattening of the Contramid® particles in the axial direction during dry compression of the tablet. The flattened particles tend to adopt their initial shape and swell more in the axial direction than in the radial. More importantly, a Contramid® tablet swells to a limit, i.e. it holds its shape, albeit bigger. This shape retention is a very important characteristic that allows for slow release of drugs. The shape retention is given by the propensity of starch to retrograde, i.e. to form double helices. These double helices are responsible for the efficient formation of a tri-dimensional network.

It has been recognized that interaction between amylopectin and amylose contribute directly to the gel properties of their mixtures (33,36,37,38). In order to obtain the necessary viscoelastic properties of a gel formed from a Contramid® tablet, a combination of important factors is considered. In a first instance, lets examine the role of chemical structures and molecular weight. Jane et al. (33) demonstrated how amylose molecular sizes and amylopectin branched chain length affected gel properties by analyzing the viscosity, clarity and strength of gels from different starch suspensions. Because high-amylose corn starches possess smaller amylopectin molecules and longer amylose chains (33), the authors concluded that a higher content of amylose in starches yielded stronger gel strength.

Water quickly penetrates the initially amorphous tablet and acts as a plasticizing agent. This plasticizing effect increases intermolecular space of free volume (39), and thus allows greater mobility to starch molecules which rapidly start reorganizing into their thermodynamically most stable conformations: double helices. While hydrogen bonding with water allows the plasticizing effect to occur, below its gelatinization temperature, hydrophobic interactions between starch molecules drive retrogradation. The reader is advised that it is often wrongly reported in literature that hydrogen bonding is the driving force behind retrogradation. In essence, hydrogen bonding allows for stabilization of the formed double helices. At room temperature, starch is insoluble in water and thus possesses intermolecular interactions (hydrophobic) and water-starch interactions (hydrophilic, plasticizing effect). If hydrogen bonding was the driving force, there would be no better molecule on earth than water to supply it and a solution would be formed, not double helices.

^{13}C Solid State NMR

To investigate retrogradation in swelling tablets, ^{13}C Solid State NMR experiments were performed. According to Veregin et al. (40) and Gidley & Bociek (41), starch crystalline and amorphous morphologies have characteristic resonance in the ^{13}C Solid State NMR spectrum. Using [13]C Cross Polarization Magic Angle Spinning Nuclear Magnetic Resonance (^{13}C CP/MAS NMR) it is thus possible to identify the conformations present in samples of starch by observing the C1 resonance in the region of 100-104 ppm. Previous studies also

90

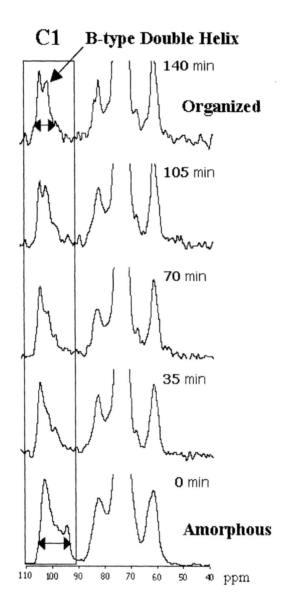

*Figure 5. Retrogradation: Disordered V-type and amorphous Contramid®
immersed in water for the indicated time converts to B-type double helix as
shown by the C1 resonance changes in the ^{13}C CP/MAS NMR spectra.*

looked at this phenomenon by observing changes of spectra for powders exposed to 100% relative humidity for various times (7,42,43). In the present study, microtablets (pellets) of Contramid® were prepared and inserted in a CP/MAS NMR probe and spectra were acquired for different swelling times. For dry tablets, the C1 region is broad and is representative of an amorphous material. Figure 5 presents spectra obtained at different swelling times. The evolution of the C1 resonance with time presented in Figure 5 indicates a transition from amorphous and V-type single helix (broad peak) to a predominant B-type double helix upon hydration of the tablet. The narrowing of the peaks corresponding to V and B phases, as well as the entire spectrum in general, indicates an increase in degree of order.

Differential Scanning Calorimetry (DSC)

Differential Scanning Calorimetry (DSC) is often used to observe and quantify thermal events in aqueous starch suspensions. The most common determination is that of gelatinization temperature which is related to the extent of retrogradation in starch suspensions that have been heated and cooled (44,45). DSC was used to investigate the calorimetric profile of swelling Contramid® tablets. The experimental set up requires the use of small pellets. The latter are swollen for two hours at room temperature in excess water, removed from water, weighed for content of water and placed in a stainless steel hermetic pan for analysis. The two hours allow enough penetration of water to start swelling of the tablet while ensuring that the tablet is not at equilibrium, meaning the retrogradation of Contramid® is not complete and should thus be observed as an exothermic event. The reference cell contains an equivalent amount of water as the one absorbed by the tablet being analyzed. Samples are first brought to 0°C at equilibrium before being ramped at 10°C/minute up to 140°C. A typical thermogram for a Contramid® tablet is presented in Figure 6 (46). It is possible to observe the retrogradation exotherm (at ca. 40°C) caused by the formation of double helices.

From this thermogram we demonstrate that heating the tablet to body temperature (37°C) speeds up the formation of double helices as shown by the retrogradation exotherm in the range of 35-50°C. Although the plasticizing effect of water is present at room temperature, we can conclude that it is optimum at a temperature close to 40°C. It is noteworthy to mention that the exotherm was not present for a tablet swollen for more than 24 hours, supporting the interpretation. As we continue heating the swollen tablet, we observe a broad endotherm around 80°C characteristic for gelatinization temperature range. This is the temperature at which all organization is destroyed; pellets lose cohesion and become a homogenous suspension in water.

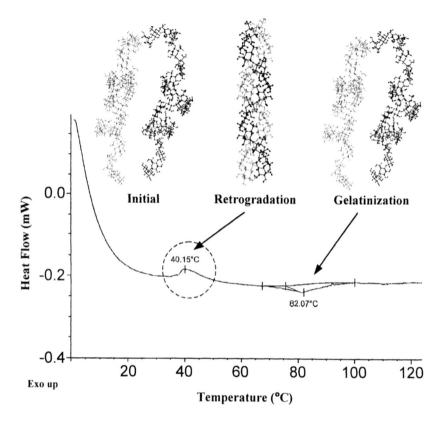

Figure 6. DSC thermogram of a Contramid® tablet swollen for 2 hours in water

In other experiments, Contramid® tablets were swollen at different temperatures (25, 37 and 55°C). After 12 hours of swelling, visual examination revealed that Contramid tablets slightly swelled at 25°C, and form a hard gel. However, the tablets split in two after a few hours due to the lack of cohesion between the swollen particles. At this temperature, the amylose chains have less mobility to self-assemble in a tri-dimensional network and maintain tablet integrity. At 37 and 55°C, tablet integrity is maintained but the gel strength of the swollen tablet, is much higher at 37°C than at 55°C. This means that the self-assembly structure of amylose that confers a high mechanical strength to the Contramid gel and maintains the tablet integrity reaches its maximum organization around 37°C, which concords with the DSC observations.

In-Situ Forming Membrane

Another interesting feature of a swelling Contramid® tablet is that a membrane is formed on the outer surface of the tablet in the first minutes of swelling. The instant retrogradation of the surface of the tablet is visible to the eye. Moussa & Cartilier observed this membrane by optical microscopy and showed that it rapidly forms when the tablet is placed in water, but very slowly thickens over time (30). What is remarkable is that the organization formed at the surface of the tablet in the first few minutes is not propagated to its core; the core maintains a coarse porous texture. Recently, Nuclear Magnetic Resonance Imaging (NMRI) has been used to observe swelling of modified high-amylose starch tablets and the same membrane formation was observed (47,48). Recent results performed with Contramid® tablets using this technique are also presented in this book (cf. Thérien-Aubin and Zhu in this book). In the present chapter we introduce x-ray tomography as a new method for investigating this heterogeneous swelling of Contramid® tablets. These studies were initiated by Marchessault and Chauve (49) at McGill University's Bone and Periodontal Research Center (Montréal, Canada). Similarly to NMRI, x-ray tomography is a non-destructive inspection technique that provides cross-sectional images in planes through a solid. It images relative electron density fluctuations in a material point by point in planes. Furthermore, using a mathematical algorithm, diffraction intensities of the x-ray beam transmitted through the tablet can yield grey level images that are directly proportional to local material density leading to a quantitative measurement of the porosity of the material (50). Using this non-invasive technique, we can image cross-sections of dry and swollen tablets of Contramid® and observe the membrane.

The relative density of planes selected approximately at mid height of both a dry (left) and 24-hour swollen tablet (right) are presented in Figure 7 (49). Tablets were made using a direct compression of 9.8 kN. The results once again demonstrate the heterogeneity of the swollen tablet and support previous observations of a membrane. It is clear that in the dry state (Figure 7, left), there is no density variation between the interior and exterior of the tablet which could explain the membrane apparition. These results further indicate that the core of swollen tablets is less dense than the outer membrane formed. While the porosity of the dry tablet is around 10%, the swollen tablet shows an average porosity of 19%. This average over the entire tablet can be divided where the core has a porosity of 34% and the membrane has a varying porosity of 2-15%. When measuring the porosity of top or bottom surfaces, one finds a value around 2% porosity. This clearly confirms that the membrane is the diffusion rate-limiting step in the delivery of drugs. This in-situ formation of an outer membrane also allows for a quasi zero-order release kinetic. This is achieved when the release rate of the drug is directly proportional to time. According to Peppas and Franson (51), a limited swelling matrix having an outer membrane controlling the release of a drug could achieve zero-order kinetic release provided that the outer membrane or polymeric layer possesses a much smaller diffusion coefficient than the inner core of that same tablet. This is readily what happens when a Contramid® tablet is placed in an aqueous environment. In the case of a continuously swelling system (e.g. HPMC), changes in both the swelling kinetics and surface area with time, renders achieving zero-order kinetics an impossible task.

We explain the existence of this membrane in two different ways: Water availability and steric hindrance. First, when the tablet is placed in water, the outer surface is exposed to an unlimited supply of water. Diffusion of water inside the tablet is governed by rapid swelling of amorphous Contramid® particles and by capillary forces caused by the interparticle micropores. Both water supplies are slowed down when the outer surface rapidly retrogrades and forms a gel with low porosity. The kinetics of diffusion of water through this newly formed membrane limits the amount of water available for retrogradation of the core in such a way that it is not sufficiently fast to allow an efficient retrogradation. Secondly, the formation of the membrane also causes the particles located inside to swell and expand with steric limitation (52). This "encasing" is also believed to play a role in the existence of the outer membrane.

Membrane structure

Lenaerts et al. (8) presented the model where double helices formed between grafted amylose on amylopectin molecules. This model of starch is supported by rheological studies of Gidley (53) that demonstrated that amylose gels are semi-crystalline networks of double helices and amorphous regions. However, this model was simple and left out important pieces of the puzzle. The linear architecture of amylose and its backbone flexibility allows multiple double helices to form at different location along the backbone, either with other amylose chains or with amylopectin as postulated by Klucinek and Thompson (37). While linear segments of amylose have a greater tendency to self-assemble

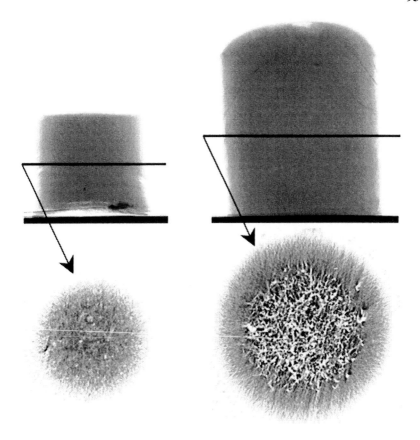

Figure 7. X-ray tomography images of dry (left) and swollen (right) Contramid® tablets. Gray levels are proportional to relative density.

in double helices (17), amylopectin may also retrograde. In the case of Contramid®, a 30% proportion of amylopectin helps in obtaining stronger gels because of the gigantic size of amylopectin molecules that allow numerous attachment points for the formation of double helices and can be represented as "anchoring points" in tridimensional network formation. These "anchoring points" allow several double helical physical crosslinks to form between amylose and amylopectin outer branches. Also, because of covalent crosslinking, amylose chains were also grafted to amylopectin molecules. These amylose chains further increase the anchoring properties of amylopectin molecules by grafting longer side chains to its usually shorter ones. From their ability to form several double helices, coupled to crosslinking, amylose chains also get entangled. The combination of all factors allows the formation of a tri-dimensional network composed of the four following components:

- Amylose-amylose and amylose-amylopectin double helices
- Amylose chain entanglement
- Amylopectin anchoring points
- Amorphous regions

Since starch particles are closely packed together by compression, the proximity further increases the self-assembling and limited swelling through retrogradation. This increase in interparticle association counterbalances the limiting effects associated to chemical grafting and crosslinking.   Figure 8 is a schematic representation of the different dynamics involved.

Drug Delivery Related Properties

Diffusion and permeability

The diffusion parameters of different drugs having various physical properties have been studied by Rahmouni et al. (52). The apparent permeability (P_{app}), partition coefficient (K) and diffusion coefficients (D_g) of drugs through the swollen Contramid® membranes were measured according to the 'lag-time' model, using the diffusion-cell technique (Figure 9).

The membranes were prepared by direct compression using a single station press (Stockes, F4), then swollen in desired buffer to equilibrium prior to use. Diffusion of solutes through hydrogels depends mainly on the solute size, the equilibrium water content of the hydrogel and its porosity (54). Factors affecting the physical properties of a gel such as degree of swelling and gel porosity may have an impact on its permeability. Compression force and particle size are known as factors that influence the porosity of dry Contramid® tablets. However, they have no significant effect on the permeation properties of the membranes obtained when swollen. The lack of a strong dependence of permeability on particle size and compression force was related to the mechanism of the transformation of Contramid® compact into a gel, as just presented.

While incorporation of hydrophilic additives such as 0.25 and 0.5% colloidal silicon dioxide or 10% hydroxypropylmethyl cellulose (HPMC) had no significant effect on the swelling and the permeation proprieties of Contramid® membranes (52), incorporation of hydrophobic additives such as magnesium stearate (0.25 and 0.5%), had a pronounced impact on these properties. This was probably due to the formation of complexes at the surface of particles between amylose chains and these agents in forms of V-type single helix. This well-known phenomenon can reduce retrogradation and thus alter the gel properties. In fact, lipidic agents are used as antistaling agents in bread and other baked goods because of this retrogradtion hindrance (55).

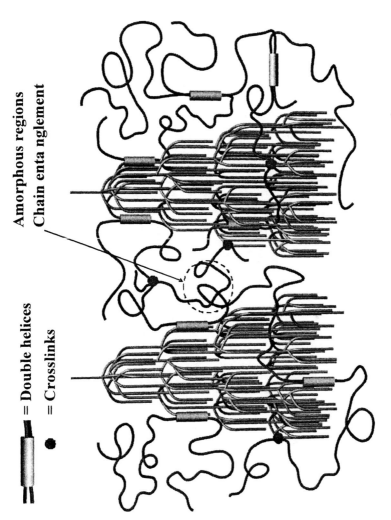

Figure 8. Schematic model representation of a Contramid® gel

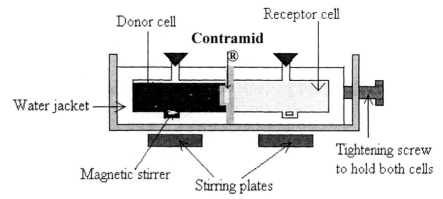

Figure 9. Diffusion cell set-up

It was also reported that P_{app} and D_g of two model drugs, rhodamine B and tramadol decreased as the membrane thickness increased (56). The reduction of P_{app} and D_g was attributed to the loss of membrane homogeneity, as it becomes thicker in the range studied, due to the membrane and core structure discussed previously. However, the authors believe P_{app} and D_g remain stable above a certain thickness where the core's permeability is much higher than that of the membrane; the membrane becoming the rate limiting diffusion step.

Furthermore, the diffusion coefficient of small solutes through Contramid® gels is greatly affected by their solubility (S_w) and to a lesser extent by their molecular weight (M_w <500 g/mol) (52). When the partition coefficient of solute-Contramid® gel was less than unity (K<1), the diffusion process occurred mainly via the 'pore' mechanism, and followed the Mackie-Meares theory. Estimation of the pore size of Contramid® gels according the Rankin equation gave a reasonable value of 22 Å. This average pore size value is in line with the findings of Ravenelle et al. (35) who reported a pore size of 16 Å from hydraulic permeability data, using the Navier-Stokes equation.

Finally, a relationship between D_g, S_w and M_w was studied according to an empirical model. Resistance to solute diffusion through Contramid® membranes was assumed to be dictated by chemical solute/gel interactions and by physical size exclusion of the gel network. Since Contramid® is a non-ionic polymer, solute/gel interactions should mainly depend on solute polarity, and thus on its S_w. The second factor, which reflects physical size exclusion by the Contramid® network, depends mainly on solute size and thus on its M_w. Good agreement was found between the experimental diffusion data and the values calculated with the proposed model.

Enzymatic degradation and resistance

Contramid® is unique in that it is based on starch, a usually highly digestible material. As mentioned above, starch is the main nutrient consumed by man. Because we have α-amylase in our gut, we are able to enzymatically degrade most starches to oligosaccharides and glucose for energy. However, there is abundant literature that discusses methods to prepare "resistant starches" (57). Eerlingen et al. reported that the higher the degree of organization, the higher the amount of double helices, the lower the enzymatic degradation (58). Hence, by partially gelatinizing starch and allowing retrogradation to occur, it is possible to reduce the enzymatic susceptibility. Generally, resistant starches are classified in four different types, depending on the conditions leading to their resistance to enzymatic digestion (57). However, as it is described below, Contramid® holds the combination of all four types in a single system:

Type 1: Physical inaccessibility (surface area, addition of protective ingredients)

Surface area is greatly reduced by forming a tablet compared to free powder. Also, because there is a shape retention property, the tablet swells to a limit, restricting enzymatic attack to the surface of the tablet. As demonstrated earlier by x-ray tomography technique, the surface porosity is very low and allows limited access to enzymatic degradation.

Type 2: Due to refractory nature of some starch granules

High-amylose corn starches have for a long time being recognized as being less susceptible to enzymatic degradation (59).

Type 3: Due to retrogradation, increase in organization degree.

The quick formation of the membrane through retrogradation (double helix formation) just discussed considerably reduces the enzymatic degradation of the tablet.

Type 4: Due to chemical modifications and crosslinking

Contramid® is also chemically functionalized. Hydrolysis of amylose and amylopectin by α-amylase and the mechanism by which the enzyme attaches itself to and degrades starch to oligo and monosaccharides has been extensively studied and is now well understood (58,60,61). This comes from the fact that to hydrolyze the α-1,4 acetal linkage, the enzyme needs to attach itself to the substrate at a specific number of binding sites (62). Because of this, chemical modifications such as crosslinking and hydroxypropylation significantly reduce Contramid®'s susceptibility to enzymatic attack in comparison to the high amylose starch starting material.

In the paper by Rahmouni et al (63), many parameters and their effect on the enzymatic degradation were evaluated. An interesting factor is the compression force. It was shown that above a compression force of 6kN, the enzymatic degradation is independent of that parameter. At lower forces, the porosity of the dry and probably swollen tablet is enough to allow more enzyme activity because of a higher surface area. Neither ionic strength nor the gastric retention time greatly affected the kinetics of enzymatic degradation.

Incorporation of gel forming polymers such as PEO and HPMC in the tablet formulation has shown to reduce degradation and thus protect the matrix from extensive erosion that would jeopardize controlled release (64). The addition of about 5-10% of HPMC for example is enough to reduce the enzymatic degradation to sastisfactory levels. This indicates that addition of PEO or HPMC does not significantly affect self-assembly of Contramid® and protects by quickly swelling in water and forming a gel that further reduces the availability of the substrate for enzymatic attack through an increase in viscosity.

Conclusion

Contramid is fabricated by altering the chemical and physical structure of starch, leading to self-assembly of an amorphous powder to a gel matrix of double helices upon swelling. An important equilibrium exists between restraining retrogradation in the dry powder and allowing it to a sufficient extent to form a viscoelastic gel. The gel formed on the surface of the tablet in the first few minutes controls the rate of water diffusion inside and out of the tablet, i.e. it allows controlled release of the drug. The intrinsic nature of high-amylose starch, the high density of the surface membrane, its low porosity, along with chemical modifications and crosslinking, significantly reduce the enzymatic degradation expected in the intestine. While highly complex and expensive, manufacturing of osmotic pumps is usually required to obtain zero-order release kinetics. With this in-situ forming rate-controlling membrane, Contramid® allows near zero-order release kinetics from a matrix tablet. Overall, Contramid® is an excellent demonstration of how one can harness the natural properties of a polysaccharide such as starch, and tune them just the right way as to obtain a high performance specialty pharmaceutical excipient.

Acknowledgements

The authors would like to thank Prof. Robert H. Marchessault and Dr. Grégory Chauve at McGill University for their help and expertise in the characterization of Contramid® and for revising this manuscript. Petr Fjurasek at McGill University's Centre for Self-Assembled Chemical Structures, CSACS, for DSC analysis and the McGill University Bone and Periodontal Research Center for the x-ray tomography analysis.

References

1. Lenaerts, V.; Dumoulin, Y.; Mateescu, M.A. *J. Cont. Rel.* **1991**, 15, 39-46.
2. Flory, P.J. *Principles of Polymer Chemistry*; Cornell University Press: Ithaca, NY, 1953.
3. Dumoulin, Y.; Mateescu, M.A.; Cartilier, L. *J. Pharm. Belgique* **1993**, 48, 150-151.
4. Dumoulin, Y.; Cartilier, L.; Predas, M.; Alex, S.; Lenaerts, V.; Mateescu, M.A. *Cont. Rel. Soc. Proc.* 21, **1994**, #1365.
5. Cartilier, L.; Moussa, I.S. *Proc. 1st World Meeting on Pharm. Biopharm. & Pharm. Tech.*, Budapest, **1995**, #241-242.
6. Szabo, P.I.; Ravenelle, F.; Hassan, I.; Preda, M.; Mateescu, M.A. *Carbohydrate Research*, **2000**, 323, 163-175.
7. Mateescu, M.A.; Lenaerts, V.; Dumoulin, Y. Canadian Patent 2 041 774, **1992**.; US Patent 5 456 921, **1995**.; *Chemical Abstract*, **1994**, 120, 226965.
8. Lenaerts, V.; Moussa, I.; Dumoulin, Y.; Mebsout, F.; Chouinard, F.; Szabo, P.; Mateescu, M.A.; Cartilier, L.; Marchessault, R.H. *Journal of Controlled Release*, **1998**, 53, 225-234.
9. Désévaux, C.; Girard, C.; Lenaerts, V.; Dubreuil, P.; *Int. J. Pharm*, **2002**, 232, 119-129.
10. Désévaux, C.; Dubreuil, P.; Lenaerts, V.; Girard, C. *J. Biomed. Mat. Res.* **2002**, 63 (6), 772-779.
11. Désévaux, C.; Dubreuil, P.; Lenaerts, V. *J. Cont. Rel.* **2002**, 82, 95-103.
12. Huneault, L.M.; Lussier, B.; Dubreuil, P.; Chouinard, L.; Désévaux, C. *J. Ortho. Res.* **2004**, 22,1351-1357.
13. Battista, O.A. *U.S. Patent* 3,146,168, **1964**.
14. Reier, G.E.; Shangraw, R.F. J. Pharm. Sci. **1966**, 55, 510-514.
15. Esnard, J.-M.; Clerc, J.; Tebbi, H.; Duchêne, H.; Lévy, J.; Puisieux, F. *Ann. Pharm. Fr.* **1973**, 31, 103-116.
16. Doelker. E. *Drug Dev. Ind. Pharm.* **1993**, 19, 2399-2471.
17. Thomas, D.J.; Atwell, W.A. *Starches* Eagan Press Handbook Series, American Association of Cereal Chemists: St-Paul, MN, 1999.
18. Shi, Y.-C.; Capitani, T.; Trzasko, P.; Jeffcoat, R. *Journal of Cereal Science*, **1998**, 27, 289-299.
19. Kasemsuwan, T.; Jane, J. *Cereal. Chem.* **1994**, 71, 282-287.
20. Buleon, A.; Colonna, P.; Planchot, V.; Ball, S. *Int J Biol Macromol.* **1998**, 23, 85-112.
21. Le Bail, P.; Morin, F.G.; Marchessault, R.H. *International Journal of Biological Macromolecules*, **1999**, 26, 193-200.
22. Le Bail, P.; Buléon, A.; Shiftan, D.; Marchessault, R.H. *Carbohydrate Polymers*, **2000**, 43, 317-328.
23. Kawada, J. ; Marchessault, R.H. *Starch/Stärke,* **2004**, 56, 13-19.
24. Wu, H.C.H.; Sarko, A. *Carbohydrate Research*, **1978**, 61, 27-40; *ibid*, **1978**, 61, 7-25.

25. Immel, S.; Lichtenthaler, F.W. *Starch/Stärke*, **2000**, 52, 1-8.
26. Jenkins, P.J.; Donald, A.M. *Polymer*, **1996**, 37, 5559-5968.
27. Lund, D. *Crit Rev Food Sci Nutr.* **1984**, 20, 249-273.
28. Gudmundsson, M. *Thermochim. Acta* **1994**, 246, 329-341.
29. Hollinger, G.; Kuniak, L.; Marchessault, R.H. *Biopol.* **1974**, 13, 879-890.
30. Moussa, I.S. ; Cartilier, L. *J. Control. Rel.* **1996**, 42, 47-55.
31. Lenaerts, V.; Beck; R.H.F.; Van Bogaert, E.; Chouinard, F.; Hopcke; R.; Desevaux; C. U.S. Patent 6,607,748, 2003.
32. Jane, J.; Xu, A.; Radosavljevic, M.; Seib, P.A. *Cereal. Chem.* **1992**, 69, 405-409.
33. Jane, J.L; Chen, J.F. *Cereal Chem.* **1992**, 69, 60-65.
34. Rahmouni, M.; Lenaerts, V.; Massuelle, D.; Doekler, E.; Leroux, J.C. *Chem. Pharm. Bull.* **2002**, 50, 1155-1162.
35. Ravenelle, F.; Légaré, A.; Buschmann, M.D.; Marchessault, R.H *Carbohydrate Polymer*, **2002**, 47, 259-266.
36. Provuori, P.; Manelius, R.; Suortti, T.; Bertoft, E.; Autio, K. *Food Hydrocolloids*, **1997**, 11, 471-477.
37. Klucinek, J.D.; Thompson, D.B. *Cereal Chem.* **1999**, 76, 282-291.
38. Boltz, K.W.; Thompson, D.B. *Cereal Chem.* **1999**, 76, 204-212.
39. Levine, H.; Slade, L. In: Franks F, ed. Water Science Reviews, Vol. 3. Cambridge, UK: Cambridge University Press; 1987.
40. Veregin, R.P.; Fyfe, C.A.; Marchessault, R.H.; Taylor, M.G. *Macromolecules*, **1986**, 19, 1030-1034.
41. Gidley, M.J.; Bociek, A.M. *Journal of the American Chemical Society*, **1988**, 110, 3820-3829.
42. Le Bail, P. Morin, F.G.; Marchessault, R.H. *Int. J. Biol. Macromol.* **1999**, 26, 193-200.
43. Shiftan, D.; Ravenelle, F.; Mateescu, M.A.; Marchessault, R.H. *Starch*, **2000**, 52, 186-195.
44. Brumovsky, J.O.; Thompson, D.B. *Cereal Chem.* **2001**, 78, 680-689.
45. Kohyama, K.; Matsuki, J.; Yasui, T.; Sasaki, T. *Carb. Polymers* **2004**, 58, 71-77.
46. Ravenelle, F.; Chauve, G.; Fjurasek, P. Marchessault, R.H. Private Communication, 2005.
47. Malveau, C.; Baille, W.E.; Zhu, X.X.; Marchessault, R.H. *Biomac.* **2002**, 3, 1249-1254.
48. Baille, W.E.; Malveau, C.; Zhu, X.X.; Marchessault, R.H. *Biomac.* **2002**, 3, 214-218.
49. Marchessault, R.H.; Chauve, G.; Private Communication, 2005.
50. Bonaldi V.M.; Garcia, P.; Coche, E.E.; Sarazin, L.; Bret, P.M. *Presse Med.* **1996**, 25, 1109-1114.
51. Peppas, N.A.; Franson, N. *J. Pol. Sci.* **1983**, 21, 983-997.
52. Rahmouni, M.; Lenaerts, .; Leroux, J.C. *S.T.P. Pharma.* **2003**, 13, 341-348.
53. Gidley, M.J. *Macromol.* **1989**, 22, 351-358.

54. Flynn, G.L.; Yalkowsky, S.H.; Roseman, T.J. *J. Pharm. Sci.* **1974**, 63, 479-510.
55. Vidal, F.D.; Gerrity, A.B. *Canadian Patent*, 1082041, **1980**.
56. Rahmouni, M.; Lenaerts, V.; Leroux, J.C. Private Communication – Labopharm Inc. Laval, Québec, Canada.
57. Thompson, D.B. *Trends Food Sci. Tech.* **2000**, 11, 245-253.
58. Eerlingen, R.C.; Jacobs, H.; Delcour, J.A. *Cereal. Chem.* **1994**, 71, 351-355.
59. Sandstedt, R.M.; Strahan, D.; Ueda, S.; Abbot, R.C. *Cereal Chem.* **1962**, 39, 123-131.
60. Colonna, P.; Leloup, V.; Buleon, A. *Eur. J. Clin. Nut.* **1992**, 46, S17-S32.
61. Jacobs, H.; Eerlinger, R.C.; Spaepen, H.; Grobet, P.J.; Delcour, J.A. *Carbohydrate Research*, **1998**, 305, 193-207.
62. MacGregor, E.a. *J. Prot. Chem.* **1988**, 7, 399-415.
63. Rahmouni, M.; Chouinard, F.; Nekka, F.; Lenaerts, V.; Leroux, J.C. *Eu. J. Pharm. Biopharm.* **2001**, 51, 191-198.
64. Chouinard, F.; Lenaerts, V. Proc 24th Int. Symp. Cont. Rel. Bioactive Mat. **1997**, 24, 265-266.

Chapter 5

Water Diffusion in Drug Delivery Systems Made of High-Amylose Starch as Studied by NMR Imaging

Héloïse Thérien-Aubin and Xiao Xia Zhu

Département de Chimie, Université de Montréal, C.P. 6128,
Succ. Centre-ville, Montréal, Québec H3C 3J7, Canada

Polysaccharides such as high-amylose starch are used as
excipients in controlled released devices for drug. The
swelling of the compressed polymer and water penetration in
the polymer matrix can be studied quantitatively by NMR
imaging. The effects of formulation, size, compression force,
and temperature on the diffusion behavior of water in tablets
of crosslinked amylose starch have been studied by NMR
imaging.

A better understanding of the process of controlled release of drugs is essential for the design and preparation of more efficient drug delivery systems. Therefore the study of the diffusion of small molecules, especially water, the swelling of polymer matrix, and the dissolution/erosion of the drug carrier is crucial. Nuclear magnetic resonance (NMR) imaging can be used to study solvent diffusion in polymeric matrixes (1-5). More recently, it has been adapted to characterize various drug delivery systems (6-14). The mechanism of drug release is based on either diffusion of drugs through the polymer matrix, or polymer degradation. It is important to study parameters related to water penetration in the polymer, swelling of polymer matrix, and the interactions in the system to improve the desing of drug release system.

The diffusion process in such systems is usually studied by gravimetric methods for solvent uptake (15) or by high performance liquid chromatography (HPLC) for the kinetics of drug released (16). These techniques and other such as optical methods (15,17,18) allows for a global characterization of the system, while NMR imaging provides localized information of the spatial distribution of the compounds of interest inside and outside the polymeric matrix.

Diffusion in polymer matrixes

Fick (19) provided the first mathematical description of diffusion. Fick's second laws of diffusion relate the diffusive flux to the concentration gradient of the diffusing species:

$$\frac{\partial C_i(\vec{r},t)}{\partial t} = D_i \nabla^2 C_i(\vec{r},t) \qquad (1)$$

where C the concentration of the diffusing species, D the diffusion coefficient, t the time, r the position and i the diffusing specie.

While Fick's laws of diffusion adequately describe the diffusion of small molecules in a solvent, diffusion in glassy polymers sometimes shows deviation from Fick's laws and the diffusion is there anomalous or non-Fickian. Diffusion in polymers is related to the properties of the polymer network and to the interactions between the polymer and the diffusing species. Alfrey (20) classified the diffusive behavior occurring in polymer matrixes according to the relation between the quantity of matter which has diffused and time :

$$d = k\,t^n \qquad (2)$$

where d is either the amount of matter having diffused or the distance covered by the diffusing front after an immersion time t, and k a parameter related to the velocity or to the diffusion. The parameter n is representative of the penetration kinetics (n is equal to 0.5 for Fickian diffusion and to 1 in the case of Case II diffusion). Intermediate values indicate an anomalous process of diffusion. According to Alfrey (*20*), Case II diffusion is relaxation-controlled whereby the relaxation of the polymer chains largely affects diffusion, the rate of relaxation being much slower than the rate of diffusion. For Fickian diffusion the relaxation of the polymer matrix does not interfere with the diffusion process since the rate of relaxation is faster than that of diffusion. Anomalous diffusion is observed when the rates of relaxation and of diffusion are rather similar. Anomalous diffusion is regarded as the superposition of Fickian and Case II diffusions (eq. 3).

$$d = k_1 \sqrt{t} + k_2 t \tag{3}$$

For the same polymer-diffusant pair, different diffusion behavior can potentially be observed (*21*) by changing either the temperature or the concentration of the diffusant.

NMR imaging

NMR imaging is widely used in medical diagnostics, providing normally the spatial mapping of proton signals. The contrast of the images can be adjusted by the difference in concentration, relaxation times, self-diffusion coefficients or flow rates. It is a non-destructive and non-invasive method since no physical slicing of the sample is required to observe the interior.

In the study of drug delivery systems, NMR imaging has been mainly used to detect the signal of the protons on mobile water molecules inside the system (*1*). NMR imaging allows for the observation at different times of the changes in the system caused by hydration and water uptake. The diffusion of drugs (*11*) may also be mapped in a water-based dissolution medium.

High-Amylose Starch Excipient

Chemically-modified high-amylose starch (CHAS) is an excipient used in the formulation of tablets and implants. When hydrated, the tablets made of modified high-amylose starch form a highly porous environment (*22*) and are very effective in the sustained release of drugs (*23*).

Water uptake in high-amylose starch excipients has been the subject of multiple studies by NMR imaging (*12-14*). The effects of temperature (*12,14*), size (*13*), compression force, and preparation method of modified starch have been investigated.

The starch used in CHAS I, II and III formulations (preparation method 1) was modified (*24*) by the gelatinization of a starch suspension in 4% NaOH, followed by crosslinking of starch by the addition of 3.25% Sodium trimeta phosphate (STMP). Starch used in CHAS IV (preparation method 2), trademark as Contramid® by Labopharm Inc. (Laval, QC, Canada), was prepared (25) by the crosslinking of starch with 0.075% of phosphorus oxychloride in an alkaline medium, followed by the functionalization of the crosslinked starch with 6% of propylene oxide and then gelatinization. Both preparations were then spray-dried and compressed into tablets.

Table 1: CHAS tablets used in NMR imaging studies

Tablet	Weight (mg)	Diameter (mm)	Thickness (mm)	Surface area (mm²)	Compr. force (± 1 kN)	Density (g/cm³)
I*	200	8.6	2.7	191	13	1.3
II*	24	4.8	1.2	54	11	1.1
III*	200	8.6	2.7	191	<10	1.3
IV**	210	9.1	3.6	234	22	0.89

* Preparation method 1 (24)

** Preparation method 2 (25)

NMR imaging studies were done after the immersion of the tablet in an excess of a 1:1 mixture of H_2O and D_2O in a NMR tube with an internal diameter of either 15 or 20 mm. A slice of either 0.5 mm or 1 mm was selected in the center of the tablet either on the radial or axial direction. The images obtained have a field of view of 20 mm and the in-plane resolution is either was 78 μm or 156 μm depending of the matrix size. For each image 4 scans were accumulated.

Swelling

Swelling is characterized by the variation of tablet dimensions. The swelling percentage S is defined as

$$S = \frac{L_t - L_0}{L_0} \times 100 \qquad (4)$$

where L_t is the thickness or the diameter after an immersion time t and L_0 the initial thickness or diameter. Upon hydration, the volume of the polymer will increase. Figures 2 shows the rapid swelling of the crosslinked starch tablet. Swelling rates are obtained with

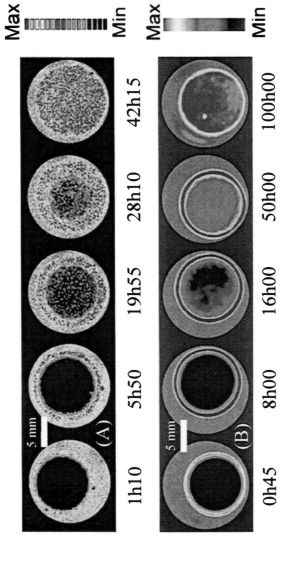

Figure 1: Radial NMR images showing the water penetration into a CHAS tablet at 37 °C as a function of immersion time. (A) CHAS I, the spatial resolution in plane is 78 μm with a slice thickness of 500 μm and (B) CHAS IV the spatial resolution in plane is 156 μm with a slice thickness of 1000 μm (adapted from reference 12)

$$S = S_{max}\left(1 - e^{-kt}\right) \tag{5}$$

where S_{max} is the maximal swelling observed at equilibrium, t the immersion time and k the rate constant describing the swelling process.

In all cases, the tablet swells to a certain extent and then stops. No degradation or erosion of the tablet is observed as in the case of other tablets, such as thoses made of microcrystalline cellulose derivatives (26).. The limited swelling of modified starch tablets is mainly due to a rearrangement of the structure of the crystalline domains. Starch in CHAS powder has a para-crystalline structure mainly composed of V-type single helix and amorphous regions. However, when the tablet is hydrated, there is conversion of the amorphous content to a B-type double helix as characterized by CP-MAS NMR spectroscopy (27). The double helical structure creates physical crosslinking points in the tablet which leads to the formation of a 3D-network in the wet tablet and restrains the swelling of the tablet.

For all tablets, swelling in thickness is larger than swelling in diameter. During the fabrication of tablets compression on the CHAS powder is applied along the axial direction (thickness). The quasi-spherical starch granules are therefore compressed into irregular discs (28). Upon hydration, stress is released and the starch granules return their spherical shape. Swelling is also affected by preparation method, which will be discussed later.

Table 2: Swelling characteristics of the CHAS tablets at 37°C

	Radial Swelling		Axial Swelling	
	S_{max} (%)	k $(10^{-5} s^{-1})$	S_{max} (%)	k $(10^{-5} s^{-1})$
CHAS I	34 ± 1	9 ± 1	56 ± 3	18 ± 5
CHAS II	48 ± 3	26 ± 1	180 ± 5	32 ± 1
CHAS IV	15 ± 1	5.8 ± 0.8	44 ± 1	15 ± 2

Water penetration

Typically, water uptake in such polymeric systems is characterized gravimetrically. In some cases, such as for CHAS tablets, gravimetric studies does not yield useful information because tablet degradation or disintegration is common (13,14) when the tablets are manipulated. In the NMR analysis, there is no agitation or abrasion of the tablet. Since the echo time used in these NMR

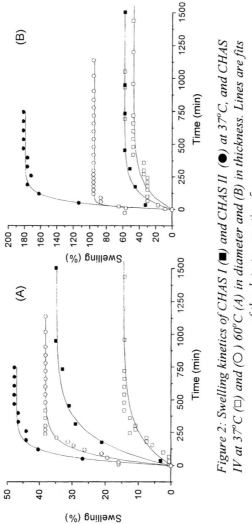

Figure 2: Swelling kinetics of CHAS I (■) and CHAS II (●) at 37°C, and CHAS IV at 37°C (□) and (○) 60°C (A) in diameter and (B) in thickness. Lines are fits of the data to equation 5.

experiments is low in comparison to the T_2 of the polymer, only the mobile water molecules are observed. Figure 3 shows the water concentration in a slice of a the CHAS tablet. These profiles are extracted from a thin slice taken in the middle of the cylinder, after a short immersion time (edges effects can be ignored as the tablet can be considered as an infinitely long cylinder).

These profiles show a decrease in water concentration toward the center of the cylinder. This type of profile is typical of Fickian diffusion; for Case II diffusion the concentration of water in the wet part would be uniform followed by a sharp decrease. However, the water penetration profile is not an absolute tool for for the description of the diffusional behavior. Other information can be obtained from these profiles for a better understanding of the mechanism of diffusion.

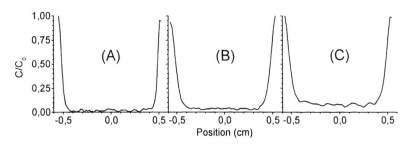

Figure 3: Water concentration profiles inside a CHAS IV tablet at 37°C after an immersion time of (A) 5 min (B) 1h30 and (C) 3h30

These profiles can be used to determine the location of the interface between the dry and wet parts of the tablet and to follow the advancing of the solvent into the tablet as depicted in Figure 4A. Diffusion behavior in CHAS IV is Fickian between 25 and 45°C and Case II at 60°C. The penetration kinetics at 37 and 45°C show Fickian behavior at the beginning followed by "non-Fickian" behavior after ca. 3h30 due to the influence of the tablet edges. The same behavior is observed at 25°C for longer immersion times. The Fickian behavior is also confirmed by NMR relative weight gain, measured by the integration over space of the NMR signals of water in the concentration profile plotted as a function of time. The results in Figure 4B clearly show that the diffusion is Fickian between 25 and 45°C and Case II at 60°C.

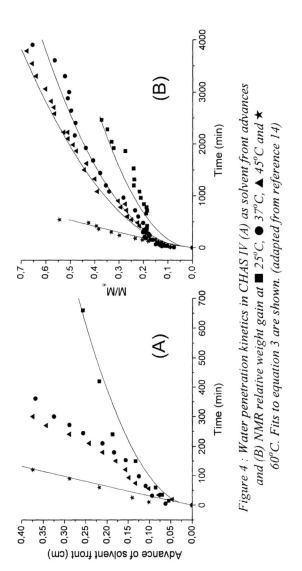

Figure 4 : Water penetration kinetics in CHAS IV (A) as solvent front advances and (B) NMR relative weight gain at ■ 25°C, ● 37°C, ▲ 45°C and ★ 60°C. Fits to equation 3 are shown. (adapted from reference 14)

Temperature effect

Temperature has an effect on both swelling and water penetration in the CHAS tablets. For CHAS IV, it has been observed (*14*) that the swelling capacity is not affected by changes in temperature between 25 and 45°C. In the case of CHAS I, the swelling is more extensive when the temperature is raised from 25 to 37°C (*12*). In both CHAS I and CHAS IV case, swelling becomes faster when the temperature was raised. An Arrhenius activation energy of 40 ± 10 kJ/mol for the swelling process was calculated from the swelling rates and can be related to the breaking of hydrogen bonds.

At 60°C, the swelling capacity is more than two times higher than at lower temperatures. When a starch-water mixture is heated, gelatinization of starch occurs The gelatinization temperature is defined as the temperature where hydrogen bonds stabilizing the double helices present in the starch are broken which induces a loss of crystallinity and a leaching of amylose out of the starch grain. This leads to the obtention of a highly viscous gel (*29*). For the chemically-modified starch such as the CHAS powder, gelatinization occurs over a broad temperature range of about 60 to 100°C (*28*). The limited extent of swelling is due to the change from amorphous to B double helixes in the crystalline domains. At 60°C, the partial loss of crystallinity leads to the formation of fewer B double helixes. Consequently, at 60°C, the 3D network restricting the swelling of the tablet is weaker and swelling is more extensive.

The diffusivity of water in the tablet increased when the temperature was raised as observed in Figure 5. In this figure the equilibrium of water penetration is reach faster at higher temperatures. To compare the rate of diffusion at different temperature, the diffusion coefficients are calculated either from mass uptake measurements or by curve fitting of the water concentration profile.

For Fickian diffusion the concentration profile inside the tablet can be described by Fick's law (equation 1). For an ideal geometry it is possible to solve equation 1. The idealized geometry of the tablet is considered as either an infinitely long cylinder or an infinitely large plane neglecting the edge effect. When the real geometry is considered, it is not possible to analytically solve equation 1. Concentration of diffusive species in an infinitely long cylinder is given by (*30*)

$$\frac{C-C_0}{C_\infty-C_0} = 1 - \frac{2}{r}\sum_{n=1}^{\infty}\frac{J_0\left(x\alpha_n\right)}{\alpha_n J_1\left(r\alpha_n\right)}e^{-D\alpha_n^2 t} \tag{6}$$

and in an infinitely large plane by (*30*)

$$\frac{C-C_0}{C_\infty-C_0} = 1 - \frac{4}{\pi}\sum_{n=0}^{\infty}\frac{(-1)^n}{2n+1}e^{\frac{-D(2n+1)^2\pi^2 t}{4l^2}}cos\frac{(2n+1)\pi x}{2l} \tag{7}$$

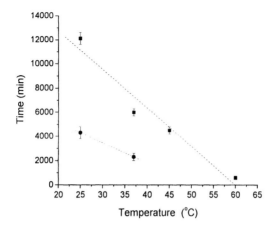

Figure 5: Immersion time to which the equilibrium is reached for ● *CHAS I and* ■ *CHAS IV.*

where C the concentration of the diffusive species after an immersion time t at position x inside the tablet, C_0 the initial concentration of the diffusive species in the tablet, C_∞ the concentration of the diffusive species once equilibrium is obtained, l the thickness of the tablet, r the radii of the tablet, J_n the Bessel function of the first kind of order n and α_n the nth root of J_0. The diffusion coefficient obtained by curve fitting are coherent with the higher diffusivity observed at higher temperatures as described by the Arrhenius equation. The activation energy obtained for the diffusion process is lower than the one obtained for the swelling process (*14*) this is mainly due to the fact that water diffusion involves few, if any, movements of the polymer matrix while swelling involves large polymer displacements.

While tablet swelling is observed, the diffusion coefficients decrease. In the starch tablets, as in the case of certain HPMC tablets (*6*), when the polymer matrix is immersed in water there is a rapid hydration followed chain relaxation to form a gel membrane at the water-tablet interface. This gel layer is held responsible for the sustained release characteristics of the starch tablets (*23*). It is mainly this gel layer that controls both water uptake and drug diffusion by the tablet (*6*). The formation of this membrane reduces the diffusion coefficient of water. Once swelling is complete and the gel layer is totally formed, there is no more decrease in the overall diffusion coefficient.

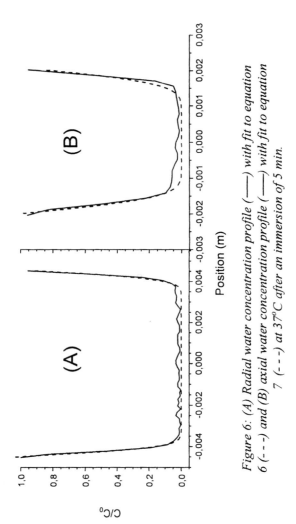

Figure 6: (A) Radial water concentration profile (——) with fit to equation 6 (- - -) and (B) axial water concentration profile (——) with fit to equation 7 (- - -) at 37°C after an immersion of 5 min.

Table 3: Diffusion coefficients of water in the CHAS tablets calculated by different means

Sample	Temperature (°C)	Initial profile fit ($10^{-11}\ m^2 s^{-1}$)	Profile fit after swelling ($10^{-11}\ m^2 s^{-1}$)	Water signal integration ($10^{-11}\ m^2 s^{-1}$)
CHAS I	25	3.9 ± 0.5	1.6 ± 0.3	1.95 ± 0.07
	37	5.4 ± 0.3	1.8 ± 0.3	2.66 ± 0.05
CHAS IV	25	3.9 ± 0.6	0.30 ± 0.02	1.1 ± 0.1
	37	7 ± 1	0.56 ± 0.06	2.11 ± 0.06
	45	8 ± 1	0.9 ± 0.1	2.60 ± 0.04
	60	16 ± 1	3 ± 1	--

Effect of preparation method

The main difference between the two types of crosslinked high-amylose starch tablets used here lies in the degree and type of crosslinking. The degree of crosslinker is much higher in CHAS I than in CHAS IV. The degree of crosslinking has an important influence on the swelling properties of the tablet; the more crosslinked starch actually swells more than the less crosslinked one (15). When starch is heavily crosslinked, it has less chain mobility and conversion from amorphous to the B double helix is difficult (15). Since the formation of the B crystalline form is held responsible for the limited swelling, fewer B helixes formed mean greater degrees of swelling. Thus, swelling of CHAS I tablets is more extensive than for CHAS IV tablets (Figure 2).

In CHAS IV tablets, water penetration is Fickian between 25 and 45°C, and non-Fickian at 60°C. For CHAS I tablets the diffusive behavior is Fickian at 25 and 37°C but the solvent front is not well defined (13) due to faster water penetration and more swelling which could lead to misinterpretation.

Equilibrium in water uptake is reached faster for CHAS I than for CHAS IV tablets (Figure 5). Therefore, the diffusivity in CHAS I should be higher. A short time after immersion, the diffusion coefficient of water is almost the same for CHAS I and for CHAS IV at the same temperature (Table 3). However, once the swelling is finished, the diffusion coefficient of water in CHAS IV decreases more than in CHAS I tablets this is related to the organization in the gel layer formed at the water-tablet interface. The efficiency of the gel layer could be related to the mobility of starch chains: in CHAS I tablets polymer mobility is larger than in CHAS IV tablets (there is formation of fewer entanglements in the gel membrane due to the higher crosslinking degree) and consequently the diffusion coefficient of water in the outer membrane of CHAS I is larger than in the gel layer of CHAS IV.

Effect of tablet size

An important factor affecting water penetration is the size of the tablet, in particular the thickness, since the compression force is applied along this direction during tablet preparation. Experiments show that a decrease of the tablet thickness leads to a larger percentage of swelling along the axial direction of the tablet (corresponding to the thickness).

Swelling of CHAS II tablets, the smallest tablets, is much larger than that of the CHAS I tablets. In both cases the compression force used is almost the same but the hardness of CHAS II tablet is lower than that of CHAS I. The greater extent of swelling can be ascribed to the lower hardness (*13*). The rate of swelling (Table 2) is higher for CHAS II tablets than for CHAS I tablets due to the larger surface/volume ratio of the CHAS II tablet. In this case, water penetration is easier in the smaller tablet since it possesses a greater relative water-tablet interface per unit volume.

Tablet size has little influence on the diffusional behavior of water in the tablet. Obviously, water uptake equilibrium is reached faster for the smaller tablets, but the diffusion coefficient of water is almost the same in both CHAS I and II once the swelling equilibrium is reached.

Effect of compression force

Different compression forces were applied during the preparation of the tablet (CHAS I and III), which results in the different hardness values (the peak force need to break the sample) of the sample. CHAS tablets produce with a lower compression force lead to weaker tablets with a lower hardness. Upon hydration, breakup of the sample is observed. Before the disintegration, a faster water penetration is observed at a mechanically weaker point in the tablet. This anisotropy of water penetration leads to an anisotropy of the swelling process, causing mechanical stress and hence possible fracture of the sample.

Conclusion

Crosslinked high-amylose starch is an innovative excipient and a better understanding of its behavior will provide ways to design more effective controlled release systems. Anisotropic swelling of the starch tablet is observed, due to the release of stress in the starch grains caused by compression forces applied along the axial direction. NMR Imaging is a useful tool to study the behavior of controlled release systems such as the starch tablets. Swelling and water uptake of the polymer have a crucial influence on the release profile. A better understanding of these processes is essential to fully characterize the diffusion process in the starch tablets. NMR imaging can provide more quantitative data on swelling and diffusion kinetics than other techniques. NMR imaging probe the interior of the tablet and therefore allows for the characterization of the diffusion behavior of small molecules within the matrix. NMR imaging studies indicate that the sustained release of drug molecules with these tablets could be ascribed to the formation of a gel layer at the water-tablet interface. This membrane reduces the rate of diffusion in the tablet.

Temperature has a large effect on water uptake in the tablets due to the phase transition of starch at certain temperatures. The diffusive behavior of water dramatically changes when the tablet is heated above the gelatinization temperature. Tablet size affects swelling since the surface/volume ratio increases when the tablet size is reduced. Formulation and fabrication of the tablets will modulate the release behavior of CHAS tablets. The NMR imaging studies provide insight into the characteristics of the polymer matrix.

Acknowledgments

The authors acknowledge the financial support from NSERC, the Canada Research Chair program and Labopharm Inc.

References

1. Weisenberger, L. A.; Koenig, J. L. *Appl. Spectrosc.* **1989**, *43*, 1117.
2. Weisenberger, L. A.; Koenig, J. L. *J. Polym. Sci. Polym. Lett.* **1989**, *27*, 55.
3. Valtier, M.; Tekely, P.; Kiene, L.; Canet, D. *Macromolecules* **1995**, *28*, 4075.
4. Malveau, C.; Beaume, F.; Germain, Y.; Canet, D. *J. Polym. Sci. Polym. Phys.* **2001**, *39*, 2781.
5. Mayele, M.; Oellrich, L. R. *Appl. Spectrosc.* **2004**, *58*, 338.
6. Rajabi-Siahboomi, A. R.; Bowtell, R. W.; Mansfield, P.; Henderson, A.; Davies, M. C.; Melia, C. D. *J. Control. Release* **1994**, *31*, 121.
7. Bowtell, R.; Sharp, J. C.; Peters, A.; Mansfield, P.; Rajabi-Siahboomi, A. R.; Davies, M. C.; Melia, C. D. *Magn. Reson. Imaging* **1994**, *12*, 361.
8. Chowdhury, M.A.; Hill, D.J.T.; Whittaker, A.K.; Braden, M.; Patel, M. P. *Biomacromolecules* **2004**, *5*, 1405.
9. Baumgartner, S.; Lahajnar, G.; Sepe, A.; Kristl, J. *Eur. J. Pharm. Biopharm.* **2005**, *59*, 299.
10. Fyfe, C. A.; Grondey, H.; Blazek-Welsh, A. I.; Chopra, S. K.; Fahie, B. J. *J. Control. Release* **2000**, *68*, 73.

120

11. Fyfe, C. A.; Blazek-Welsh, A. I. *J. Control. Release* **2000**, *68*, 313.
12. Baille, W. E.; Malveau, C.; Zhu, X. X.; Marchessault, R. H. *Biomacromolecules* **2002**, *3*, 214.
13. Malveau, C.; Baille, W. E.; Zhu, X. X.; Marchessault, R. H. *Biomacromolecules* **2002**, *3*, 1249.
14. Thérien-Aubin, H.; Baille, W. E.; Zhu, X. X.; Marchessault, R. H. *Biomacromolecules* **2005**, *in press.*
15. Moussa, I. S.; Cartilier, L. H. *J. Control. Release* **1996**, *42*, 47.
16. Sangalli, M. E.; Maroni, A.; Zema, L.; Busetti, C.; Giordano, F.; Gazzaniga, A. *J. Control. Release* **2001**, *73*, 103.
17. Bussemer, T.; Peppas, N. A.; Bodmeier, R. *Eur. J. Pharm. Biopharm.* **2003**, *56*, 261.
18. Gao, P.; Meury, R. H. *J. Pharm. Sci.* **1996**, *85*, 725.
19. Fick, A. E. *Annalen der Physik und Chemie* **1855**, *94*, 59.
20. Alfrey, T. J.; Gurnee, E. F.; Lloyd, W. G. *J. Polym. Sci. Polym. Sym* **1966**, *12*, 249.
21. Hopfenberg, H. B.; Frisch, H. L. *J. Polym. Sci. Polym. Lett.* **1969**, *7*, 405.
22. Ravenelle, F.; Marchessault, R. H.; Legare, A.; Buschmann, M. D. *Carbohydr. Polym.* **2001**, *47*, 259.
23. Lenaerts, V.; Moussa, I.; Dumoulin, Y.; Mebsout, F.; Chouinard, F.; Szabo, P.; Mateescu, M. A.; Cartilier, L.; Marchessault, R.H. *J. Control. Release* **1998**, *53*, 225.
24. Dumoulin, Y.; Carriere, F.; Ingenito, A. WO Patent 98/35992, **1998**.
25. Lenaerts, V.; Beck, R. H. F.; Van Bogaert, E.; Chouinard, F.; Hopcke, R.; Desevaux, C. WO Patent 2002/002084, **2002**.
26. Lerk, C. F.; Bolhuis, G. K.; de Boer, A. H. *J. Pharm. Sci.* **1979**, *68*, 205.
27. Shiftan, D.; Ravenelle, F.; Mateescu, M. A.; Marchessault, R. H. *Starch/Staerke* **2000**, *52*, 186.
28. Le Bail, P.; Morin, F. G.; Marchessault, R. H. *Int. J. Biol. Macromol.* **1999**, *26*, 193.
29. Whistler, R. L.; BeMiller, J. N.; Paschall, E. F. *Starch : Chemistry and Technology*, 2nd ed.; Academic Press: Orlando, 1984.
30. Crank, J. *The Mathematics of Diffusion*, 2nd ed.; Clarendon Press: Oxford, 1979.

Chapter 6

Cross-Linked Starch Derivatives for Highly Loaded Pharmaceutical Formulations

Mircea Alexandru Mateescu, Pompilia Ispas-Szabo, and Jérôme Mulhbacher

Department of Chemistry and Biochemistry, Université du Québec à Montréal, C.P. 8888, Succ. A, Montréal, Québec H3C 3P8, Canada

Starch derivatives were obtained from Crosslinked High Amylose Starch (HASCL) by substitution with Carboxymethyl (CM-), Aminoethyl (AE-) or Acetate (Ac-) groups. The new polymers are able to generate ionic or neutral networks involved in the control of drug release. Surprisingly, it was found that the drug loading capacity of the new derivatives was markedlly higher compared to the unsubstituted HASCL. The HASCL derivatives ensured a close to linear release for 16-24h and they were proposed as excipients for oral solid dosage forms. Dissolution kinetics and mechanistic studies (related to swelling and diffusion aspects) allowed a better understanding of the physical and molecular phenomena controling the drug delivery from these novel matrices.

121

Starch modification by crosslinking was first aimed to prevent its retrogradation (1) contributing thus to the stabilization of the amylaceous preparations used in the food industry. Crosslinked high amylose starch (HASCL), a hybrid plant product was suggested in the late seventies as a material for size-exclusion chromatography (2) and as a specific stationary phase for affinity chromatography separation of alpha amylase (3) and haemoglobin (4) as well as for specific retention of macrophage cells (5). Although crosslinked, the HASCL still remains a substrate of alpha-amylase, with the mention that the rate of amylolysis decreases with the increase of the crosslinking degree (6). This led to the use of CrossLinked (CL)-Amylose as substrate for selective and fast determination of alpha-amylase (6,7). Formulated as a tablet test (Iodocrom®), it allowed a fast (5-10 min), single step diagnostic of acute pancreatitis (8). Carboxymethyl (CM-) and diethylaminoethyl (DEAE-) crosslinked high amylose starch were proposed, also in the late seventies, as chromatographic materials for biochemical separations (9).

Ten years later, in the early nineties, cross-linked high amylose starch was introduced as an excipient with a high potential for pharmaceutical formulation of many therapeutical molecules (12). A particular feature of HASCL, was the non-monotonous dependency of the drug release times with increasing crosslinking degree (cld). At high cross-linking, the hydration is so strong that the material acts as a powerful disintegrant - Liamid® (15-16). Under the same treatment conditions, when the polymer is subjected only to physical treatment (gelatinization, drying) without chemical reticulation, the uncross-linked high amylose starch was found to generate a faster release, caused by the tablet capping and final disintegration. Maximal release time was obtained for a narrow interval of low cld, whereas at high crosslinking, a fast decrease of release time was found. Considering the percentage of crosslinker used to react with 100 g of polymer as a conventional crosslinking degree, the best drug release time in vitro was obtained for HASCL-6 (10). This particular behaviour (11,13,14) can be explained by the fact that only a low cld will allow enough chain flexibility for self-assembling and stabilization by physical forces, mostly, hydrogen bonding.

Abbreviations: CM-HASCL-6, carboxymethyl high amylose starch cross-linked 6*; AE-HASCL-6, aminoethyl high amylose starch cross-linked 6*; Ac-HASCL-6, acetate high amylose starch cross-linked 6*; *cld, cross-linking degree (expressed in grams of bifunctional agent used to cross-link 100 g of polymers).

These structural aspects induce the formation of the matrix which, in fact, controls the drug release. At higher cld this self-stabilization by hydrogen bonding is hindered and hydroxyl groups are free to hydrate and trigger the polymer swelling.

The HASCL behaves as a hydrogel at low cross-linking and the drug liberation is governed by swelling only or by swelling and diffusion mechanisms. This control was found for a drug loading of 20% or moderately higher while at a loading higher than 30 %, the control of the release is lost and tablets loose their integrity.

As mentioned before, despite crosslinking, the HASCL excipient is still susceptible to alpha-amylase action. This behavior led to a new concept known as Enzymatically Controlled Drug Release (ECDR), applicable for drugs exhibiting particularly long release times or incomplete releases (i.e. due to the drug low solubility or to interactions with starch matrices). The drug liberation can be accelerated by an enzymatically-controlled release, based on the addition of alpha-amylase to the formulation (17,18).

During the last decade, we developed a new series of crosslinked high amylose starch obtained by further derivatization through carboxymethylation, aminoethylation and acetylation (19,20). The first reason for derivatization was to reduce the susceptibility of HASCL to intestinal amylolysis which improves the tablet stability and the drug release time. Surprisingly, it was found that the drug loading capacity was markedly higher compared to the unsubstituted HAS-CL. It is worth noting also , a lower susceptibility (21) to alpha-amylase attack of the polymeric derivatives.

We are now reviewing these new series of crosslinked starch derivatives, discussing their high loading capacity and several mechanistic aspects allowing a better understanding of the processes controlling the drug release from these novel matrices.

Derivatization of crosslinked high amylose starch

It was of interest to follow the behavior of HASCL derivatized with cationic, anionic or less polar groups, and to correlate the matrix modification with their release control properties and to evaluate possible interactions with

124

charged and uncharged drugs. Can these modified matrices modulate the release of drugs through their polar (positive or negative) or non polar global groups and can they generate a slower release, particularly in formulations with a high drug loading ? (22,23).

Hydroxyl groups were shown to play an important role in the organization of the network and consequently in the control of the drug release (10-11). In terms of interacting forces, the carboxylic or amino groups present in CM- and AE- starch derivatives can be involved, for instance, in new hydrogen bonds with the hydroxyl groups of the matrix. On the other hand, the drug characteristics (i.e. molecular size, conformation, ionization, etc.) can affect its diffusion through the matrix. When the gel and the drug are ionized, attracting or repelling forces between them will increase or decrease the diffusion rate.

For the present studies, three types of derivatives were synthesized from HASCL-6 by substitution of hydroxylic groups with cationic (carboxymethyl: CM), anionic (aminoethyl: AE) or less polar acetate (Ac) groups (Scheme I).

Synthesis of CM-HASCL-6 – was realized as described by Mulhbacher et al. (19). After neutralization and washing the remaining wet gel paste was finally dried with pure acetone to obtain the CM-HASCL in powder form.

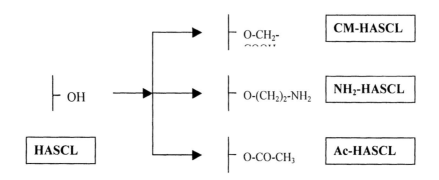

Scheme 1. Schematical presentation of High Amylose Starch and its derivatives.

Synthesis of AE-HASCL-6:
The single-step procedure was almost the same as for CM-HASCL, with the mention that the HAS was first crosslinked and then derivatized with 85 g of

chloroethylamine hydrochloride (9) at a pH between 9 and 10, to generate AE-HASCL.

Synthesis of Ac-HASCL-6:
The one step procedure was similar to that for CM-HASCL-6, with cross-linking and then treatment with 142 mL of acetic anhydride (9) to yield the Ac-HASCL derivatives.

For all derivatives here presented, the degree of substitution can be varied by using different ratios of : substitution reagent / HASCL polymer.

Controlled release properties and dissolution kinetics

It was shown that these new polymeric excipients were able to control the release over 20 h from monolithic tablets loaded with 20 to 60% (w/w) drug. Using different drugs as model tracers, such as Acetaminophen (uncharged molecule), Acetylsalicylic acid (carrying acidic group) and Metformin (carrying a basic group), it was found that the release of the ionic drugs from ionic polymeric matrices (CM-HASCL and AE-HASCL) could be partially controlled via the ionic interaction established between pendant groups of polymeric matrix and drugs. The substitution degree of HASCL derivatives can also be varied in order to modulate the release time (19) of the drug. At 20% acetaminophen loading, no major differences between release times from various derivatives were found (Fig. 1a). For all tested formulations, the release time was estimated as the duration required for 90% of drug to be released in to receiving media, (phosphate buffer 0.05M at pH 7). At higher acetaminophen loading (40% and 60% drug), the CM-HASCL, AE-HASCL and Ac-HASCL derivatives were able to control the release over 16 to 22 h, while for tablets based on non-modified HASCL, 90% release time was much lower: only 2-6 h (Fig. 1b,c). No erosion, no sticking and no flotation of the tabletsbased on derivativzed HASCL were observed during the dissolution tests (19).

Despite the common trend of all derivatives to provide a good control for 16 to 22 h, their general behavior during drug dissolution was different from one derivative to another. The CM-HASCL exhibited a significant swelling capacity where the carboxylic groups possibly play a role, modulating the water access and contributing thus to the drug release. The AE-HASCL tablets have shown remarkable mechanical properties during drug release and less swelling capacity. The amino groups are probably involved in hydrogen bonding, enhancing thus tablet stability and controlling the drug release through a more compact structure. The Ac-HASCL derivatives contained less polar acetate groups and weaker hydrogen stabilisation, but exhibited also good mechanical properties.

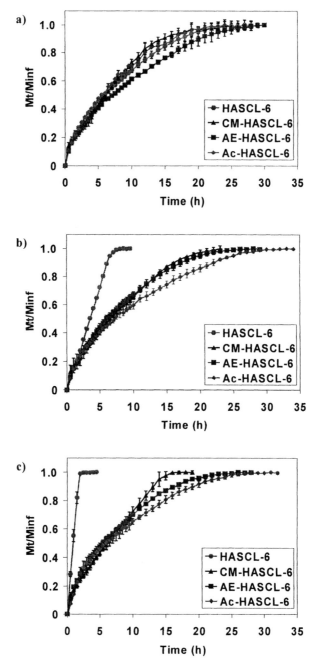

Figure 1: The release of acetaminophen from tablets based on HASCL-6 and derivatives, containing 20% (a), 40% (b) and 60% (c) drug (from Mulhbacher et al, J. Control. Release 2001, with permission).

Longer release times from Ac-HASCL tablets can be explained in this case by a limited water access into the matrix due to the less hydrophilic character of the polymeric excipient.

Acetyl-salicylic acid formulated in highly loaded dosages with the ionic (CM-HASCL or AE-HASCL) or with the less polar (Ac-HASCL) matrices clearly generates longer release times than highly loaded of Metformin (containing amino function), for which only the CM-HASCL matrix allowed longer release times (19). It should be noted that Metformin is a highly soluble (chlorohydrate) molecule and that the required daily dosagee is high enough (approx. 1g). CM-HASCL can be therefore, an excipient of choice for Metformin slow-release.

In an attempt to elucidate if AE-HASCL and CM-HASCL matrices could exert ionic interactions with incorporated drugs and thus modulate their release, additional dissolution tests were conducted with two model molecules: phenylacetic acid (136 Da) and benzylamine (107 Da) having similar molecular size and conformation, except the ionic groups: carboxyl in the first case and amino for the second. Data from Table I clearly indicate that ionic interactions exist between AE-HASCL (with amino groups) and Phenylacetic acid (with a relatively strong carboxylic group) giving longest release (24h). Similar interactions occur between CM-HASCL (with carboxylic groups) and Benzylamine (with a basic amino group) yielding longest release (24h).

Table 1: The influence of charged drugs on the 70% and 90% dissolution times from neutral, anionic and cationic HASCL-6 derivative matrices.

Matrices	Benzylamine Release time (h)		Phenylacetic acid Release time (h)	
	70 %	90%	70 %	90%
Ac-HASCL-6	4	9	8	17
AE-HASCL-6	4	10	15	24
CM-HASCL-6	10	23	10	18

In fact, the CM-HASCL generated a 90% release time of 23 h for benzylamine and 18 hours for phenylacetic acid whereas with AE-HASCL the 90% release time was about 24 hours for phenylacetic acid and only 10 hours for benzylamine. The Ac-HASCL, with no ionic charges (considered as control) generated 9 hours for benzylamine and of 17 hours for phenylacetic acid.

The substitution degree appears to have an important effect on the release time. For the three tracers, the 90% release times from the CM-HASCL and Ac-

HASCL tablets increase when the substitution degrees increase, and this for each drug loading (20, 40 and 60%). In the case of tablets based on AE-HASCL, release kinetics of Acetaminophen and Metformin are similar to those from CM- and Ac- derivatives. In the case of Acetyl-salicylic acid tracer, a higher substitution degree of AE-HASCL generates longer release times (19). Unexpectedly, the dissolution of Acetylsalicylic acid (Fig. 2), was longest at higher drug loading (normally higher loading generates a faster release). This behavior suggests an interaction of Acetyl salicylic acid with the CM-HASCL-6 matrix.

In conclusion, each type of derivative can generate optimal release times for each of the drugs tested. The Ac-HASCL allows the best release time for Acetaminophen that is the less polar drug tested. The AE-HASCL induced the longest release time for Aspirin (carboxylic groups) and CM-HASCL ensured the longest release for Metformin (amino groups). Therefore, the release control of the ionic drugs from CM-HASCL and AE-HASCL matrices could be modulated by ionic interaction. Furthermore, all HASCL derivatives represent novel excipients allowing a good control of the release of drugs from high dosage formulations. For each type of drug, there can be an optimal choice of polymeric starch derivative that generates the best release time as a function of its molecular characteristics.

Mechanistic studies I. Swelling properties

Aspects of the swelling properties of HASCL, CM-HASCL, AE-HASCL and Ac-HASCL matrices in relation with the pH and ionic strength of the dissolution media were recently reported (25) and largely discussed in terms of equilibrium swelling ratio, swelling velocity and values of n exponent, as defined by Khare and Peppas (22). The equilibrium swelling ratio for HASCL and Ac-HASCL after 24h were the lowest, followed in increasing order by those of AE-HASCL and CM-HASCL (Fig. 3a). The swelling ratio of the CM- derivative was the highest. The Ac- derivative and HASCL presented almost the same swelling ratio, despite the fact that the acetyl groups are less polar and, in addition, can hinder the hydrogen association of the amylose, creating thus some amorphous region in the matrix which could be filled and swollen by water. Higher equilibrium swelling ratio values were found for AE- and CM-derivatives with more hydrophilic functional groups. The swelling ratio of the HASCL and its Ac- and AE- derivatives did not change with the pH increase from 1.2 (gastric) to 7 (intestinal), whereas for CM-HASCL the swelling was pH dependent, markedly increasing with pH increase (Fig 3a). The swelling ratio of each polymeric material increased at pH 10 (Fig. 3a), probably due to the fact that hydroxyl groups of high amylose starch and derivatives, still involved in hydrogen bonds (chain-chain) at pH 1-7, begin to be deprotonated at values between pH 7-10.

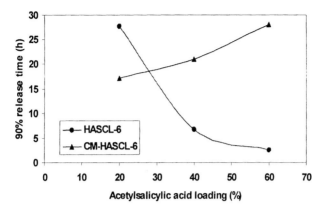

Figure 2: The 90% release time of acetylsalicylic acid from HASCL-6 and CM-HASCL-6 tablets depending on the drug loading.

Surprisingly, in the case of AE-HASCL tablet, there was no decrease in the swelling ratio with the increase of pH, as it would be expected for basic polymers in gel slurry (22). This seems to be an argument in favor of the hypothesis of stronger hydrogen bonding occurring in tablet forms (10-11).

At pH 7, the ionic strength seems to have no effect on the swelling ratio of HASCL and its Ac- and AE- derivatives, whereas for CM-HASCL the increase of ionic strength leads to a marked decrease of the swelling ratio. This can be due to the higher initial hydration volume of carboxyl groups, which would be then in competition for the water uptake with the salt present in the surrounding media with high ionic strength, inducing thus the CM-HASCL swelling decrease.

The initial swelling velocity of HASCL and its Ac- and AE- derivatives was not affected by the ionic strength and the pH, except for high pH 10 (Fig. 3b-c). The CM-HASCL initial swelling velocity was found to decrease with the increase of ionic strength and acetaminophen loading (not shown) and to increase with the increase of pH, as in the case of the equilibrium swelling ratio (Fig. 3a).

The values of exponent n, calculated from the swelling kinetics, presented the same type of dependency upon the pH and ionic strength as the swelling ratio at equilibrium. Differently from the decreasing swelling ratio and the initial swelling velocity, the n exponent was found to increase slightly at higher acetaminophen loading. The n exponent values for the Ac-HASCL (n ≈ 0.55) and for HASCL (n ≈ 0.57) suggested almost a Fickian profile of swelling kinetics whereas AE-HASCL (n ≈ 0.62) corresponded rather to anomalous swelling kinetics. The n exponent of CM-HASCL was of 0.7 (anomalous profile)

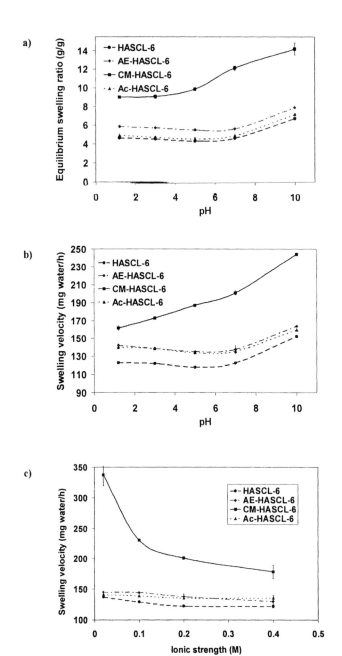

Figure 3: Variation of the equilibrium swelling ratio depending on the pH (a) and swelling velocity variation with the pH (b) and ionic strength (c).

in conditions allowing a low swelling and grew to 0.8 (closer to linear profile) in a medium ensuring a higher swelling.

The values obtained for the n exponent from swelling were markedly different from those obtained from the drug release kinetics at loading of 10% or 20% acetaminophen (Table II), suggesting that the release was controlled not only by the swelling but also by another mechanism (*ie* drug diffusion). The swelling studies were conducted at lower loading (10% or 20%) and using reduced tablet mass (200 mg instead of 500mg) in order to ensure conditions of monodimensional penetration of water (the ratio thickness/diameter should be 1/10) required to allow an appropriate mathematical treatment (25).

Table 2: The n exponent obtained from the swelling and from the release kinetics of tablets containing 10% and 20% acetaminophen.

Matrices	n Exponent			
	Swelling (6h)		*Release (6h)*	
	10 % drug	20% drug	10 % drug	20% drug
HASCL	0.643 ± 0.008	0.678 ± 0.005	0.945 ± 0.029	0.921 ± 0.017
Ac-HASCL	0.621 ± 0.018	0.621 ± 0.009	0.885 ± 0.030	0.984 ± 0.054
AE-HASCL	0.647 ± 0.007	0.664 ± 0.034	0.971 ± 0.071	1.018 ± 0.019
CM-HASCL	0.728 ± 0.022	0.725 ± 0.005	0.925 ± 0.038	0.965 ± 0.085

Mechanistic studies II. Diffusion

Aspects of the diffusion characteristics of various tracers through HASCL, CM-HASCL, AE-HASCL and Ac-HASCL matrices were recently discussed (26). The study followed the variation of permeability dependency on polymeric membrane properties (i.e swelling behavior which will play a role in the membrane porosity, tortuosity, etc.) and on solute characteristics (molecular weight, ionic charge, etc.). Thin tablets (100 mg; 12.72 mm diameter and 1.23 mm thickness) were first hydrated in the appropriate buffer for 24 h and then placed as membranes between the two compartments of the diaphragm-cell apparatus (27). The donor cell was filled with the appropriate buffer containing the diffusant whereas the acceptor cell was filled with the same buffer without diffusant. The same method was used by another group interested in permeation of different tracers trough a swollen cross-linked amylose starch membrane (28).

In agreement with the previous swelling data, the increase of the ionic strength exerted no significant effect on the acetaminophen permeability for HASCL, AE-HASCL and Ac-HASCL matrices, except for CM-HASCL which was sensitive to ionic strength changes (Fig. 4a). It is supposed that a decrease of the swelling volume of the matrix would reduce the drug permeability due to a more compact, less porous matrix. Unexpectedly, in the particular case of the CM-HASCL, a higher permeability of acetaminophen at lower swelling (induced by increasing ionic strength) was found possible (Fig. 4a,b). Similar results were obtained with acetylsalicylic acid as diffusant.

A high swelling volume increase could enhance the water regain (water entrapped into the gel) and consequently the local viscosity within the tablet, which would induce a harder penetration of the acetaminophen (Fig. 4b). Scanning electron microscopy analysis (Fig. 4c,d) showed a more compact structure and smaller pore size for tablets previously swollen for 24 hs at higher ionic strength.

For instance, the size of the pores appears two times greater for the CM-HASCL previously swollen in 0.2 M than in 0.4 M ionic strength buffer. It is worthnoting that, prior to lyophilization, the CM-HASCL tablets swollen in 0.2 M ionic strength buffer were larger than those swollen in 0.4 M (25), whereas once lyophilized, they had the same size irrespective of the buffer in which they were swollen. At low ionic strength there is more water associated with the highly swollen hydrogel. The increase of ionic strength signifies more hydrated ions, higher fluidity and an easier diffusion of active molecules (Fig. 4b). The high swelling capacity of the CM-derivative could affect the partition coefficient of the acetaminophen which was slightly lower (1.07) compared to those obtained for HASCL and its amino and acetate derivatives (about 1.33). In the case of CM-derivative, the interactions between acetaminophen and the polymer were "diluted" by the higher water content present in the membrane (26).

The effect of molecular weight of diffusant on the permeability was also studied. Using polyvinyl-pyrollidone (PVP, Kollidon M.W. 2.5 – 1250 kDa, BASF) as tracers, the permeability coefficients were higher in the case of CM-HASCL than for HASCL, Ac-HASCL and AE-HASCL. When plotted on a log-log scale, the experimental values of the permeability coefficient for the five PVP grades through membranes based on each starch derivative, fit well with a power trendline (26). These data are in agreement with those of Ju et al. (28) regarding the diffusion coefficients of polymers in the layer adjacent to a swollen matrix.

For starch derivatives, the diffusion was shown to depend on the molecular weight of the diffusant, whereas the partition coefficients were affected by the affinity of the diffusant for the polymeric matrices and for the dissolution medium modulated by attractive or repulsive ionic interactions. The fact that for the HASCL, Ac-HASCL and AE-HASCL matrices no major differences were found in terms of diffusion and permeability measured at various ionic strengths is important, showing the versatility of these matrices (possibility to use any of them in function of the required formulation characteristics). This also shows the stability of the release profile, irrespective to the site of delivery in the intestinal tract. Furthermore, diffusion and partition studies could represent a useful tool to evaluate eventual drug - excipient interactions, important for the formulation design of various active agents. These results can also be useful for the pharmaceutical formulation of high molecular weight drugs.

Conclusions

The new excipients based on cross-linked high amylose starch derivatives can modulate the release of drugs in relation with their positive or negative global charge and better control of their release, in case of high load dosages, particularly when ionic or hydrophobic drug-matrix interactions occur. Their substitution degree can also be varied to control the drug's kinetic profile over 24 h. Furthemore, hydration would induce a reorganization of the polymeric chains, generating even more stable networks able to control the drug release. The swelling behavior and the diffusion characteristics of various tracers through such tablets based on starch derivatives was not significantly affected by the pH or ionic strength of the surrounding media, except for the carboxymethyl derivative which exhibited a markedly higher equilibrium swelling ratio and a pH dependency.

Acknowledgements

Jérôme Mulhbacher was the recipient of a University – Industry graduate studentship from FQRNT with participation of Labopharm Inc. Both contributions are gratefully acknowledged.

References

1. Wurzburg, O. B.; Preparation of starch derivatives. *US* 2935510/1960.
2. Serban, M.; Shell, H.D.; Mateescu, M. A. Preparation and properties of new amylose based carriers for exclusion chromatography. *Rev. Roum. Biochim.* **1975**, 12, 187-191.
3. Schell H.D., Mateescu M.A., Bentia T., Jifcu A. Alpha-amylase purification and separation from glucoamylase by affinity chromatography on cross-linked amylose. *Anal. Letters* **1981**, 14, 1501-1514.
4. Jacques, W.; Mateescu, M. A. Affinity chromatography of hemoglobin on Cross-linked Amylose. *Anal. Letters.* **1993**, 26, 875-886.

134

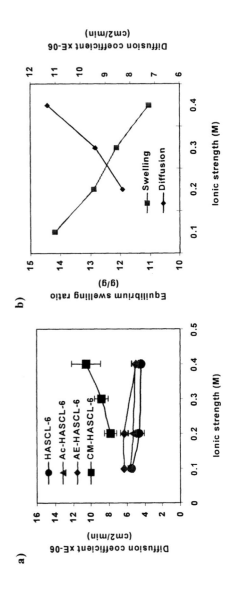

a)

b)

135

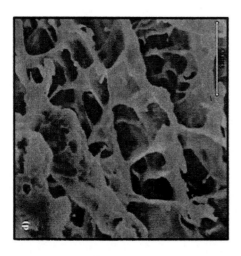

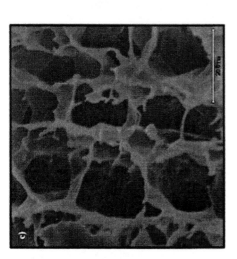

Figure 4: Effect of the ionic strength on the Diffusion coefficient of acetaminophen through HASCL and its derivatives (a), and on the CM-HASCL swelling and permeability to acetaminophen (b); Scanning Electron Microscopy (magnitude x1500) of lyophilised CM-HASCL-6 previously swollen in a buffer with: 0.2 M (c) and 0.4 M (d) ionic strength.

5. Desmangles, R.; Flipo, D.; Fournier, M; Mateescu, M.A. Fast separation of macrophages by retention on cross-linked amylose and release by enzymatic amylolysis of the chromatographic material. *J. Chromatogr.* **1992**, 584, 121-128.

6. Mateescu, M.A.; Schell, H.D.; Enache, E.; Valsanescu, T.; Bentia, T.; Petrescu, A.; Zarchievici, V.; Rotaru, C. Selective determination of alpha-amylase with cross-linked substrate tablets. *Anal. Letters* **1985**, 18, 79-91.

7. Mateescu, M.A.; Schell, H.D. A new amyloclastic method for the selective determination of alpha-amylase. Carbohydr. Res. **1983**, 124, 319-323.

8. Valsanescu, T.; Mateescu, M.A. "All-reagent" tablet-test and method for rapid and selective alpha-amylase iodometric determination. *Anal. Biochem.* **1985**, 146, 299-306; *Romanian Patent 85456/1984).*

9. Schell, H.D.; Serban, M.; Mateescu, M. A.; Bentia, T. Acid and basic amylose ionic exchangers. *Rev. Roum. Chim.* **1978**, 23, 1143-1147.

10. Dumoulin, Y.; Alex., S.; Szabo, P.;. Cartilier, L.; Mateescu, M. A. Cross-linked amylose as matrix for drug controlled release. *Carbohydr. Polym.* **1998**, 32, 361-370.

11. Ispas-Szabo, P.; Ravenelle, F.; Hassan, I.; Preda, M.; Mateescu, M.A. Structure-properties relationship in cross-linked high amylose starch for use in drug controlled release. *Carbohydr. Res.* **2000**, 323, 163-175.

12. Lenaerts, V.; Dumoulin, Y.;. Mateescu, M.A. Cross-linked amylose tablets for linear slow - release of theophylline. *J. Control. Release* **1991**, 15, 39-46.

13. Lenaerts, V.; Moussa, I.; Dumoulin, Y.; Mebsout, F.; Chouinard, F.; Szabo, P.; Mateescu, M. A.; Cartilier, L.; Marchessault, R. H. Cross-Linked High Amylose Starch for Controlled Release of Drugs: Recent Advances. *J. Control. Release* **1998**, 53, 225-234.

14. Ravenelle, F.: Rahmouni, M.; Contramid®: Self-assembly of high-amylose starch for controlled drug delivery. *Chapter 4 (the present book).*

15. Dumoulin, Y.; Clément, P.; Mateescu, M. A.; Cartilier, L. Cross-linked amylose as a new binder/disintegrant in compressed tablets. *S.T.P. Pharma. Sciences* **1994**, 4, 329-335.

16. Cartilier, L.; Mateescu, M. A.; Dumoulin, Y.; Lenaerts, V. Cross-linked amylose as a binder/disintegrant in tablets. US 5 616 343/1997.

17. Mateescu, M.A.; Dumoulin, Y.; Cartilier, L.; Lenaerts, V. Cross-linked polyhydroxylic material for enzymatically controlled drug release. *PCT: WO94/02121; US Patent 5 603 956 / 1997.*

18. Dumoulin, Y.; Cartilier, L.;.Mateescu, M.A. Cross-Linked amylose tablets containing alpha-amylase: Enzymatically Controlled Drug Release (ECDR) system. *J. Control. Release* **1999**, 60, 161-167.

19. Mulhbacher, J.; Ispas-Szabo, P.; Lenaerts, V.; Mateescu, M.A. Cross-Linked high amylose starch derivatives as matrices for controlled release of high drug loadings. *J. Control. Release* **2001**, 76, 51-58.

20. Lenaerts, V.; Chouinard, F.; Mateescu, M. A.; Ispas-Szabo, P. Cross-linked high amylose starch having functional groups as a matrix for the slow release of pharmaceutical agents. *US 6 419 957/2002*.
21. Mulhbacher, J.; McGeeney, K.; Ispas-Szabo, P.; Lenaerts V.; Mateescu M.A. Modified high amylose starch for immobilization of uricase for therapeutic application. *Biotechnol. Appl. Biochem.* **2002**, 36, 163-170.
22. Khare, A. R.; Peppas, N. A. Swelling/deswelling of anionic copolymer gels. *Biomaterials.* **1995**, 16, 559-567.
23. Berger, M.R.; Mayer, J.M.; Flet, O.; Peppas, N.A.; Gurny, R. Structure and interactions in covalently and ionically crosslinked chitosan hydrogels for biomedical applications. *Eur. J. Pharm. Biopharm.* **2004**, 57, 19-34.
24. Mateescu, M. A.; Lenaerts, V.; Dumoulin, Y. Use of cross-linked amylose as a matrix for the slow release of biologically active compounds. *US patent 5456921/1995*.
25. Mulhbacher, J.; Ispas-Szabo, P.; Mateescu, M.A. Cross-Linked high amylose starch derivatives for drug . II. Swelling properties and mechanistic study. *Int. J. Pharm.* **2004**, 278, 231-238.
26. Mulhbacher, J.; Mateescu, M.A. Cross-Linked high amylose starch derivatives for drug release. III. Diffusion properties.. *Int. J. Pharm.* **2005**, 297, 22-29.
27. Peppas, N.A.; Wirght, S.L. Drug diffusion and binding in ionizable interpenetrating networks from poly(vinyl alcohol) and poly(acrylic acid). *Eur. J. Pharm. Biopharm.* **1998**, 46, 15-29.
28. Rahmouni, M.; Lenaerts, V.; Leroux, J-C. Drug permeation through a swollen cross-linked amylose starch membrane. *S.T.P. Pharma Sci.* **2003**, 13, 341-348.
29. Ju, R.T.C.; Nixon, P.H.; Patel, M.V. Diffusion coefficients of polumer chains in the disffudion layer adjacent to a swollen hydrophilic matrix. *J. Pharm. Sciences* **1997**, 86, 1293-1298.

Hyaluronan

Chapter 7

Hyaluronan: Investigations into the Mode of Action of Hyaluronan in Topical Drug Delivery

Marc B. Brown, Stuart A. Jones, Weijiang He, and Gary P. Martin

Pharmaceutical Sciences Research Division, King's College London, 150 Stamford Street, London SE1 9NH, United Kingdom

Hyaluronan (HA) is a polyanionic, polysaccharide ubiquitous in mammals where it has a protective, structure stabilising and shock-absorbing role. The unique viscoelastic nature of HA along with its biocompatibility and non-immunogenicity has led to its use in a number of clinical applications. The ability of HA to localise therapeutic agents in the superficial layers of the skin has led to the commericalisation of Solaraze®, a 3% diclofenac in 2.5% HA gel, for the topical treatment of actinic keratoses (AKs), the third most common skin complaint in the US. However, the means by which HA enhances topical drug delivery remains unclear. The data described in this study demonstrate that HA offers promotes the delivery and localisation of drugs in the skin. In addition, it would appear that the molecular weight and concentration of HA may be critical for such drug delivery properties.

Introduction

Hyaluronic acid was discovered in bovine vitreous humour by Meyer and Palmer in 1934 [1]. It is most frequently referred to as hyaluronan (HA) due to the fact that it exists *in vivo* as a polyanion and not in the protonated acid form. HA is ubiquitous in that it is distributed widely in vertebrates and is present as a component of the cell coat of many strains of bacteria (e.g. *A. Streptococcus*) [2]. Commercially produced HA is isolated either from animal sources, within the synovial fluid, umbilical cord, skin, and rooster comb, or from bacteria through a process of fermentation or direct isolation. The molecular weight of HA is heavily dependent on its source however, refinement of these isolation processes has resulted in the commercial availability of numerous molecular weight grades extending up to a maximum of 5,000,000 daltons [3]. Extensive studies on the chemical and physicochemical properties of HA and its physiological role in humans, together with its versatile properties, such as its biocompatibility, nonimmunogenicity, biodegradability and viscoelasticity, have proved that it is an ideal biomaterial for cosmetic, medical and pharmaceutical applications. Several of HA's important physicochemical properties are molecular weight dependent and therefore discrete differences in function over the wide range of commercially available molecular weights enables HA to be used in a diverse set of applications. In addition, chemical modification of HA can produce a more mechanically and chemically robust material which still retains its biocompatibility and biodegradability. Methods detailing the synthetic transformation of HA and the resulting applications in cosmetic, medical and pharmaceutical applications have been extensively reviewed previously [4-8] and are not discussed in further detail here.

Physical, Chemical and Biological Properties of HA

The primary structure of HA comprises an unbranched linear chain with repeating units of of D-glucuronic acid and N-acetyl-D-glucosamine, linked together through alternating $\beta_{1,3}$ and $\beta_{1,4}$ glycosidic bonds. Hydrophobic faces exist within the secondary structure of HA, formed by the axial hydrogen atoms of about 8 CH groups on the alternating sides of the molecule. Such hydrophobic patches, energetically favour the formation of a meshwork-like β-sheet tertiary structure as a result of molecular aggregation [9]. The tertiary structure is stabilised by the presence of inter-molecular hydrogen bonding [10]. The hydrophobic and hydrogen bonding interactions allow large numbers of molecules to aggregate leading to the formation of molecular networks (matrices) of HA. HA in aqueous solution has been reported to undergo a transition from Newtonian to non-Newtonian flow with increasing molecular weight, concentration or shear rate.

In addition, the higher the molecular weight and concentration of HA, the higher the viscoelasticity the solutions possess [11,12]. The viscoelasticity of HA in aqueous solution is pH-dependent and effected by the ionic strength of the environment in which it is placed [13].

HA in combination with other glycosaminoglycans (GAGs) such as dermatan sulphate, chondroitin sulphate and keratin sulphate are prominent in tissues such as the skin. HA is found to exist together with protein cores (aggrecans) to which the other GAGs are also attached. The most important property of these molecules is their ability to bind to water and this induces the proteoglycans to become hydrated such that a gel-like system is formed.

HA is found in almost all vertebrate organs, but most abundantly in the extracellular matrix of soft connective tissues. HA is particularly abundant in mammalian skin where it constitutes a high fraction of the extracellular matrix of the dermis [14]. The estimated total amount of HA in human skin has been reported to be 5 g, about a third of the total amount of HA believed to be present within the entire human body.

Cosmetic, Medical and drug delivery applications of HA

HA has been extensively utilised in cosmetic products because of its viscoelastic properties and excellent biocompatibility. Application of HA containing cosmetic products to the skin is reported to moisturise and restore elasticity to the skin thereby achieving an anti-wrinkle effect, albeit no rigorous scientific proof exists to substansiate this claim. HA-based cosmetic formulations or sunscreens may also be capable of protecting the skin against UV irradiation due to the free radical scavenging properties of HA [15]. In addition, HA, either in a stabilised form or in combination with other polymers, is used as a component of commercial dermal fillers (e.g. Hylaform®, Restylane® and Dermalive®) in cosmetic surgery. It is reported that the long-term injection of such products into the dermis, can reduce facial lines and wrinkles with fewer side effects and better tolerability compared to the use of collagen [16]. The main side effect may be an allergic reaction, possibly due to impurities present in HA [17]. Lin et al. (1994) [18] have investigated the feasibility of using HA as an alternative implant filler material to silicone gel in plastic surgery and HA is commonly used as growth scaffold in surgery, wound healing and embryology. In addition, administration of purified high molecular weight HA into orthopaedic joints has been reported to restore the desirable rheological properties and alleviate some of the symptoms of osteoarthritis [19]. The success of the medical applications of HA has led to the production of several successful commercial products which have been extensively reviewed previously [8].

HA has been extensively studied in ophthalmic, nasal and parenteral drug delivery. In addition, more novel applications including, pulmonary, implantation and gene delivery have also been suggested [8]. In the majority of drug delivery applications HA acts either as a mucoadhesive, to retain the drug at its site of action/absorption, or as an *in vivo* release/absorption modifier, to improve drug bioavailability.

Administration of therapeutics to the skin using HA

Topical delivery for the treatment of skin disorders offers numerous potential advantages over systemic therapies, such as those involving the use of oral or parenteral products. These include avoidance of hepatic first-pass metabolism, improved patient compliance and ease of access to the absorbing membrane, i.e. the skin. In addition, by directly administering the drug to the pathological site, any adverse effects associated with systemic toxicity can be minimised. However, the targeted delivery of drugs for the treatment of topical disorders is not trivial since percutaneous drug absorption is a partition-diffusion process. First the drug partitions from the formulation into the stratum corneum and then, after diffusing across the stratum corneum, it partitions into and diffuses across the epidermis. This partition diffusion process is repeated as the drug moves to deeper skin layers.

The physiological function of the stratum corneum, the outermost and non-viable layer of the skin, is to act as a protective barrier for the body and as such, it is particularly effective at preventing the permeation of hydrophilic molecules, including some drugs, into deeper skin layers where viable cells are located. The excellent barrier properties of the stratum corneum means that in order to achieve therapeutic levels of many drugs additional methods of chemical or physical penetration enhancement are often required. However, the use of such methods can promote the transport of drug all the way through the skin into the systemic circulation. Entry into the blood stream can be desirable if systemic effects are required (as with transdermal patches) but, if the target receptor for the drug lies within the epidermis, it increases the chances of undesirable side effects. Consequently, in topically applied formulations there is a need for a system which aids drug permeation through the stratum corneum whilst avoiding further penetration all the way across the deeper layers of the skin (the dermis) and into the blood stream. Obviously, such a delivery system using traditional formulation approaches is difficult to achieve. Relatively few investigations have been conducted into the development of drug delivery technologies that facilitate targeting of drugs to the superficial layers of the skin, although one such system which is receiving increasing attention is HA [20-22]. Solaraze®, a 3% diclofenac in 2.5% HA gel for the topical treatment of actinic keratoses (AKs), has recently received regulatory approval in the US, Canada and Europe [23]. AK is the third most common skin complaint in the US and results from excessive sun exposure. AK lesions are most common on the scalp, face, and/or dorsum of the hands in light skinned persons aged ≥50 years, particularly those with a history of occupational exposure to the sun. AK are now considered to be synonymous with squamous cell carcinoma (SCC) in situ with growing evidence that AK and SCC lie on a clinical, histological, cytological and molecular continuum [24]. However, a mechanism of action to explain the topical delivery properties of HA remains to be elucidated. Thus, the aim of this study was to compare the effect of molecular weight and concentration of HA, in comparison to other glycosaminoglycans and pharmaceutically relevant polysaccharides, on the percutaneous penetration of diclofenac and ibuprofen.

Materials and Methods

Materials

Sodium hyaluronate (viscosity average molecular weight (Mv) 600,000) was kindly donated by SkyePharma (London, UK). Sodium hylauronate (viscosity average Mv 137,000) was prepared by autoclaving a solution of high Mv HA at 121°C for 110 min. All other materials were obtained from Sigma (Poole, UK). All solvents used for chromatography were analytical grade.

Partition cells were made on site at King's College London and were constructed from polypropylene in two parts. The lower part provided a base to support the skin sample during the partition studies, whilst the upper part, comprising a hollow threaded cylinder, was employed to constrain the skin sample *in situ*. The Franz cells comprised a sampling port, a donor compartment (internal diameter 0.85 cm) and receptor compartment (volume 1.80 ml) which were fixed together using a clip. The Franz cells were also made on site from glass.

Methods

Skin preparation

Human skin was obtained from patients undergoing elective abdominoplastic surgery. This procedure recieved full approval from King's College London Research Ethics Committee and informed consent was obtained from the donors who were females, aged between 25-40 y. Excised skin was frozen at -20°C within 2 h of surgery and stored until use. Full thickness skin sections (dermis and epidermis layers) were prepared by carefully removing subcutaneous fat and other debris using forceps and scalpel. For the epidermal sheet experiments individual portions of skin were immersed in water at 60°C for 45 s. The skin was then placed flat, dermis side down, on a corkboard and the epidermis (comprising SC and viable epidermis) was gently removed from the underlying dermis using forceps and floated onto the surface of cold water. All skin samples were then cut into small circular pieces using a template of similar dimensions to the partition cell or Franz cell. Skin from the same section was used for each drug study enabling comparison of the formulation effect for the same drug but not between drugs.

Skin-partitioning studies

Donor solutions containing drug (10 μg/ml for diclofenac, 20 μg/ml for ibuprofen) and 1% w/w polysaccharide were prepared in deionised water (controls were drug in deionised water alone, DW) and allowed to hydrate for 24 h. The skin was mounted in the partition cells with the epidermis side up, and the unit was then placed into an air tight 50 ml glass jar containing 20 ml of the donor solution such that only the epidermal surface was exposed to the solution. The bottle was then transferred to a water bath (32°C) and shaken for 48 h. Preliminary experiments showed that this time was required for equilibration to be established. Samples were then removed and assayed for drug content by HPLC. The amount of drug which partitioned into the skin was measured indirectly by the loss of compound from the donor solution. The percentage of drug partitioning into the skin was calculated from the concentration of compound in the donor solution before and after equilibrium. Control experiments were performed with no skin in the partition cell to determine any loss due to adsorption to the partition cell or glass. Results are expressed as mean $\pm$ s.d for n = 4 experiments.

Franz cell studies

Small sections of full thickness skin for the dermal deposition studies or epidermal sheet for the diffusion studies were mounted, stratum corneum side up, in the Franz cells. The receptor compartment was filled with pH 7.4 phosphate buffer after which the diffusion cell was placed on a stirring plate in a water bath maintained at 32°C. The receptor fluid was continuously stirred using a small teflon-coated magnetic bar to ensure complete mixing of the receptor fluid. The Franz cells were allowed to stand for 12 h in order to equilibrate the skin and also to remove any visceral debris that may have remained on the dermal side of the skin. After equilibration, the receptor fluid was changed by replacing with fresh buffer and any air bubbles which had accumulated inside the receptor compartment or at the skin/receptor interface, were removed by tilting the cell.

In the dermal deposition experiments 50 µg of each formulation containing 1.75% w/w diclofenac or 20% w/w ibuprofen in 1% w/w polysacharides (or deionised water) unless otherwise stated, was applied to the surface of the skin (n = 4 for each formulation) and rubbed in with a circular motion using a pre-weighed glass rod. Throughout the 48 h of the experiment, the receiver fluid was stirred to ensure homogeneity, and was also maintained at the same level as the skin. After 48 h, the experiment was terminated and the skin removed from the diffusion cell to enable the mass balance study to be performed in order to quantify the amount of drug on the skin, in the skin and in the receiver fluid. The extraction techniques, which involved homogenisation, collagenase digestion, ultracentrifugation and solvent extraction, were all validated and proved to be 100% efficient in recovering diclofenac and ibuprofen from all of the sample matrices. Drug concentration was measured using HPLC analysis.

The diffusion experiments were performed using similar methodology to the dermal deposition studies apart from 10 µg of a saturated solution of each drug in the specified HA concentration being used. In addition, during the 48 hours of the experiment, at appropriate time intervals, 0.2 ml of the receptor fluid was withdrawn and replaced by the the same volume of fresh buffer solution. Drug concentration of each of the removed samples was measured using HPLC analysis.

HPLC analysis of drug concentration

The HPLC analysis of diclofenac and ibuprofen was performed using a 15 cm x 4.6 mm I.D. Spherisorb RP-C18 (5 µm) column (Hichrom Ltd., Reading, Berkshire, UK), with a 10 mm C18 guard column (S5ODS2-10C5) (Hichrom Ltd., Reading, Berkshire, UK) using a CM 4000 pump and a CI-4100 integrator connected to a SpectroMonitor 3100 UV detector (all LDC Analytical, Florida, USA). Analysis was performed isocratically with a mobile phase comprising 73% phosphate buffer (45 mM, pH 7) and 27% acetonitrile: tetrahydrofuran (7:3 v/v). All samples were made up in mobile phase, containing an internal standard (20 µg/ml). For diclofenac analysis, ibuprofen was used as the internal standard, and vice versa. An injection volume of 100 µl was used, the samples were eluted at a flow rate of 1.10 ml/min and monitored at 273 nm.

Results

Partition studies

The effects of 1% w/w polysaccharide concentrations on the skin partitioning of ibuprofen is shown in figure 1. Similar results were found with diclofenac (data not shown). The results demonstrate that all three glycosaminoglycans: HA, Chondroitin sulfate (CS) and heparin (HP) significantly increased the partitioning of diclofenac and ibuprofen into the skin compared to the aqueous control. The extent of the increase induced by HA however was greater than that for both CS and HP at the same w/w concentrations. In contrast, NaCMC was found not to increase the partitioning of either diclofenac or ibuprofen into skin ($p > 0.05$).

Franz cell studies

Dermal deposition studies

The amount of drug on the surface of the skin and diffused through to the receptor compartment was determined 48 h after application of 1.75% w/w diclofenac or 20% w/w ibuprofen all formulated in 1% w/w polysaccharide formulations. None of the polysaccharides had a significant effect on the amount of diclofenac remaining on the skin surface compared to the control (data not shown). However, HA significantly reduced the amount of ibuprofen remaining ($p < 0.01$) on the skin surface compared to the control, a phenomena not observed for the other polysaccharides. In the receptor compartment, significantly less diclofenac and ibuprofen was found in the receptor fluid when formulated in 1% w/w HA compared to the other polysaccharides (data not shown).

The amount of diclofenac in the skin 48 h after application of all vehicles tested (1% w/w) is shown in figure 2 and similar results were also obtained for ibuprofen. In general, the amount of either drug delivered into the skin was observed to obey the following trend: HA>CS>NaCMC≅DW. Significantly more diclofenac and ibuprofen was delivered into the skin when applied in HA or CS, when compared to the other polysaccharides and the aqueous control. In addition, it was found that HA delivered significantly more drug to the skin when compared to CS ($p < 0.05$).

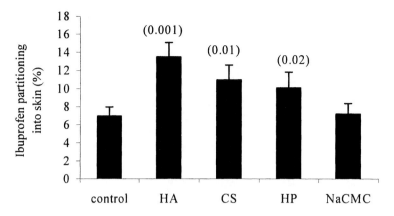

Figure 1 . The effect of 1% w/w polysaccharides on the dermal partitioning of 20 µg/ml ibuprofen, n=4, mean±SD (P-value of significance compared to control).

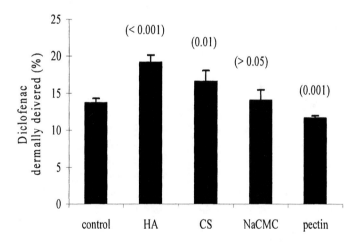

Figure 2. The effect of 1% w/w polysaccharide on the percentage of the applied dose of diclofenac (1.75% w/w) remaining in the skin after 48 h, n=4, mean±SD (P-value of significance compared to control).

150

Diffusion studies

The current Solaraze formulation contains high molecular weight HA at a concentration of 2.5% w/w thus the effect of both HA concentration and the molecular weight were investigated on the diffusion of diclofenac across human epidermal sheet and the results are depicted in figures 3 and 4 respectively. The data supported those obtained in the dermal deposition studies and demonstrate that for localised skin delivery a high molecular weight and concentration of HA is required.

Discussion

It is apparent that the tendency to promote the localisation of the drugs within the skin was more apparent when HA was incorporated in the topical vehicle that when other GAGs or other pharmaceutically relevant polysaccharides were utilised. It also would appear that the ceffects attributable to HA are concentration and molecular weight dependent.

Despite the above data an exact mechanism of action to explain the topical delivery properties of HA, remains to be elucidated. It is known that the degree of hydration of the stratum corneum influences skin permeability. Increased hydration opens the compact substance of the stratum corneum by loosening the dense, closely packed cells thus increasing the permeability to many drugs [25]. Hydrophobic patches and oleaginous formulations are based on such a principle in that they occlude the skin, inhibiting transepidermal water loss and increasing stratum corneum hydration resulting in enhanced percutaneous delivery of the formulated drug. Polysaccharides, in general, are regarded as moisture-control agents [26] which are the reason for their incorporation in many topical and cosmetic preparations. However, the skin hydration properties of HA are considered to be much higher than other polysaccharides because of its considerable capacity to bind water, especially at high concentration and molecular weight [27]. Thus, the topical application of HA may result in increased hydration of the stratum corneum and lead to the enhanced topical delivery of a concomitantly applied drug across this skin barrier.

Such properties, however, do not account for drugs formulated in HA being retained in the skin with little systemic absorption, as indicated previously. Browne *et al* (1999) [28] have reported the surprising movement of HA into the keratin, epidermal and dermal layers of mouse and human skin, with radiolabelled HA apparently being absorbed rapidly from the surface of the skin and into the epidermis. There are three possible factors, which may be involved in this behaviour. First, the presence of underlying skin HA receptors may direct the localisation of applied HA. Second, the specific structure of hydrated HA and the presence of a hydrophobic area may enable the absorption of this macromolecule across membranes. In turn, the localisation of HA can then

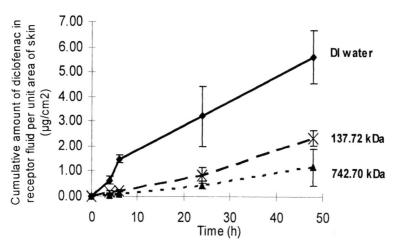

Figure 3. The effect of HA concentration on the diffusion of diclofenac across epidermal sheet over 48 h, n=3, mean±sd.

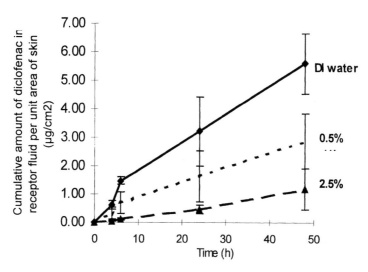

Figure 4. The effect of HA molecular weight on the diffusion of diclofenac across epidermal sheet over 48 h, n=3, mean±sd.

influence the topical delivery of the drug. Finally, the most favoured explanation, the increased hydration of the surface layers of the skin in the presence of HA not only enhances drug absorption across the stratum corneum, but also facilitates the retention of drug within the more hydrated epidermal layers (possibly by exposure of potential drug binding sites) and in doing so decreases drug diffusion into the lower skin layers. The latter mechanism is based on a more general HA effect and does not necessarily rely on the co-localisation of drug and HA.

Conclusions

Hyaluronan is a naturally occurring polysaccharide which is biocompatible and non-immunogenic, rendering it an ideal formulation aid. The results in this study demonstrate that the inclusion of hyaluronan as a vehicle excipient offers clear potential in the dermal delivery and localisation of drugs although the mode of action still remains to elucidated. Such localisation would be desirable for the dermal use of many drugs e.g. cytotoxics, corticosteroids, anaesthetics, retinoids, sex hormones, antifungal, antibacterial, antiparasitic and antiviral agents for a variety of skin disorders.

References

1. Meyer, L.; Palmer, J. *Biol. Chem.* **1934**, *107*, 629-634.
2. Laurent, T. C. In Structure of hyaluronic acid; Balazs, E. A., Ed.; Chemistry and the Molecular Biology of the Intracellular Matrix; Academic Press: London, 1970; pp 703-732.
3. Milas, M.; Rinaudo, M.; Roure, I.; Al Assaf, S.; Phillips, G. O.; Williams, P. A. *Biopolymers* **2001**, *59*, 191-204.
4. Larsen, N. E.; Balazs, E. A. *Adv. Drug. Dev. Rev.* **1991**, *7*, 279-293.
5. Denlinger, J. L. In Hyaluronan and its derivatives as viscoelastics in medicine; Laurent, T. C., Ed.; The Chemistry, Biology and Medical Applications of Hyaluronan and Its Derivatives; Portland Press: London, 1998; pp 235-242.
6. Band, P. A. In Hyaluronan derivatives: chemistry and clinical applications; Laurent, T. C., Ed.; The Chemistry, Biology and Medical Applications of Hyaluronan and Its Derivatives; Portland Press: London, 1998; pp 33-42.
7. Lapcik, L.; Lapcik, L.; De Smedt, S.; Demester, J.; Chabrecek, P. *Chem. Rev.* **1998**, *98*, 2663-2684.
8. Brown, M. B.; Jones, S. A. *J. Eur. Acad. Dermatol. Venereol.* **2005**, *In Press*.

9. Laurent, T. C.; Fraser, J. R. E. *FASEB J.* **1992**, *6*, 2397-2404.
10. Scott, J. E.; Heatley, F. *Proc. Natl. Acad. Sci. U. S. A.* **1999**, *96*, 4850-4855.
11. Gribbon, P.; Heng, B. C.; Hardingham, T. E. *Biochem. J.* **2000**, *350*, 329-335.
12. Gibbs, D. A.; Merrill, E. W.; Smith, K. A.; Balazs, E. A. *Biopolymers* **1968**, *6*, 777-791.
13. Kobayashi, Y.; Okamoto, A.; Nishinari, K. *Biorheology* **1994**, *31*, 235-244.
14. Juhlin, L. *J. Intern. Med.* **1997**, *242*, 61-66.
15. Trommer, H.; Wartewig, S.; Bottcher, R.; Poppl, A.; Hoentsch, J.; Ozegowski, J. H.; Neubert, R. H. H. *Int. J. Pharm.* **2003**, *254*, 223-234.
16. Bergeret-Galley, C.; Latouche, X.; Illouz, Y. G. *Aesthetic Plast. Surg.* **2001**, *25*, 249-255.
17. Friedman, P. M.; Mafong, E. A.; Kauvar, A. N. B.; Geronemus, R. G. *Dermatol. Surg.* **2002**, *28*, 491-494.
18. Lin, K.; Bartlett, S. P.; Matsuo, K.; Livolsi, V. A.; Parry, C.; Hass, B.; Whitaker, L. A. *Plast. Reconstr. Surg.* **1994**, *94*, 306-315.
19. Balazs, E. A.; Denlinger, J. L. *J. Rheumatol.* **1993**, *20*, 3-9.
20. Brown, M. B.; Forbes, B.; Martin, G. P. In The use of hyaluronan in topical drug delivery; Kennedy, J., Phillips, G. O., Williams, P. A., Hascall, V., Eds.; Hyaluronan: Biomedical, Medical and Clinical Aspects; Woodhead Publishers: Cambridge, 2002; pp 249-256.
21. Liao, Y. H.; Brown, M. B.; Martin, G. P. *J. Pharm. Pharmacol.* **2001**, *53*, 549-554.
22. Brown, M. B.; Marriott, C.; Martin, G. P. In A study of the transdermal drug delivery properties of hyaluronan; Willoughby, D. A., Ed.; Hyaluronan in Drrug Delivery; Royal Society of Medicine Press: 1995; pp 53-71.
23. Jarvis, B.; Figgitt, D. P. *American Journal of Clinical Dermatology* **2003**, *4*, 203-213.
24. Schartz, R. A. *Dermatological Surgery* **1997**, *23*, 1009-1019.
25. Bucks, D.; Maibach, H. I. Marcel Dekker: New York, 1999; pp 81-108.
26. Whistler, R. L. In Introduction to industrial gums; Whistler, R. L., BeMiller, J. N., Eds.; Industrial Gums: Polysaccharides and Their Derivatives; Academic Press: San Diego, 1993; pp 1-3.
27. Cowman, M. K.; Liu, J.; Hittner, D. M.; Kim, J. S. In Hyaluronan Interactions: salt, water, ions'; Laurent, T. C., Ed.; The Chemistry, Biology and Medical Applications of Hyaluronan and Its Derivatives; Portland Press: London, 1998; pp 17-25.
28. Brown, T. J.; Alcorn, D.; Fraser, J. R. E. *J. Invest. Dermatol.* **1999**, *113*, 740-746.

Chapter 8

Biomedical Applications of Hyaluronic Acid

Samuel J. Falcone, David Palmeri, and Richard A. Berg

FzioMed, Inc., 231 Bonetti Drive, San Luis Obispo, CA 93401

Hyaluronic acid (HA), also named hyaluronan, is a high molecular weight polysaccharide found in the body in pericellular matrices, various bodily fluids, and in specialized tissues such as the vitreous humor of the eye and cartilage. Hyaluronic acid possesses both biological activities and physical properties that add to the uniqueness of this ubiquitous polysaccharide. The uniqueness of HA and its importance both biologically and physically apparently accounts for its identical structure when synthesized in bacteria, birds, and mammals. Because of this property, HA has been purified from chickens or bacteria for use as a biomaterial in medical devices in humans or other mammals. Hyaluronic acid is unique because of its viscoelastic and hydrodynamic properties, its assembly into extracellular and pericellular matrices, and its effects on cell signaling. Its use as a biomaterial has been driven largely by its physical properties and viscoelastic behavior. The biological properties of HA and its fragments have largely been ignored in its use as a biomaterial. As HA is more widely used and studied, these properties are becoming increasingly apparent and important in its use in medical devices.

Hyaluronic acid (HA) is a component of the extracellular matrix and is a ubiquitous substance found abundantly in nature. Many types of cells synthesize HA; it interacts with other constituents (proteins) of the extracellular matrix including the cell surface to create the supportive and protective structure around the cells. HA has also been shown to provide extracellular signals to cells related to locomotion and gene expression. It is a constituent of all body fluids and tissues, and it is found in higher concentrations in the vitreous humor of the eye, hyaline cartilage, and the synovial fluid. It was initially isolated from the vitreous of the eye.

Structurally, it is a polydisaccharide containing D-glucuronic acid and D-N-acetylglucosamine with repeating β(1-3), β(1-4) saccharide linkages. Three isozymes of hyaluronan synthases have been identified in bacterial and animal cells (for review, see reference(*1*)). These enzymes produce HA of variable molecular weights depending on the tissue. It is a most unusual polymer in that the same polysaccharide is produced in bacteria, birds, and mammals although the molecular weights may differ depending on its source. The highest molecular weight HA is found in cartilage. Hyaluronic acid being an extracellular and pericellular polymer is catabolized by receptor-mediated endocytosis and lysosomal hydrolysis in various tissues (*2*) or after transport to lymph nodes (*3*). Hyaluronidases are broadly distributed enzymes involved in tissue invasion and remodeling, (*4*) and are found in plasma (*5*).

Of special interest is that HA can be degraded by and fragmented to smaller oligosacharides via reactive oxygen (*6*) such as that produced by inflammatory cells, supporting a role for HA fragments in wound healing and inflammation (*7*) (for review of HA homeostasis in the body see reference (*1*)).

Hyaluronic acid has been purified in quantity from cartilage and bacteria owing to the relative abundance of HA in those materials. Once HA was purified its rheological properties as a viscoelastic polymer became apparent. The physical properties of HA point to its pivotal role in the extracellular and pericellular matrix in tissues. Hyaluronic acid, in solution, is viscoelastic and the rheological properties depend on concentration, ionic strength, pH, and molecular weight (*8, 9*). These rheological properties have been exploited for use in some biomedical applications. Solutions of high molecular weight HA are viscous, cohesive, lubricious, and hydrophilic. These properties have found use as viscoelastic adjuncts in ophthalmic surgery, viscosupplementation in the synovium of arthritic joints, and as covalently bound lubricious hydrophilic coatings for medical devices, e.g., stents, catheters (*10,11*). For medical applications that require extended residence time in situ, HA can also be crosslinked by a variety of chemical methods. Cross-linking HA dramatically increases the elastic component of the overall modulus of HA causing the polymer gel to behave more like an elastic solid and less like a viscous fluid in response to deformation. Crosslinked HA products have found use as surgical adhesion barriers, synovial viscosupplements, and more recently as materials for augmentation of the dermis (*12-14*).

This review will focus on the biological and physical properties of HA that have been successfully exploited for its use as a biomaterial.

Biological Properties of Hyaluronic Acid

Hyaluronic acid is found both in the extracellular matrix between cells in tissues and associated with cell surfaces, in the pericellular matrix. Furthermore, HA is not an inert material. It is recognized specifically by several proteins in the extracellular matrix that give rise to important biological functions (*15*).

The first two proteins discovered to interact with HA are the link protein and the protein aggrecan found in the extracellular matrix of cartilage (*16*). The association of these two proteins with HA contributes to the formation of proteoglycan aggregates in cartilage, which are responsible for the bulk elastic and hydrodynamic properties of cartilage. A second type of interaction is with receptor proteins found on cell surfaces (for review see (*17*)). Two of these proteins are CD 44, a receptor found on many cell types (*18*) and RHAMM, a receptor for hyaluronan-mediated motility (*19*).

CD 44 is a transmembrane receptor that connects the pericellular matrix on the outside of cells with cytoskeletal proteins (*17*). The close association of HA with the cell pericellular matrix suggests an explanation for the role of CD44 in leukocyte migration, neoplasia, and in wound healing (*17*). In order for cells to migrate in tissues they must be capable of disrupting connections to the extracellular matrix. As cells migrate they are exposed to HA that is present in the extracellular matrix. A protein that is also implicated in cell mobility is RHAMM which is present in several intracellular compartments and is also exported to the cell surface where it interacts with HA (*17*). RHAMM was shown to be involved in fibroblast locomotion (*20*) and cell mobility (for review, see reference (*17*)).

Since HA is found in pericellular matrix it is not surprising that cell migration, mobility and wound healing may be influenced by HA. Of interest is the finding that fragments of HA, presumably resulting from its degradation, are more active in interaction with these cell surface receptors than native intact HA(21) leading to a role for fragments of HA in acute (*22*) and chronic inflammation (*7*). HA fragments can induce genes coding for inflammatory mediators (*22, 23*). These observations suggests that HA fragments, resulting from degradation of high molecular weight HA, are capable of cell signaling through interaction of receptor-mediated control pathways, followed by alterations in cell mobility and gene expression. Support for this pathway is that fragments of HA are capable of inducing cytokine expression in macrophages (*24*).

The finding that HA interacts specifically with proteins in the body serving both structural and cell signaling functions points to a potentially complex response by tissue when HA is injected into the body in the form of a medical device. For example, as HA is degraded it may have effects on leukocyte mobility and therefore inflammation and wound healing.

Physical Properties of Hyaluronic Acid

HA Structure in Solution and Rheology

The structure of uncrosslinked HA in dilute and concentrated solution has been extensively studied (for a review see reference (25)). The original work envisioned that uncrosslinked HA is a high molecular weight unbranched polysaccharide that behaves as a stiffened random coil in solution. The molecule occupies a large hydrated volume and shows solute-solute interactions at unusually low concentration. Since HA is a polyelectrolyte, the solution properties are greatly affected by ionic strength (25). The conformation of HA in solution has been studied at neutral pH and at physiological concentration, using nuclear magnetic resonance and circular dichroism. These studies support a model incorporating dynamically formed and broken hydrogen bonds that contribute to the semi-flexibility of the polymer chain (26).

Hyaluronic acid that is not crosslinked is water soluble, rapidly resorbed, and has a short residence time in situ that limits its utility for use in biomedical applications. HA in solution is subject to degradation via ultrasound, UV irradiation, thermal, and free radicals (25). Uncrosslinked HA in solution behaves as a pseudoplastic shear thinning fluid and the zero shear viscosity (η_o) correlates to the solution concentration multiplied by the molecular weight (27). Figure 1 describes the relationship of the log η_o to the log of the (concentration *molecular weight) for a series of HA solutions prepared from HA of three molecular weights, 1800 kg mol^{-1}, 680 kg mol^{-1}, and 350 kg mol^{-1}, at several concentrations, between 10-90mg/ml, in phosphate buffered saline. Also included in Figure 1 are seven commercial preparations of uncrosslinked HA used as medical devices. The data for these HA medical products are included in Table 1. The plot of the log of the solution η_o vs. the log of the solution (concentration*MW) is linear. This means that the viscosity of an HA solution can be controlled by adjusting polymer molecular weight and/or the solution concentration.

The data in Figure 1 also imply that the properties of two uncrosslinked HA solutions of a given viscosity can be quite different. Consider two HA solutions both with a η_o of ~170 Pas. Solution A is a 16mg/ml solution of HA 1800 kg mol^{-1} and solution B is a 70mg/ml solution of HA 350 kg mol^{-1}. The η^* at low frequency, 0.0628 rad/s, low deformation rates, or low shear is 172 Pas for both materials. The viscoelastic properties of these two solutions are quite different and this is shown in Figure 2. This Figure describes the complex viscosity (η^*) and the storage viscosity (η''), the elastic component of the complex viscosity vs. frequency for HA solutions A and B. At low frequency both solutions have the same η^*; as the frequency is increased, the η^* of the high molecular weight HA solution decreases more rapidly than the η^* of lower molecular weight HA. At high frequency, the η^* of the concentrated low molecular weight HA is greater than 10 times higher than the dilute solution of high molecular weight HA. The data in Figure 2 also show that the elastic component of the η^*, η", at low frequency for the solution of high molecular weight HA is much higher than that of low molecular weight HA. At low deformation rates, the high molecular weight HA solution with an entangled or aggregated chain structure responds elastically to deformation. Under these conditions the concentrated low molecular weight solution responds in a primarily viscous manner. The high molecular weight HA solution, up to a frequency of ~ 1 rad/s, is more elastic than the solution of low molecular weight HA. Both materials have the same viscosity at low deformation rates, but solution A is a dilute entangled cohesive solution ideally suited for bulk removal during eye surgery, while solution B is a concentrated non-cohesive tissue adhesive material possibly better suited for tissue coating and lubrication medical applications. The higher viscosity and elasticity of low molecular weight HA at high frequency also indicate that it may be a better tissue coating and lubricating material under conditions of high deformation rates.

Cohesive Properties of Hyaluronic Acid: Cohesion-Dispersion Index (CDI)

The cohesive nature of HA is of prime importance to its use as a viscoelastic adjunct in eye surgery (28). The more cohesive the material is the more likely it is to be readily removed from the anterior chamber of the eye during cataract surgery. Cohesion is a result of intermolecular entanglement of the HA polymer chains that imparts a bolus-like behavior to the viscoelastic solution. A quantitative dynamic aspiration method that measures the cohesion of viscoelastic agents has been developed (29). The principle of the technique is to measure the amount of sample aspirated with increasing vacuum applied to the sample; vacuum levels of 127, 254, 381, 533, and 711 mm of Hg were used. At each vacuum level, the weight of material aspirated through a 0.5mm pipette tip in two seconds is measured. The data are plotted as the percent aspirated vs. vacuum level and the slope of the steepest portion of the curve is determined. The cohesive-dispersive index (CDI) is equal to the slope at steepest portion of the curve. Materials that are more cohesive are removed in bulk at a lower vacuum than materials of lesser cohesiveness. Table I list the CDI results for a series of HA viscoelastic agents that have been used in eye surgery. From these data it was concluded that the cohesive nature of HA increases with increasing molecular weight (29). The study was extended to include the effect of concentration and molecular weight on the cohesive properties of uncrosslinked HA (27). It was found that the CDI correlates very well to solution concentration

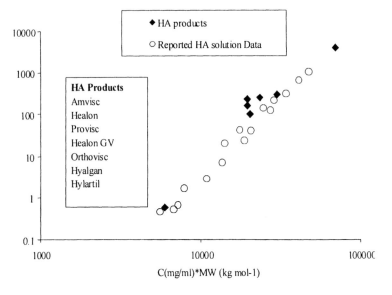

Figure 1. The zero shear viscosity (η_o) of uncrosslinked HA materials as a function of MW and concentration. Open circles are data from Falcone et al.(27) and filled diamonds are data from Poyer et al.(29)

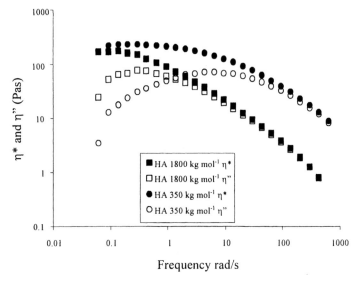

Figure 2. The complex modulus (η^*) and storage modulus (η'') vs. frequency for two HA solutions of different molecular weight and concentration; 1.6% HA 1800 kg mol^{-1} and 7% HA 350 kg mol^{-1} in phosphate buffered saline.

of HA, and both sets of data are plotted in Figure 3. The results of this study also confirmed that the cohesiveness of HA increases with increasing molecular weight and that the cohesive nature of HA correlates to the complex viscosity at high frequency.

The cohesive properties of HA polymeric solutions are a function of η_o, but the zero shear viscosity does not accurately predict the cohesive behavior of the solution. The cohesive properties of uncrosslinked HA preparations were found to be proportional to the dynamic rheological properties of the HA polymer solutions. The cohesive properties, CDI, of the HA solutions were found to correlate well to the dynamic viscosity at high frequency, 62.8 rad/s. Figure 4 describes the relationship of the log of the CDI to the log of the dynamic viscosity, η' for a series of HA solutions prepared from HA of three molecular weights, 1800 kg mol^{-1}, 680 kg mol^{-1}, and 350 kg mol^{-1}, at several concentrations, between 10-90 mg/mL, in phosphate buffered saline. Also included in Figure 4 are three commercial preparations of uncrosslinked HA used as medical devices: Healon, Provisc, and Viscoat. For both series, the CDI correlates well to the η' at high frequency. For this data, the CDI decreases as the η' increases and the cohesive nature of uncrosslinked HA correlates to the high frequency viscous component of the viscoelastic material. High molecular weight HA has less viscosity and stiffness and elastic response at high deformation rates. These rheological properties produce a more cohesive entangled structure that rapidly and easily flows through a small orifice, thus yielding a high CDI. The data presented here reveal that the CDI correlates well with the dynamic viscosity at high frequency in dynamic viscoelastic testing.

From the rheological properties of high molecular weight HA, it appears to be ideally suited for a wide variety of medical applications. Under conditions of low shear, the applications include viscoelastic gels for eye surgery in the anterior chamber of the eye or as a temporary replacement for the vitreous humor where a gel-like material is required. Their use in ocular surgery is improved by their high viscosity at low shear rates in the anterior chamber of the eye, whereas their low high shear viscosity facilitates their injection into and removal from the eye through a small-bore needle.

Studies of the synovial joint have indicated that HA is responsible for the viscoelasticity of synovial fluid (28). Analysis of the synovial fluid of an osteoarthritic joint has shown that, compared to a healthy joint, the molecular weight and viscosity of the HA are reduced (30, 31). Hyaluronic acid would be expected to function well as a shock absorber under low shear in the resting knee. However, owing to the shear thinning nature of high molecular weight HA and its cohesiveness, high molecular weight HA would be expected to be a less effective lubricant than low molecular weight HA. This observation may account for some of the discrepancy in results of various formulations of HA in the treatment of osteoarthritis, OA (32, 33) where it is not clear that the function of HA in reducing pain in joints is due to a lubricating effect, a shock absorber effect, or a biological effect.

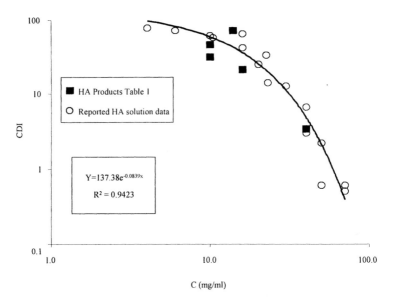

Figure 3. The cohesiveness of uncrosslinked HA vs. the solution concentration. This data is taken form two independent studies. Filled squares, data from Table I. Open circles, data from Falcone et al.(27)

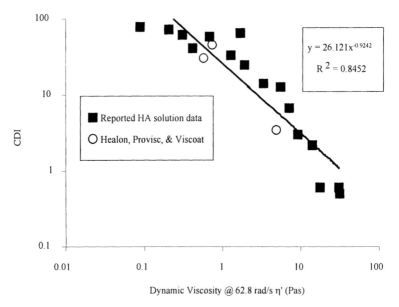

Figure 4. The log of the CDI vs. the log of the dynamic viscosity, η', for uncrosslinked HA solutions and products. Filled squares, data from Falcone et al.(27)

Crosslinked Hyaluronic Acid

Hyaluronic acid has been chemically covalently crosslinked to provide HA derivatives that effectively have an increased molecular weight, improved residence time in the body, and provide different viscoelastic properties than uncrosslinked HA (*11, 33*) (see references in Table II). Typically, crosslinking of HA is done by adding the crosslinker to HA in solution to make a solid gel and the crosslinked gel is dried and broken into particles and screened for different sizes. These crosslinked particles are suspended in aqueous solution to form the products listed in Table I. The molecular weight and material properties now depend on crosslink density and particle size (*34*). The gel suspension of crosslinked HA also has different rheological properties than uncrosslinked HA. In dilute solution, the modulus at low frequency of crosslinked HA is higher that uncrosslinked HA and this indicates that a network structure forms for the crosslinked materials. At higher concentrations, when the entangled network has formed, the steady shear viscosity or elastic modulus at high frequency for crosslinked HA is not higher than that of uncrosslinked HA (*33*). The increase in molecular weight of crosslinked HA increases elastic component of the viscoelastic properties at low frequency which are generally thought to improve efficacy of HA containing formulations for a given medical indication, e.g., viscosupplementation for treatment of arthritis and dermal skin augmentation (*30, 34-36*). The covalently crosslinked HA products are gel-like in structure whereas uncrosslinked HA forms dilute to concentrated polymer solutions depending on the molecular weight and concentration of HA (*33*). Figure 5 describes the effect of frequency on the elastic and viscous modulus for two uncrosslinked and two crosslinked HA products. The uncrosslinked HA products, Healon and Provisc, both exhibit a transition from a predominately viscous regime to a predominately elastic regime at ~ 0.5 rad/sec and ~ 25Pa. This behavior is typical of an entangled high molecular weight HA concentrated solution (*27*). These modulus data support data in Table I indicating that Healon and Provisc are of the same solution concentration and same approximate molecular weight. The modulus data for the crosslinked products Restylane and Hylaform in Figure 5 support a gel-like structure. For these materials, the elastic modulus is considerably higher than the loss modulus across the entire frequency range. It is also noteworthy that the elastic modulus for Restylane is ~10 higher than that of Hylaform and the loss modulus is >10 higher over the entire frequency range tested.

Table II list the cross-linkers that are used for each of the HA crosslinked products. Hylaform, Hylaform Plus, and Synvisc are crosslinked with divinyl sulfone (DVS). The DVS crosslinked HA materials can be prepared with varying amounts of cross-link density in the HA matrix by changing the mole ratio of DVS to HA in the crosslinking reaction. The cross-linking reaction is conducted as a single phase in acidic or basic medium and addition of DVS to HA at high pH forms an insoluble gel matrix. After neutralization, the insoluble

Table I. Products derived from HA and their properties

Product	C(mg//ml)	MW (kg mol^{-1})	η_o (Pa-s)	% Elasticity @ 6.2 rad/s	CDI
Eye viscoelastic adjuncts					
Amvisc Plus	16	2000	100		21.4
Healon	10	1900	300	69	31.2
Healon GV	14	5000	3000		72.3
Healon 5	23	4000	7000		
Provisc	10	2300	256	69	46
Viscoat*	40*	>500*	72	46	3.4
Injectables for Osteoarthritis					
Orthovisc	15	1300	160	56	
Hyalgan	10	700	.06		
Hylartil (Healon)				67	
Artzal				27	
Synvisc**	8	6000	1400	78	
Skin Augmentation					
Restylane**	20		23320	82	16.7
Perlane**	20		7396	84	
Hylaform**	4.5-6.0		14970	90	12.4
Hylaform Plus**	4.5-6.0		10270	95	
Juvederm 24**	24	3000	1436	65	
Juvederm 24 HV**	24	3000	6857	75	
Juvederm 30**	24	3000	3462	71	
Anti adhesion products					
Hyalobarrier**	40		652	65	2.7
Intergel**	5		284	59	38

* HA + chondroitin sulfate: HA of low molecular weight (500 kg mol^{-1}<) at 40mg/ml.

** Crosslinked HA product

CDI values for eye viscoelastic adjuncts taken from Poyer et al. (29)

% elasticity @ 6.2 rad/s taken from Balazs (30)

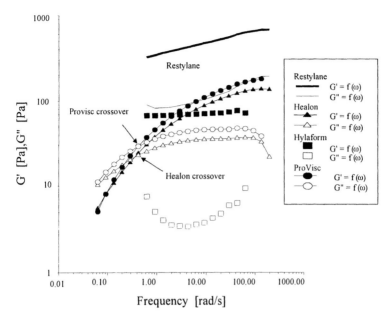

Figure 5. The elastic (G') and loss (G") modulus vs. frequency for two crosslinked, Restylane and Hylaform, and two uncrosslinked, Healon and Provisc, HA products.

Table II. Cross-linking Methods for HA

Product	Cross-linker	Reference
Intergel	Ferric iron	US 5,532,221
Hyalobarrier	Internal esterification	US 5,676,964
Orquest HA	Oxidation	US 6,303,585
Restylane	1,4-butanediol diglycidylether	US 5,827,937
Perlane	1,4-butanediol diglycidylether	US 5,827,937
Hylaform	Vinyl sulfone	US 4,500,676
Hylaform Plus	Vinyl sulfone	US 4,500,676
Juvederm 24	1,4-butanediol diglycidylether	US 6,685963
Juvederm 24HV	1,4-butanediol diglycidylether	US 6,685963
Juvederm 30	1,4-butanediol diglycidylether	US 6,685963
Synvisc	Vinyl sulfone	US 4,500,676
N/A	Multifunctional succinimidyl esters of polyethylene glycol	US 5,470,911

composition is isolated and dried, and the dehydrated powder can be completely re-hydrated in saline solution. The crosslink density controls the rheological properties and the heat resistance of the gel matrix. Varying the mole ratio of DVS to HA and the percent solids in solution can control the viscosity of the resultant Hylan from 2-185 Pas before sterilization. More importantly, changing the cross-link density can also control the viscosity after terminal sterilization. Using a DVS:HA mole ratio of 0.6-1, virtually no decrease in viscosity is seen after steam sterilization of the formulation.

Restylane is a biopolymer gel consisting of hyaluronan from Streptococcus equi or S. zooepidemicus that is crosslinked with the di-functional di-glycidyl ether of 1,4 butane diol. The crosslinking reaction can be carried out in basic medium to promote the formation of ether linkages as the cross-links. The cross-link density is kept low, <1.0%, which after neutralization and evaporation affords a biocompatible viscoelastic gel. This formulation is reported to have optimal viscoelastic properties for augmenting the dermis (*36*). Perlane is also crosslinked with the di-glycidyl ether of 1,4 butane diol (see Table II).

Juvederm 24, 24HV, and 30 are all crosslinked insoluble networks of HA for dermal augmentation. These materials are covalently crosslinked with butanediol diglycidyl ether by a proprietary process that allows a high solids formulation of crosslinked HA that still has good injectability. The difference between these materials is the crosslink density and the size of the crosslinked particles in the viscoelastic gel.

Hyalobarrier is an anti-adhesion gel consisting of autocrosslinked HA. In this case, the crosslinking chemistry is an inter- and intramolecular esterification promoted by several reagents. Using this technique, it was envisioned that the in situ residence time of the HA formulation could be increased without introducing toxic crosslinking agents that could be leached out of the crosslinked polymer matrix (*12*).

HA Products and % Elasticity

Several studies concerning HA products have referred to a rheological property called the percent elasticity (*30, 35*). Percent elasticity is calculated as $(G')*100/(G'+G'')$ and is reported as the proportion of elasticity in an HA formulation. The higher the percent elasticity the more viscoelastic the HA formulation. These studies have also suggested that the efficacy of a HA preparation is related to the viscoelasticity of the formulation. The efficacy of both viscosupplementation and dermal augmentation are reportedly improved as the percent elasticity of the formulation increases. This has been reported for HA products that are uncrosslinked and crosslinked. The percent elasticity vs. frequency for several HA preparations used in augmenting the dermis are shown in Figure 6. The percent elasticity for all of these formulations is relatively

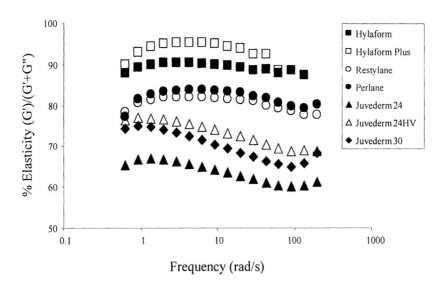

Figure 6. The change in percent elasticity with frequency for several HA products currently indicated for augmentation of the dermis

constant over the entire frequency range tested. The Juvederm products have the lowest percent elasticity and the Hylaform products have the highest elasticity with Restylane and Perlane in between. For the crosslinked HA products, which are used for augmentation of the dermis, the relationship of the percent elasticity of the formulation to efficacy of the formulation is still in question.

The percent elasticity at a frequency of 1 Hz and the η_o for several HA products are listed in Table I. The percent elasticity at 1 Hz is plotted against the η_o for all of these HA products in Figure 7. The data are presented for all products in Table I. that have both percent elasticity@ 6.2 rad/s and η_o values listed. For all these products, including both uncrosslinked and crosslinked HA, the percent elasticity correlates in a logarithmic manner to the η_o.

Biomedical Uses of Hyaluronic Acid

Hyaluronic acid has a number of physical properties that vary depending on molecular weight, concentration and whether it is uncrosslinked, crosslinked, or formed into particles. Table III lists some of the properties that are expressed by various forms of HA. Each application depends on a specific combination of physical and biological properties.

HA in Ophthalmology

Hyaluronic acid is contained in ocular fluid. In vitro experiments have shown that HA provides a protective effect on damaged animal endothelium or human corneal endothelium after intraocular lens implantation (37). HA was introduced into ophthalmic surgery to take advantage of its physical property of high viscosity, and its presence in the vitreous body of the eye. For viscoelastic adjuncts used in eye surgery, high concentrations, 10-70 mg/ml of high molecular weight HA, 2000-5000 kg mol^{-1} have been used. The notable exception is Viscoat, which is a mixture of relatively low molecular weight HA, < 500 kg mol^{-1} and chondroitin sulfate. These viscoelastic formulations are uncrosslinked and characterized as being highly entangled concentrated HA polymer solutions with η_o of ~100-7000 Pas. These materials are formulated to be highly cohesive so that they can be injected and removed entirely as a bolus before and after eye surgery (29).

Rheumatoid or Osteoarthritic Tissues

In the normal undamaged joint, hyaluronic acid pervades the surface layer of the articular tissues and diffuses into the synovial space to lubricate the joint at low deformation rates such as resting or walking. Hyaluronic acid is reported

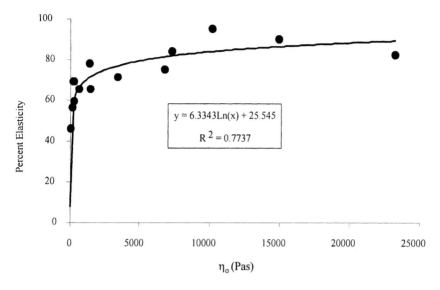

Figure 7. Plotted here is the percent elasticity vs. the η_o for all HA products in Table I with both values listed.

Table III. Medical Indications and Formulation Properties for Hyaluronic Acid Biomaterials.

Ophthalmology (uncrosslinked)	OA (uncrosslinked)	Dermal and OA (crosslinked)*
Cohesive	Not cohesive	Not cohesive
Not tissue adhesive	Tissue adhesive	Not tissue adhesive
Poor residence time in body	Poor residence time in body	Space filling
Animal and bacterial sources	From animal and bacterial sources	Non-uniform and Lumpy
Viscoelastic adjuncts for eye surgery	Pseudoplastic shear thinning fluids. Viscosity controlled by solution concentration and polymer molecular weight	Gel-like rheology Crosslinked particles

* covalent crosslinked

to prevent mechanical damage when the joint is operated at high deformation rates (running) by acting as a shock absorber (*38*). Hence, the protective, lubricating, and stabilization of cells and tissue layers of a joint by HA are due to its viscoelastic properties (*38*). In a joint damaged by arthritis, it is reported that the molecular weight and concentration of HA are decreased and the subsequent resultant decrease in the viscoelastic rheological properties of HA lead to arthritic complications, e.g., pain, inflammation, and limited joint mobility (*13*). Thus intra-articular injections of HA in patients suffering from arthritis may benefit by improving the rheological properties of degraded synovial fluid. The HA injectable preparations used for treatment of osteoarthritis can be uncrosslinked or crosslinked HA derivatives. In one study, the correlation of pain reduction to the elastoviscous properties of Synvisc, Orthovisc, and Hyalgan was determined (*30*). The elastoviscous properties of hyaluronan or hylan (crosslinked HA) are determined by the solution concentration and molecular weight. As the concentration and/or molecular weight are increased, the elastoviscous properties are increased and the analgesic effect of the formulation improves (*39*). Table I lists some commercial formulations of HA for use in osteoarthritis and their physical properties.

The function of HA injected as a viscosupplement is unclear. It has been reported to serve as a viscoelastic supplement or a lubricant. It has been difficult, however, to explain the effect on pain completely as a viscous supplement. Recently, it has been shown that HA also may have analgesic effect by protecting noncioreceptors in the joint pointing to a possible biological role for HA in joint tissues (*30*). Also, since HA is a large polyanion, it has been suggested that it may function as a scavenger of low molecular weight inflammatory or pain modulators by binding small molecules for removal from the joint space or to facilitate Macrophage uptake in this manner to detoxify the joint of factors that bind to the injected HA (*30*).

Dermal Augmentation

Hyaluronic acid derivatives (hylans) have also been used more recently in dermal filler formulations. Since HA has been used in ophthalmology and orthopedics for 20 years, scientists in aesthetic medicine considered using HA in cosmetic medicine. Three of the HA commercial products used for cosmetic surgery and their physical and rheological properties are listed in Table I. For use in dermal filler formulations, HA is crosslinked which prolongs residence time at the site of injection and increases the elastoviscous properties. The zero shear viscosity values for the crosslinked HA products, Restylane, Hylaform, and

Hylaform Plus, are all much higher than the zero shear viscosity values listed for the uncrosslinked HA commercial preparations. In addition, the percent elasticity for the dermal filler formulations has been increased to 82-95% at 1 Hz. The cross-linkers used for Restylane, Hylaform, and Hylaform Plus, are included in Table II. Restylane is crosslinked with the di-glycidyl ether of 1,4 butane diol and Hylaform and Hylaform Plus are crosslinked with divinyl sulfone.

The use of HA in dermal augmentation has prompted several studies attempting to understand some observed tissue reactions (40). Currently, it is unknown whether tissue reactions purported for HA result from the polysaccharide itself, chemical modifications due to crosslinking, or from contaminating proteins that may have not been completely removed during purification (41). Also, antibodies associated with HA have been suggested to be due to determinants on the polymer itself although they may be directed against contaminants (35). As hyaluronic acid is used in the dermis, there is also the possibility that degradation products may exert a biological activity in receptor-mediated response to inflammation or tissue damage (7, 21, 22).

Anti Adhesion

Two crosslinked HA commercial preparations, Hyalobarrier and Intergel, used in anti adhesion are included in Table I. The effectiveness of HA in adhesion prevention depends on the ability of the polymer to be tissue adherent and coat tissues that may come into contact with each other (42). Intergel is composed of Hyaluronic acid crosslinked with ferric iron. It has been studied in human clinical trials, although it is no longer on the market in the US (43).

As with all applications of HA in medical devices, it is important to evaluate the tissue reaction in every specific indication where HA is injected into the body since fragments of HA resulting from degradation are known to be involved in inflammation (7, 22).

Summary and Conclusions

Hyaluronic acid is unique in that it has both biological and physical functions in the body. Its relative ease of purification in quantity and its similarity in many species has made it a popular biomaterial as it is available and biocompatible. Its viscoelastic property has driven the use of HA as a fluid, gel, or gel particles in specific biological applications where replacement of fluids or gels is required to restore a lost function. Since HA has important biological effects in addition to its physical properties in the body, a clear understanding of its biological effects is essential when considering its use in a specific biomaterial application.

References

1. Tammi, M. I.; Day, A. J.; Turley, E. A., Hyaluronan and homeostasis: a balancing act. *J Biol Chem* **2002**, 277, (7), 4581-4.
2. Fraser, J. R.; Laurent, T. C.; Laurent, U. B., Hyaluronan: its nature, distribution, functions and turnover. *J Intern Med* **1997**, 242, (1), 27-33.
3. Prevo, R.; Banerji, S.; Ferguson, D. J.; Clasper, S.; Jackson, D. G., Mouse LYVE-1 is an endocytic receptor for hyaluronan in lymphatic endothelium. *J Biol Chem* **2001**, 276, (22), 19420-30.
4. Csoka, T. B.; Frost, G. I.; Stern, R., Hyaluronidases in tissue invasion. *Invasion Metastasis* **1997**, 17, (6), 297-311.
5. Frost, G. I.; Csoka, A. B.; Wong, T.; Stern, R., Purification, cloning, and expression of human plasma hyaluronidase. *Biochem Biophys Res Commun* **1997**, 236, (1), 10-5.
6. Uchiyama, H.; Dobashi, Y.; Ohkouchi, K.; Nagasawa, K., Chemical change involved in the oxidative reductive depolymerization of hyaluronic acid. *J Biol Chem* **1990**, 265, (14), 7753-9.
7. Horton, M. R.; McKee, C. M.; Bao, C.; Liao, F.; Farber, J. M.; Hodge-DuFour, J.; Pure, E.; Oliver, B. L.; Wright, T. M.; Noble, P. W., Hyaluronan fragments synergize with interferon-gamma to induce the C-X-C chemokines mig and interferon-inducible protein-10 in mouse macrophages. *J Biol Chem* **1998**, 273, (52), 35088-94.
8. Gibbs, D. A.; Merrill, E. W.; Smith, K. A.; Balazs, E. A., Rheology of hyaluronic acid. *Biopolymers* **1968**, 6, (6), 777-91.
9. Gatej, I.; Popa, M.; Rinaudo, M., Role of the pH on hyaluronan behavior in aqueous solution. *Biomacromolecules* **2005**, 6, (1), 61-7.
10. Verheye, S.; Markou, C. P.; Salame, M. Y.; Wan, B.; King, S. B., 3rd; Robinson, K. A.; Chronos, N. A.; Hanson, S. R., Reduced thrombus formation by hyaluronic acid coating of endovascular devices. *Arterioscler Thromb Vasc Biol* **2000**, 20, (4), 1168-72.
11. Balazs, E. A.; Denlinger, J. L., Clinical uses of hyaluronan. *Ciba Found Symp* **1989**, 143, 265-75; discussion 275-80, 281-5.
12. Belluco, C.; Meggiolaro, F.; Pressato, D.; Pavesio, A.; Bigon, E.; Dona, M.; Forlin, M.; Nitti, D.; Lise, M., Prevention of postsurgical adhesions with an autocrosslinked hyaluronan derivative gel. *J Surg Res* **2001**, 100, (2), 217-21.
13. Balazs, E. A., Viscosupplementation for treatment of osteoarthritis: from initial discovery to current status and results. *Surg Technol Int* **2004**, 12, 278-89.

14. Manna, F.; Dentini, M.; Desideri, P.; De Pita, O.; Mortilla, E.; Maras, B., Comparative chemical evaluation of two commercially available derivatives of hyaluronic acid (hylaform from rooster combs and restylane from streptococcus) used for soft tissue augmentation. *J Eur Acad Dermatol Venereol* **1999**, 13, (3), 183-92.

15. Toole, B. P., Hyaluronan and its binding proteins, the hyaladherins. *Curr Opin Cell Biol* **1990**, 2, (5), 839-44.

16. Hascall, V. C.; Heinegard, D., Aggregation of cartilage proteoglycans. I. The role of hyaluronic acid. *J Biol Chem* **1974**, 249, (13), 4232-41.

17. Turley, E. A.; Noble, P. W.; Bourguignon, L. Y., Signaling properties of hyaluronan receptors. *J Biol Chem* **2002**, 277, (7), 4589-92.

18. Aruffo, A.; Stamenkovic, I.; Melnick, M.; Underhill, C. B.; Seed, B., CD44 is the principal cell surface receptor for hyaluronate. *Cell* **1990**, 61, (7), 1303-13.

19. Hardwick, C.; Hoare, K.; Owens, R.; Hohn, H. P.; Hook, M.; Moore, D.; Cripps, V.; Austen, L.; Nance, D. M.; Turley, E. A., Molecular cloning of a novel hyaluronan receptor that mediates tumor cell motility. *J Cell Biol* **1992**, 117, (6), 1343-50.

20. Turley, E. A.; Austen, L.; Vandeligt, K.; Clary, C., Hyaluronan and a cell-associated hyaluronan binding protein regulate the locomotion of ras-transformed cells. *J Cell Biol* **1991**, 112, (5), 1041-7.

21. Termeer, C. C.; Hennies, J.; Voith, U.; Ahrens, T.; Weiss, J. M.; Prehm, P.; Simon, J. C., Oligosaccharides of hyaluronan are potent activators of dendritic cells. *J Immunol* **2000**, 165, (4), 1863-70.

22. Horton, M. R.; Shapiro, S.; Bao, C.; Lowenstein, C. J.; Noble, P. W., Induction and regulation of macrophage metalloelastase by hyaluronan fragments in mouse macrophages. *J Immunol* **1999**, 162, (7), 4171-6.

23. West, D. C.; Kumar, S., Hyaluronan and angiogenesis. *Ciba Found Symp* **1989**, 143, 187-201; discussion 201-7, 281-5.

24. McKee, C. M.; Penno, M. B.; Cowman, M.; Burdick, M. D.; Strieter, R. M.; Bao, C.; Noble, P. W., Hyaluronan (HA) fragments induce chemokine gene expression in alveolar macrophages. The role of HA size and CD44. *J Clin Invest* **1996**, 98, (10), 2403-13.

25. Lapcik, L., Jr.; Lapcik, L.; De Smedt, S.; Demeester, J.; Chabrecek, P., Hyaluronan: Preparation, Structure, Properties, and Applications. *Chem Rev* **1998**, 98, (8), 2663-2684.

26. Cowman, M. K.; Matsuoka, S., Experimental approaches to hyaluronan structure. *Carbohydr Res* **2005**, 340, (5), 791-809.

27. Falcone, S.; Palmeri, D.; Berg, R., Rheometric and Cohesive Properties of Hyaluronic Acid. *Submitted* **2005**.

28. Goa, K. L.; Benfield, P., Hyaluronic acid. A review of its pharmacology and use as a surgical aid in ophthalmology, and its therapeutic potential in joint disease and wound healing. *Drugs* **1994**, 47, (3), 536-66.

29. Poyer, J. F.; Chan, K. Y.; Arshinoff, S. A., Quantitative method to determine the cohesion of viscoelastic agents by dynamic aspiration. *J Cataract Refract Surg* **1998**, 24, (8), 1130-5.

30. Balazs, E. A., Analgesic effect of elastoviscous hyaluronan solutions and the treatment of arthritic pain. *Cells Tissues Organs* **2003**, 174, (1-2), 49-62.

31. Lee, H. G.; Cowman, M. K., An agarose gel electrophoretic method for analysis of hyaluronan molecular weight distribution. *Anal Biochem* **1994**, 219, (2), 278-87.

32. Himeda, Y.; Yanagi, S.; Kakema, T.; Fujita, F.; Umeda, T.; Miyoshi, T., Adhesion preventive effect of a novel hyaluronic acid gel film in rats. *J Int Med Res* **2003**, 31, (6), 509-16.

33. Milas, M.; Rinaudo, M.; Roure, I.; Al-Assaf, S.; Phillips, G. O.; Williams, P. A., Comparative rheological behavior of hyaluronan from bacterial and animal sources with crosslinked hyaluronan (hylan) in aqueous solution. *Biopolymers* **2001**, 59, (4), 191-204.

34. Bothner, H.; Wik, O., Rheology of hyaluronate. *Acta Otolaryngol Suppl* **1987**, 442, 25-30.

35. Micheels, P., Human anti-hyaluronic acid antibodies: is it possible? *Dermatol Surg* **2001**, 27, (2), 185-91.

36. Narins, R. S.; Brandt, F.; Leyden, J.; Lorenc, Z. P.; Rubin, M.; Smith, S., A randomized, double-blind, multicenter comparison of the efficacy and tolerability of Restylane versus Zyplast for the correction of nasolabial folds. *Dermatol Surg* **2003**, 29, (6), 588-95.

37. Lee, K. Y.; Mooney, D. J., Hydrogels for tissue engineering. *Chem Rev* **2001**, 101, (7), 1869-79.

38. Kelly, M. A.; Kurzweil, P. R.; Moskowitz, R. W., Intra-articular hyaluronans in knee osteoarthritis: rationale and practical considerations. *Am J Orthop* **2004**, 33, (2 Suppl), 15-22.

39. Gomis, A.; Pawlak, M.; Balazs, E. A.; Schmidt, R. F.; Belmonte, C., Effects of different molecular weight elastoviscous hyaluronan solutions on articular nociceptive afferents. *Arthritis Rheum* **2004**, 50, (1), 314-26.

40. Sasaki, M.; Miyazaki, Y.; Takahashi, T., Hylan G-F 20 induces delayed foreign body inflammation in Guinea pigs and rabbits. *Toxicol Pathol* **2003**, 31, (3), 321-5.

41. Lowe, N. J.; Maxwell, C. A.; Lowe, P.; Duick, M. G.; Shah, K., Hyaluronic acid skin fillers: adverse reactions and skin testing. *J Am Acad Dermatol* **2001**, 45, (6), 930-3.

42. Peck, L. S.; Goldberg, E. P., Polymer Solutions and Films as Tissue-Protective and Barrier Adjuvants. In *Peritoneal Surgery*, ed.; diZerega, G. S., 'Ed.'^'Eds.' Springer: New York, 2000; 'Vol.' p^pp 499-520.

43. Johns, D. B.; Keyport, G. M.; Hoehler, F.; diZerega, G. S., Reduction of postsurgical adhesions with Intergel adhesion prevention solution: a multicenter study of safety and efficacy after conservative gynecologic surgery. *Fertil Steril* **2001**, 76, (3), 595-604.

Chitin and Chitosan

Chapter 9

Chitosan Nanoparticles for Non-Viral Gene Therapy

Julio C. Fernandes[1], Marcio José Tiera[2], and Françoise M. Winnik[3]

[1]Orthopedic Research Laboratory, Hôpital du Sacré-Cœur de Montréal,
Université de Montréal, Montréal, Québec H4J 1C5, Canada
[2]Departamento de Química e Ciências Ambientais, UNESP-Universidade
Estadual, Paulista, Brazil
[3]Faculty of Pharmacy and Department of Chemistry, Université
de Montréal, Montréal, Québec, Canada

The cationic polysaccharide chitosan has been widely used for
non-viral transfection in vitro and in vivo and has many
advantages over other polycations. Chitosan is biocompatible
and biodegradable and protects DNA against DNase
degradation. However following administration the Chitosan-
DNA polyplexes must overcome a series of barriers before
DNA is delivered to the cell nucleus. This paper describes the
most important parameters involved in the chitosan-DNA
interaction and their effects of on the condensation, shape, size
and protection of DNA. Strategies developed for chitosan-
DNA polyplexes to avoid non-specific interaction with blood
components and to overcome intracellular obstacles as the
crossing of the cell membrane, endosomal escape and nuclear
import are presented.

Introduction

Basic Concepts

Gene therapy involves the introduction of exogenous genes into target cells where production of the encoded protein will occur. In the case of acquired or inherited genetic disorders, this enables the replacement of a missing or defective gene, leading to normal cell function. Moreover, the introduction of DNA encoding antigenic proteins into cells has been studied for immunotherapy of cancer and viral infections (1,2). It is also considered a promising approach for tissue engineering, especially in the field of orthopedics, where the delivered gene could be used as a short-term local enhancer for repair of injured musculoskeletal tissue (3). DNA is delivered to cells either by the viral or nonviral systems. Viral vectors include retroviruses, adenoviruses, adeno-associated viruses, herpes simplex virus and lentivirus (4). Naked DNA, cationic liposomes, cationic lipids and polymers, as well as DNA/cationic liposome/polycation complexes are utilized in the nonviral approach (5–7).

Although viral systems have demonstrated higher transfection efficiency compared to non-viral vectors, they suffer from a number of drawbacks that severely hinder their use in vivo (8). Such limitations include their rapid clearance from the circulation, the reduced capacity to carry a large amount of genetic information and the associated risks of toxicity and immunogenicity, which limits the possibility of subsequent administrations. The main advantages of the non-viral method resides in its being a safer alternative, demonstrating no immunogenicity, negligible toxicity, having the ability to carry large therapeutic genes and a reduced production cost (9,10). For these reasons, there is an increased interest in the development of a safe and efficient non-viral gene delivery system that can circumvent the limitations seen with the viral approach. However to effectively produce a therapeutic effect a non viral vector must overcome multiple biological barriers, ie, i) avoid degradation by Dnase in the extracellular (as well as intracellular) environment, ii) interact effectively with the cell surface in order to be internalized by endocytosis or pinocytosis , iii) the endocytosed vector has to protect the DNA promoting the escape from endosome/ lysosomes, iv) DNA has to reach the nucleus by diffusion within the cytosol v) Entry of DNA in the nucleus and gene expression followed by mRNA being transported out the nucleus vi) Protein translation from mRNA. (Figure 1). In the following sections we will discuss all these steps presenting alternatives to overcome these barriers by employing chitosan as a non-viral DNA carrier.

Chitosan : Structure and General properties

Chitosan (a $(1 \rightarrow 4)$ 2-amino-2-deoxy-β-D-glucan is obtained by alkaline deacetylation of chitin. Chitin, a polysaccharide found in the exoskeleton of

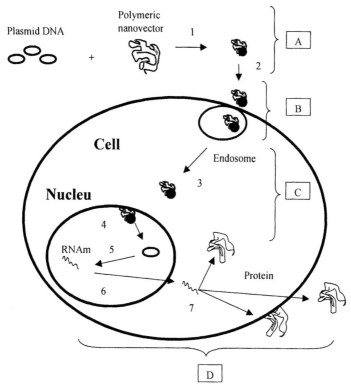

Figure 1: Schematic gene therapy mechanism. (A) Bioavailability and extracellular trafficking (1) gene-vector complexation; (B) internalization and endocytosis by the cell membrane; (C) intracellular trafficking (2) uptake of the vector complex into intracellular endolysosome; (3) DNA-chitosan release from the endosome into the cytoplasm; (4) Internalization of the complex into the nucleus; (D) gene expression; (5) DNA dissociation from the vector; (6) ARNm transcription from the gene; (7) protein translation from ARNm. The protein can be secreted out of the cell, be released into the cytoplasm, or fixed onto the membrane. (Adapted from Mansouri et al, (10))

crustaceans and insects (*11*) (Figure 2). It is a copolymer of N-acetyl-D-glucosamine and D-glucosamine, which behaves as a weak base. The pKa value of the D-glucosamine residue of about 6.2–7.0. Chitosan is insoluble in water at neutral and alkaline pH values. In acidic medium, the amine groups will be positively charged, conferring to the polysaccharide a high charge density (*12*). Chitosan excels in enhancing the transport of drugs across the cell membrane. Its cationic polyelectrolyte nature provides a strong electrostatic interaction with mucus, negatively charged mucosal surfaces and other macromolecules such as DNA (*12,13*).

Chitosan is a non-toxic biodegradable polycationic polymer with low immunogenicity (14). It is a good candidate for gene delivery system because when positively charged, it can be complexed with negatively charged DNA (15,16). Chitosan can effectively bind DNA and protect it from nuclease degradation (17,18). Moreover chitosan/DNA complexation takes place without sonication and does not require the use of organic solvents, therefore minimizing possible damage to DNA during complexation. DNA-loaded chitosan microparticles were found to be stable during storage (19). The application of DNA–chitosan nanospheres has advanced in vitro DNA transfection research and data have been accumulating that shows their usefulness for gene delivery (20,21).

Condensation, Shape, Size and Protection of DNA-Chitosan Polyplexes

The Chitosan-DNA interaction is driven mainly by the electrostatic interaction between the amino groups of chitosan and the charged phosphate groups of DNA (22, 23) and, although different methodologies have been published on the preparation of chitosan nanoparticles, for gene therapy, the complex coacervation has been the most employed procedure. In general the encapsulation process is entirely performed in aqueous solution at low temperatures preserving the bioactivity of the plasmid DNA. Stable complexes are formed only when chitosan is added in molar excess relative to DNA, with zeta potential values between 10 and 20 mV, depending upon the degree of excess. However it is well documented that molecular weight, charge ratio (+/-) and pH are important parameters in providing the needed protection, as well as in determining the nanoparticles shape. Transmission electronic microscopy measurements have provided evidence that chitosan of 8 kDa condenses plasmid into toroids and rod shaped particles whose sizes were estimated about 66 nm by light scattering measurements (24). Although condensation of the DNA by the chitosan is of great importance, the plasmid DNA must remain intact to assure its functionality once inside the cell. A widely used method to monitor the DNA condensation and the effect of such conditions on the integrity of the plasmid DNA is that of gel electrophoresis. Chitosan at a concentration of 0.02%, used for synthesis of the complexes, is viewed in Figure 3 (20). The intact DNA, before complexation, is seen in lane 2, while complexes of chitosan-DNA of molecular weights 150, 400 and 600 kDa were loaded in lanes 3, 5 and 7, respectively. The DNA in these three lanes is unable to migrate and remains in the gel loading wells, indicating a strong attachment of the DNA to the chitosan. No unbound DNA is seen in these lanes. Following digestion with chitosanase and lysozyme, the integrity of plasmid DNA can be evaluated (viewed in lanes 4, 6, and 8). The plasmid is intact and shows no signs of degradation. This suggests that the coacervation process does not affect the integrity of the condensed DNA. The size and morphology of the chitosan-DNA nanoparticles

181

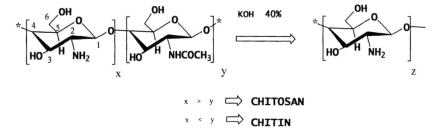

x > y ⇨ **CHITOSAN**

x < y ⇨ **CHITIN**

Figure 2: Chemical structure of chitin and chitosan and the deacetylation process to produce the highly deacetylated product. In the 2-amino-2-deoxy-d-glucopyranose ring is shown the commonly used numbering for the carbon atoms.

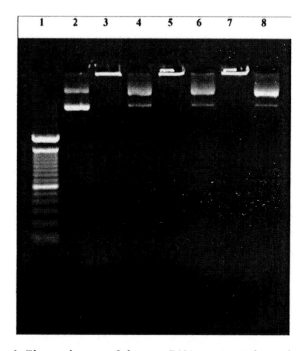

Figure 3. Electrophoresis of chitosan-DNA nanoparticles to determine plasmid integrity following synthesis. Nanoparticles were digested with chitosanase and lysozyme after synthesis and the released DNA was visualized with ethidium bromide. Lane 1: Molecular weight marker; lane 2: VR1412 plasmid DNA; lane 3: nanoparticles composed of 150 kDa chitosan; lane 4: lane 3 + digestion; lane 5: nanoparticles composed of 400 kDa chitosan; lane 6: lane 5 + digestion; lane 7: nanoparticles composed of 600 kDa chitosan; lane 8: lane 7 + digestion. (Reproduced with permission from ref. 20)

can be vizualised by electron microscopy. The nanoparticles are embedded in Epon, sectioned and the DNA stained with uranyl acetate and lead citrate, giving it a darker appearance (Figure 4). The micrographs demonstrate a homogenous distribution of DNA within the particles. On the other hand, atomic force microscopy (AFM) revealed the size and morphology of the synthesized particles. The complexes made employing a chitosan 150 kDa appear spherical with a mean size that is inferior to 100 nm (Figure 5).

Recently Danielsen et al. (25-26) have confirmed by AFM that polyplexes made by mixing plasmid DNA with chitosan from 10 to 200 kDa yielded a blend of toroids and rods. The ratios between the fractions of toroids and rods were observed to decrease with increasing acetylation degree (FA) of the chitosan indicating that the charge density of chitosan, proportional to $(1 - FA)$, is important in determining the shape of the compacted DNA. The amount of chitosan required to fully compact DNA into well-defined toroidal and rodlike structures were found to be strongly dependent on the chitosan molecular weight, and thus its total charge. A higher charge ratio (+/-) was needed for the shorter chitosans, showing that an increased concentration of the low degree of polymerization (DP) chitosan could compensate for the reduced interaction strength of the individual ligands with DNA. The stability of these DNA–chitosan complexes was studied after exposure to heparin and hyaluronic acid (HA) using atomic force microscopy (AFM) and ethidium bromide (EtBr) fluorescence assay. Studies of the polyplex stability when challenged by HA showed that whereas HA was unable to dissociate the complexes, the degree of dissociation caused by heparin depended on both the chitosan chain length and the amount of chitosan used for complexation(26).

Liu et al. have showed that the charge density is an important parameter by studying the formation of polyplexes in different pH values (27). Working with 5kDa chitosan 99% deacetylated their results indicated that upon interacting with chitosan, the DNA molecules saved a B conformation, and the binding affinity of chitosan to DNA was dependent on pH of media. At pH 5.5, highly charged chitosan had a strong binding affinity with DNA; whereas in pH 12.0 medium, only weak interactions existed. However the authors have reported that no typical toroid patterns were observed, which was attributed the strong compaction of DNA caused by highly charged chitosan.

In a recent report Kiang et al. have showed that in fact the electrostatic interaction with the chitosan chain is very important to bind efficiently DNA. They reported that the charge (+/_) ratio to achieve complete DNA complexation increases for chitosans having smaller deacetylation degrees, which in turn affects the stabilization of the particles, affecting the transfection in vitro and in vivo (28). However other interactions as hydrogen bonding and hydrophobic interactions can not be neglected since hydrophobic interactions have been also detected in aqueous solution of the polyelectrolyte itself, leading to formation of polymer chain aggregates (29,30). Therefore it is reasonable to consider that the formation of the polyplexes may result from a combined electrostatic-hydrophobic driving force leading to the packing of Chitosan-DNA polyplexes.

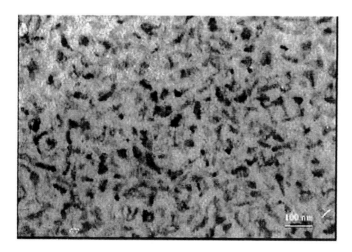

Figure 4. Transmission electron microscopy of chitosan-DNA nanoparticles.
Homogenous distribution of DNA is seen following staining
with uranyl acetate and lead citrate. Bar = 100 nm. . (Reproduced with
permission from ref. 20)

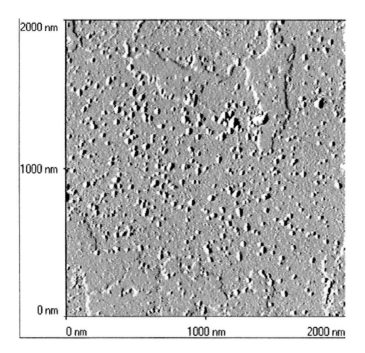

Figure 5. Atomic force microscopy analysis of complexes formed
between chitosan and DNA. (Reproduced with permission from ref. 20)

Koping Hoggard et al. (*31,32*) prepared monodisperse oligomers of chitosan fully deacetylated (6-, 8-,10-, 12- 14- and 24-mer), with very low polydispersity, and ultrapure chitosan (UPC) of 154 kDa. Depending on chain length of chitosan, charge ratio and buffer properties, chitosan-DNA complexes appeared as different physical shapes such as coils, soluble globules, soluble aggregates, precipitated globules and precipitated aggregates. It was shown that only UPC and 24-mer chitosans could form stable complexes with DNA, and 24-mer was more efficient in mediating gene expression in vitro and in vivo than was UPC. Sato et al (*33*) have also found that molecular mass of chitosan, pH of medium, and serum concentration are very important to promote transfection efficiency. Working on chitosans samples whose average molecular were 15, 52, and 100 kDa, they found that transfection efficiency mediated by chitosan of 100 kDa was less than that by chitosan of 15 and 52 kDa, but clearly indicating that dependence with cell lines was also observed. However Bozkir et al. (*34*) have suggest that formulations with high molecular weight (HMW) chitosan can be an effective non-viral method of gene vector in animal studies. The authors have studied the influence of two different preparation (the solvent evaporation method and the complex coacervation method) methods on the encapsulation of a model plasmid with chitosan. Protection of encapsulated pDNA offered by these nanoparticles from nuclease attack was confirmed by assessing degradation in the presence of DNase I, and the transformation of the plasmids with incubated nanoparticles were examined by β-galactosidase assay. The results obtained showed that pDNA existed as a mixture of both supercoiled (84.2%) and open circular (15.8%) forms, and that formulation prepared by the complex coacervation method protected the supercoiled form of pDNA effectively. There weren't any significant changes in nanoparticle size and zeta potential values at pH 5.5 for a period of 3 months, but differences in particle sizes were observed after lyophilization with a cryoprotective agent. The efficiency of nanoparticles mediated transformation to *Escherichia coli* cells was significantly higher than naked DNA or poly-l-lysine (PLL)-DNA polycation complexes. The transfection studies were performed in COS-7 cells. A 3-fold increase in gene expression was produced by nanoparticles as compared to the same amount of naked plasmid DNA (pDNA).

The degree of acetylation is expected to affect the Chitosan-DNA interactions since the charge density is decreased, i.e., the number amino groups is decreased providing to the polymer chain an increased hydrophobic character. (*28*). Various workers aiming to increase the transfection have also conducted the trymethylation (quaternization) of the amino groups (*35-38*). Trimethyl chitosan (TMO) derivatives of 40% (TMO-40) and 50% (TMO-50) degrees of quaternization were synthesized and examined for their transfection efficiencies in two cell lines: COS-1 and Caco-2 (*35,36*). Results showed that quaternized chitosan oligomers were able to condense DNA and form complexes with a size ranging from 200 to 500 nm. The results indicate that serum proteins marginally affected transfection efficiency of the trimethylated chitosan/DNA complexes, which might indicate that dissociation of the complex is also minimal. This has been attributed to non-pH-dependent positive charges that quaternized chitosan oligomers bear in contrast to chitosan, which at neutral pH is poorly positively charged (*36*). However serum proteins in the culture medium had a dramatic decreasing effect on the transfection efficiency of DOTAP lipoplexes.

The protection of DNA-chitosan nanoparticles is also important since the carrier must be capable of surviving in the bloodstream for a significant length of time, so that it can reach the target tissue. Therefore circulation time is a most prominent and fundamental factor for reaching the target cells in successful gene delivery, especially in cases of intravenous injection (*39*). Cationic polymer/DNA complexes also show short plasma circulation times with rapid hepatic uptake and accumulation or deposit in organs such as the skin and intestine. Aggregation of cationic particles, such as cationic polymer/DNA complexes, can occur following interaction with blood components. Albumin is known to bind to cationic particles. As a particulate drug carrier, Ch-DNA nanoparticles are susceptible to be taken up by macrophages of the mononuclear phagocytic system which recognize them as a foreign body material. This occurs as a result of opsonization process which leads to their covering by plasma proteins and glycoproteins (*40*). A representative flow cytometric assessment of nanoparticles uptake by macrophages is shown in Figure 6a (*41*). The internalization of FITC labeled Ch-DNA nanoparticles by THP-1 macrophages occurs within the first hour (Fig. 6b) and increased at 24 h (*41*).

The results can also be accessed by fluorescence microscopy and FITC labeled Ch-DNA nanoparticles are seen in the THP-1 macrophages within 1 h of incubation (Figures. 7a–d) (*41*). The facility of macrophages to internalize nanoparticles could play an important role in their spreading and distribution (*42*). Consequently, several attempts aiming at the development of therapeutic strategies using viral and non-viral systems for macrophage-associated pathogenesis have been reported (*43–45*).

However for gene therapy the opsonization represents a major biological barrier to the delivery of DNA using condensed particles. Therefore it is necessary to prevent unwanted interactions between particles and the dynamic environment of the blood circulation by introducing hydrophilic surfaces to the cationic particles.The modification of chitosan by introducing hydrophilic polymers may create a cloud of hydrophilic chains at particle surface, which may avoid the interaction with proteins and phagocytes prolonging the circulation time in the bloodstream (*46*). Many hydrophilic groups can be attached to the chitosan backbone or to pre-formed nanoparticles to improve the solubility in water (*47-51*). Nanospheres synthesized by salt-induced complex coacervation of cDNA and gelatin and chitosan were evaluated as gene delivery vehicles (*52*). These nanospheres were subsequently modified by introducing PEG_{5000} chains. The attaching of these groups has avoided the aggregation of the particles during the lyophilization and the storage during one month did not alter the properties of the nanospheres. Chitosan-DNA nanospheres were effective in tranfecting 293 cells but not HeLa Cells and the tranfection efficiency was not

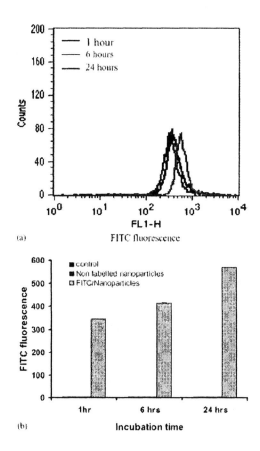

Figure 6. The kinetics of FITC labeled Ch-DNA nanoparticles uptake by THP-1 was evaluated by flow cytometry after 1, 6 and 24 h (a). The FITC fluorescence was detected for the cells which were incubated with labeled nanoparticles. Cells alone and those which were incubated with non-labeled nanoparticles were used as controls. In these cases, no signal was detected. A histogram representation of the fluorescence as a function of incubation time is shown in the bottom diagram (b). (Reproduced with permission from ref. 41)

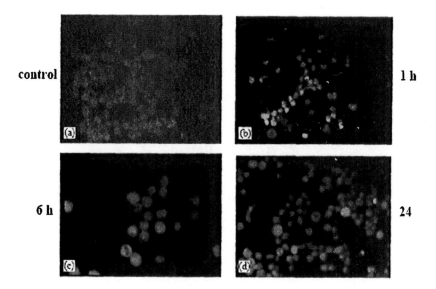

Figure 7. After Ch-DNA nanoparticle FITC labeling, their uptake by THP-1 macrophages is showed by fluorescence microscopy following different incubation periods (a) control, (b) after 1 h, (c) after 6 h and (d) after 24 h. (Reproduced with pemission from ref. 41).

affected by PEG derivatization. Poly(vinyl pyrrolidone) was also grafted on galactosylated chitosan- (GCPVP) and showed improved physicochemical properties over the unmodified chitosan (53). The binding strength of GCPVP 10K/DNA was superior to that of the GCPVP 50K/DNA one, which was attributable to its higher flexibility due to the smaller size. However Dnase I protection of GCPVP 10K/DNA complex was inferior to that of the GCPVP 50K/DNA one. The DNA-binding property showed mainly depended on the molecular weight of chitosan and composition of PVP.

Although the attaching of hydrophilic groups decreases the toxicity conferring improved physicochemical characteristics, the internalization process may be inhibited. The coating with PEG has been showed to decrease markedly the transfection efficiency of the polyplexes (54,55). This has been attributed to the neutral surface charge of the pEGylated polyplexes, which prevents non-specific adhesion to the cell surface (54). However this approach can be useful when nanoparticles are derivatized with cell-specific targeting ligands, what will be discussed later in the chapter.

Cellular uptake

Internalization Contact and Crossing of the Cell Membrane

The charge (+/-) ratio of the polyplexes has been also indicated as important to effectively proceed with the transfection process. Assuming that DNA wrapped in inter-polyelectrolyte complexes is well protected from DNase degradation, the following step is to reach its target, the cell. Although the mechanism of internalization contact and crossing of the cell membrane is not fully understood, it is well accepted that the Chitosan-DNA polyelectrolyte complex exhibiting a net positive charge binds to negatively charged cell membrane (56). Therefore a net positive charge is fundamental in the process and the level of transfection is showed to increase with the charge ratio (+/-) reaching a maximum and decreasing at higher stochiometries. It has been reported by Ishi et al that the level of transfection with plasmid/chitosan complexes was found to be highest when charge (+/-) ratio was between 3 and 5, and transfection medium contained 10% serum at pH 7.0 (57). Lee et al. using chitosan of low molecular weight found the most efficient transfection was obtained at a charge (+/-) ratio 3:1 (58), while MacLaughlin et al. have reported that the highest level of expression was obtained at a ratio of 2:1 complex made with chitosan 102kDa (24). However it must be considered that the transfection efficiency of chitosan-DNA nanoparticles is also cell type dependent and the ratio chitosan:plasmid must be controlled to obtain the appropriate particle size aiming to maximize the transfection (25).

The particle size is also recognized as a key parameter since it may affect the blood circulation time and the cellular uptake. It has been reported that polycation-DNA gene delivery systems mostly enter the cell by endocytosis or pinocytosis, having a size requirement inferior to 100 nm (25,59). However the experimental results available are contradictory and reasonable transfection efficiencies were reported for particles having sizes varying from 100 nm (21) to 2.0 micrometers (60). Recently on basis of AFM images Liu et al. have proposed that Chitosan-DNA complexes in the range of several hundreds of nanometers are transferred into the cell mainly via endocytosis (61).

Cell Surface Receptors Mediating the Endocytosis

It is well known that some kinds of saccharide play important roles in biological recognition on cellular surfaces. Liver parenchymal cells exclusively express large numbers of asialoglycoprotein receptors that strongly bind with galactose (62,63). The asialoglycoprotein receptor (ASGP-R) is known to be present only on hepatocytes at a high density of 500,000 receptors per cell, and retained on several human hepatoma cell lines (64,65). The asialoglycoprotein receptor system can not only bind galactose-containing ligands, but can also internalize them within membrane-bond vesicles or endosomes (66). Once a ligand binds to the galactose receptor, the ligand–receptor complex is rapidly internalized and the receptor recycles back to the surface (67,68), allowing high binding capacity and efficient uptake of galactosylated ligands by liver cells. Murata et al. have suggested that the utilization of a pre-quaternized and subsequently galactosylated chitosan could be used to transfect cells HepG2 cells (68). Park et al. have used this specificity to synthesize galactosylated chitosan (GC) aiming to transfect HepG2 human hepatoblastoma cell line and HeLa human cervix epithelial carcinoma cells(69-71). The particle sizes for the DNA complexes using GC 13K and GC 18 K/ show tendency to decrease with increasing charge ratio of GC to DNA and had a minimum value around 240 and 100 nm, respectively, at the charge ratio of 5. Cytotoxicity study showed that GC prepared by the water-soluble chitosan had no cytotoxic effects on cells. The results showed that the transfection efficiency into HepG2, which has asialoglycoprotein receptors (ASGP-R), was higher than that into HeLa without ASGP-R. Attaching of poly(ethyleneglycol) to galactosylated chitosan (GPC) was also performed in another study of the same group in order to improve the stability in water and enhanced cell permeability (72). In this case it was reported that GCP/DNA complexes were only transfected into Hep G2 having asialoglycoprotein receptors (ASGR), indicative of specific interaction of ASGR on cells and galactose ligands on GCP. The transfection with GCP/DNA complexes was subsequently investigated by confocal laser scanning microscopy using primary hepatocytes and HepG2 human hepatocarcinoma cell line (73). The more efficient transfection of the complex occurred in the human-derived HepG2 cells than in primary hepatocytes. Erbacher et al. (21) synthesized lactosylated-modified chitosan derivatives (having various degrees of substitution) and tested their transfection efficiencies in many cell lines. The in vitro transfection was found to be cell-type dependent. HeLa cells were efficiently transfected by this modified carrier even in the presence of 10% serum, but neither chitosan nor lactosylated chitosans have been able to transfect HepG2 and BNL CL2 cells. Gao et al. (74) have used highly purified low molecular chitosan (LMWC) to attach lactobionic acid (LA) bearing galactose group. A series of galactosylated-LMWC (gal-LMWC) having different groups contents were obtained. The tranfection efficiency was evaluated in human hepatocellular carcinoma cell (HepG2), L-02. SMMC-7721, and human cervix adenocarcinoma (Hela) cell lines in vitro. The transfection of gal-LMWC/DNA complex showed a very selective transfection to hepatocyte and the efficiency was showed to increase with the improvement of the galactosylation degree.

Transferrin is other interesting targeting ligand to attach to the chitosan backbone. Transferrin receptors are found on the surface of most proliferating cells and, in elevated numbers, on erythroblasts and on many tumors where they have been linked to drug resistance (75). Leong et al. have reported the conjugation of PEG to pre-formed nanoparticles of gelatin and chitosan using different procedures (52,19). Tranferrin was bonded to nanospheres utilizing a sulfhydryl derivative obtained from protein reaction with 2-iminothyalone, while the PEG conjugation was performed utilizing PEG derivatives containing the succinimidyl attached to end of backbone. For the chitosan carrier the plain nanospheres were reported as effective as the PEG-chitosan modified nanospheres. However the conjugation of PEG to the surface of the nanospheres minimized any aggregation in solution (19). The KNOB protein was also conjugated to the nanoparticles using a bis-succinimidyl PEG derivative. The conjugation of KNOB protein to chitosan-DNA nanoparticles was reported to increase 130 –fold the transfection efficiency in HeLa cells and several folds in HEK293 cells. Chitosan has been also investigated for its ability to form polymeric targeted vesicle drug carriers. Glycol chitosan modified by attachment of a strategic number of fatty acid pendant groups (11-16 mols%) assembles into unilamellar polymeric vesicles in the presence of cholesterol (76-78). These polymeric vesicles were found to be biocompatible and haemocompatible and capable of entrapping water-soluble drugs aiming to target receptors overexpressed in some tumors. Recently it was reported the *in vivo* biological evaluation of doxorubicin formulated in tranferrin targeted polymeric vesicles made from palmitoylated glycol chitosan (GCP). The transferrin conjugated vesicles showed a statistically significant uptake advantage when compared to the nontargeted vesicles (*79*).

The receptor for the folic acid is other approach that can be used to target tumors since cancerous cells divide rapidly and need folic acid to DNA synthesis. This morphological phenomenon is best noted by up-regulation of membrane folate binding protein expression for a subsequent increase in folate internalization. The conjugation of folic acid has been studied for poly(ethyleneinimine)(*80-82*),poly-L-lysine (*54*) and poly(dimethylaminomethyl methacrylate) (*55*). In general it is observed that the attaching of folates to the distal termini of PEG-modified polymers greatly enhances the transfection activity of the corresponding DNA complexes over the polyplexes containing only PEG-modified polyplexes. The enhancements observed at all charge (+/-) ratios can be blocked partially by co-incubation with free folic acid, which suggest the involvement of folate receptor in gene transfer. Folate conjugation, therefore, presents a potential strategy for tumor-selective targeted gene delivery. Recently Mansouri et al. (*83*) have reported the first conjugation with chitosan. The nanoparticles obtained for folic acid-modified chitosan (FA-CH) showed a very low toxicity and the coacervation process utilized to obtain the FA-CH-DNA did not affect the integrity of the utilized plasmid. Currently these new nanoparticles are being tested in transfection studies.

Release from Endosomes

After the internalization the following crucial step in gene delivery with cationic polymers is the escape of the polymer/DNA complexes from the endosome. The inefficient release of the DNA/polymer complex from endocytic vesicles into the cytoplasm is indicated as one of the primary causes of poor gene deliver. In this respect the approach is to enhance endosomal escape by using cationic polymers with a pK_a slightly below the physiological pH. The endosomal escape is believed take place through the mechanism named "proton sponge", and the importance of this step has been recognized and reviewed recently by Cho et al. (*84*). This hypothesis has been proposed to explain the high transfection activity of polyethyleneinimine PEI. (*85*). It has been demonstrated that PEI has buffering capacity over a broad pH range hence, once PEI-based polyplexes are present in the endosome, they can absorb protons that are pumped into this organelle. Because of repulsion between the protonated amine groups, swelling of the polymer occurs. Moreover, to prevent the build up of a charge gradient due to the influx of protons, an influx of Cl⁻ ions also occurs. The influx of both protons and Cl⁻ ions increases the osmolarity of the endosome and causes water absorption. Combination of swelling of the polymer and osmotic swelling of the endosome leads to a destabilization of the endosome and release of its content into the cytoplasm (86). Subsequently, transport to and uptake in the nucleus as well as dissociation of the polyplex has to occur before transcription of the DNA take place. However Funhoff et al. (87) working with a series of synthetic polymers indicated that the endosomal escape is not always enhanced by polymers buffering at low pH. In this respect the research focusing chitosan is scarce but it is believed that protonation of the free amine groups of chitosan must contribute to unpack the DNA delivering it in the cytoplasm. Ishii et al. (57) have indicated that when the transfection is performed at pH 6.5 chitosan cannot release from endosome, due to the fact that chitosan is drastically protonated when pH of medium is altered from pH 7 to 6. Recently, to enhance the transfection efficiency of chitosan, water-soluble chitosan (WSC) was coupled with urocanic acid (UA) and the transfection efficiency was investigated (88). The authors reported that the transfection efficiency of chitosan into 293T cells was much enhanced after coupling with UA and increased with an increase of UA contents in the urocanic acid-modified chitosan. Kiang et al (86) have used poly(propyl acrylic acid) (PPAA) as an approach to enhance the release of endocytosed drugs into the cytoplasmic compartment of the cell. By incorporating this polymer in chitosan nanoparticles the release of plasmid DNA from the endosomal compartment was enhanced. In vitro transfection studies confirmed that the incorporation of PPAA into the chitosan-DNA nanoparticles enhanced gene expression in both HEK293 and HeLa cells compared to chitosan nanoparticles alone. The dose and time at which PPAA was incorporated during the complex formation affected the release of DNA and transfection efficiency. The authors have suggested that the PPAA triggered membrane disruption resulting in the release of DNA from the endosomal compartment.

Nuclear Transport

The transport of the complexes through the cytoplasm to the nucleus is poorly understood. Ishii et al. have used plasmid/chitosan complexes made with FITC-labeled plasmid and Texas Red-labeled chitosan to observe cellular localization(57). Their results show that at 4 h post-incubation, the fluorescence of the two molecules had accumulated in the nucleus and remained in the same localization at 24 h post-incubation. On the other hand, FITC-plasmid/(poly-l-lysine or lipofectin) complexes did not display fluorescence images even at several hours post-incubation. The authors have interpreted these results as an indication that plasmid/chitosan complexes have high nuclear accumulation and resistance to digestion by cellular and nuclear enzymes. However this suggests that the chitosan may not separate from DNA prior the nuclear entry, as observed with polyethyleninimine (89-91). Modifications of chitosan with nuclear localization signal peptides (92-94) or the inclusion in the plasmid of nucleotide sequences with affinity for cellular proteins, such as transcription factors, may be a good strategy to increase the nuclear import, since they are recognized to enhance the transfection efficiency (92,95,96). After nuclear import the vector unpacking play an important rule in the transfection efficiency. In this respect studies with poly-l-lysine(97,98) have showed that the polycation molecular weight is very important, therefore must be considered in future studies employing chitosan.

Conclusion

In recent years gene transfer technology has evolved at an increasingly fast pace. Its projected applications in medicine range from gene therapy of inherited diseases and cancer to arthritis. Cationic polymers, such as chitosan (Ch), are promising candidates for DNA transport in non-viral delivery systems. The enabling characteristics of Ch-nanoparticles include biocompatibility, multiple ligand affinity, and a capacity of taking up large DNA fragments, while remaining small in size (~100 nm). These vectors show good biocompatibility, but low transfection efficiency. Multiple strategies to improve transfection rates have been proposed to take advantage of chitosan uptake by cells, including low molecular weight chitosan, membrane receptor targeting and internalization.

Non-viral gene delivery methods should also be considered in future therapeutics due to pharmacoeconomic concerns. Industrial costs of producing viral-based systems strongly favor non-viral vectors ("one kg of chitosan will always cost less than one kg of virus"), with comparable safety and efficacy.

In conclusion, chitosan-based gene delivery shows promising and reproducible results in the hands of several different research groups. Second- and third generation nanovectors should overcome efficacy problems in the near future.

Acknowledgements

This study was supported in part by a research grant from The Arthritis Society of Canada. Dr Fernandes holds a clinician scientist scholarship of the Fonds de la recherche en sante du Quebec. Dr Tiera holds a post-Ph D scholarship from the CAPES program from Brazil.

References

1. Vile, R.G.; Russell, S.J.; Lemoine, N.R. Cancer gene therapy: hard lessons and new courses. *Gene Ther.* **2000**,7, 2–8.
2. Okuda K.; Ihata A.; Watabe, S.; Okada, E.; Yamakawa, T.; Hamajima, K, Yang J.; Ishii, N.; Nakazawa, M.; Okuda, K.; Ohnari, K.; Nakajima, K.; Xin, K.Q. Protective immunity against influenza A virus induced by immunization with DNA plasmid containing influenza M gene. *Vaccine* **2001**, 19, 3681–3691.
3. Bonadio, J.; Smiley, E.; Patil, P.; Goldstein, S. Localized, direct plasmid gene delivery in vivo: prolonged therapy results in reproducible tissue regeneration. *Nat. Med.* **1999**, 5, 753–759.
4. Oligino, T. J.; Yao, Q.; Ghivizzani, S.C.; Robbins, P.; Vector systems for gene transfer to joints. *Clin. Orthop.* **2000**, (379Suppl), S17–30.
5. Zelphati, O.; Nguyen, C.; Ferrari, M.; Felgner, J.; Tsai Y.; Felgner, P.L.;Stable and monodisperse lipoplex formulations for gene delivery. *Gene Ther.* **1998**, 5, 1272–1282.
6. Park, I. K.; Kim, T.H.; Park, Y.H.; Shin, B.A.; Choi, E.S.; Chowdhury, E.H.; Akaike, T.; Cho, C.S. Galactosylated chitosan-graft-poly(ethylene glycol) as hepatocyte-targeting DNA carrier. *J. Control. Rel.* **2001**,76,349–362.
7. Gao, X.; Huang, L. Potentiation of cationic liposome-mediated gene delivery by polycations. *Biochemistry* **1996**,35,1027–1036.
8. Luo, D.; Saltzman, W.M. Synthetic DNA delivery systems. *Nat. Biotechnol.* **2000**, 18, 33–37.
9. Romano, G.; Michell P.; Pacilio, C.; Giordano, A. Latest developments in gene transfer technology: achievements, perspectives, and controversies over therapeutic applications. *Stem Cells* **2000**, 18, 19–39.
10. Mansouri, S.; Lavigne, P.; Corsi, K.; Benderdour, M.; Beaumont, E.; Fernandes, J.C.; Chitosan-DNA nanoparticles as non-viral vectors in gene therapy: strategies to improve transfection efficacy. *Eur. J. Pharm. Bioph.*, **2004**, 57, 1-8.
11. Romoren, K.; Thu, B.J.; Evensen, O. Immersion delivery of plasmid DNA. II. A study of the potentials of a chitosan based delivery system in rainbow trout (Oncorhynchus mykiss) fry. *J. Controlled Release* **2002**, 85, 215–225.
12 Hejazi, R.; Amiji, M. Chitosan-based gastrointestinal delivery systems. *J. Controlled Release* **2003**, 89, 151–165.
13. Fang, N.; Chan, V.; Mao, H.-Q.; Leong, K.W. Interactions of phospholipid bilayer with chitosan: Effect of molecular weight and pH *Biomacromolecules* **2001**, 2, 1161–1168.
14. Muzzarelli, R. A. A.; in: R. Belcher, H. Freiser (Eds.), *Chitosan in Natural Chelating Polymers; Alginic acid, Chitin, and Chitosan*, Pergamon Press, Oxford, 1973, pp. 144–176.

15. Liu, W.; Sun, S.; Cao, Z.; Zhang, X.; Yao, K.; Lu W. W.; Luk K.D.K., An investigation on the physicochemical properties of chitosan/DNA polyelectrolyte complexes *Biomaterials* **2005**, *2*, 2705-2711.

16. Richardson, S.C.W.; Kolbe, H.V.J.; Duncan, R. Potential of low molecular mass chitosan as a DNA delivery system: biocompatibility, body distribution and ability to complex and protect DNA. *Int. J. Pharm.* **1999**, 178, 231–243.

17. Cui, Z.; Mumper, R. J.; Chitosan-based nanoparticles for topical genetic immunization. *J. Controlled Release* **2001**,75, 409–419.

18. Illum, L.; Jabbal-Gill, I.; Hinchcliffe, M.; Fisher, A. N. ; Davis, S.S. Chitosan as a novel nasal delivery system for vaccines. *Adv. Drug Deliv. Rev.* **2001**, 51, 81–96.

19. Leong, K.W.; Mao, H.-Q.; Truong-Le, V.L.; Roy, K.; Walsh, S. M.; August, J.T. DNA-polycation nanospheres as non-viral gene delivery vehicles. *J. Controlled Release* **1998**, 53, 183–193.

20. Corsi, K.; Chellat, F.; Yahia, L.; Fernandes, J.C. Mesenchymal stem cells, MG63 and HEK293 transfection using chitosan-DNA nanoparticles. *Biomaterials* **2003**, 24, 1255–1264.

21. Erbacher, P.; Zou, S.; Bettinger, T.; Steffan, A.-M.; Remy, J.-S. Chitosan-based vector/DNA complexes for gene delivery: biophysical characteristics and transfection ability. *Pharm. Res.* **1998**, 15, 1332–1339.

22. Huang, M.; Fong, C-W.; Khor, E.; Lim, L-Y. Transfection efficiency of chitosan vectors: Effect of polymer molecular weight and degree of deacetylation, *J. Controlled Release* **2005**, *106*, 391-406.

23. Berger, J.; Reist, M.; Mayer, J. M.; Felt, O.; Gurny, R. Structure and interactions in chitosan hydrogels formed by complexation or aggregation for biomedical applications, *Eur. J. Pharm. Biopharm.* **2004**, *57,* 35-52.

24. MacLaughlin , F. R.; Mumper, J.; Wang, J.; Tagliaferri, J. M.; Gill, I.; Hinchcliffe, M.; Rolland, A. P. Chitosan and depolymerized chitosan oligomers as condensing carriers for in vivo plasmid delivery. *J. Controlled Release* **1998**, *56*, 259–272.

25. Danielsen, S.; Varum, K. M.; Stokke, B. T. Structural analysis of chitosan mediated DNA condensation by afm: influence of chitosan molecular parameters. *Biomacromolecules,* **2004,** *5*, 928-936

26. Danielsen, S.; Strand, S.; Davies, C. L.; Stokk, B.T. Glycosaminoglycan destabilization of DNA-chitosan polyplexes for gene delivery depends on chitosan chain length and GAG properties., *Biochim. Biophys. Acta (BBA)-Gen. Subj.* **2005** , *1721*, 44-54.

27. Liu, W.; Sun., S.; Cao Z.; Zhang, X.; Yao, K.; Lu, W. W.; Luk, K .D. K. An investigation on the physicochemical properties of chitosan/DNA polyelectrolyte complexes, *Biomaterials,* **2005**, *26*, 2705-2711.

28. Kiang, T.; Wen, J.; Lim, H. W.; Leong, K. W. The effect of the degree of chitosan deacetylation on the efficiency of gene transfection. *Biomaterials* **2004**, *25*, 5293–5301.

29. Schatz, C.; Pichot, C., Delair, T.; Viton, C.; Domard, A.; Static light scattering studies on chitosan solutions: from macromolecular chains to colloidal dispersions. *Langmuir* **2003**, *19*, 9896-9903.

30. Philippova, O. E.; Volkov, E. V.; Sitnikova, N. L.; Khokhlov, A. R. Two types of hydrophobic aggregates in aqueous solutions of chitosan and its hydrophobic derivative. *Biomacromolecules* **2001**, *2*, 483-490.

31. Koping-Hoggard, M.; Melnikova Y. S.; Varum, K. M.; Lindman, B.; Artursson, P.; Relationship between the physical shape and the efficiency of oligomeric chitosan as a gene delivery system in vitro and in vivo *J. Gene Med.* **2003**, *5*,130–141.

32. Koping-Hoggard, M.; Varum, K. M.; Issa M.; Danielsen, S.; Christensen, B. E.; Stroke, B. T.; Artusson, P. Improved chitosan-mediated gene delivery based on easily dissociated chitosan polyplexes of highly defined chitosan oligomers. *Gene Ther.* **2004**, *11*, 1441-1452.

33. Sato, T. ; Ishii, T. ; Okahata, Y. In vitro gene delivery mediated by chitosan. Effect of pH, serum, and molecular mass of chitosan on the transfection efficiency *Biomaterials* **2001**, *22*, 2075-2080.

34. Bozkir, A. ; Saka, O. M.; Chitosan-DNA nanoparticles: Effect on DNA integrity, bacterial transformation and transfection efficiency *J. Drug Targeting* **2004**, *12*, 281-288.

35. Thanou, M. M. ; Kotze, A.F. ; Scharringhausen, T. ; Lueben, H.L. ; de Boer, A.G. ; Verhoef, J.C. ; Junginger, H.E. Effect of degree of quaternization of *N*-trimethyl chitosan chloride for enhanced transport of hydrophilic compounds across intestinal Caco-2 cell monolayers. *J. Controlled Release* **2000**, 64 , 15–25.

36. Thanou, M.; Florea, B. I.; Geldof, M.; Junginger, H. E.; Borchard, G.; Quaternized chitosan oligomers as novel gene delivery vectors in epithelial cell lines. *Biomaterials* **2002**, *23*, 153–159.

37. Thanou, M.; Florea, B. I.; Junginger, H. E.; Borchard, G.; Quaternized chitosan oligomers as gene delivery vectors in vitro *J. Controlled Release* **2003**, *87*, 294-295.

38. Jansma, C. A. ; Thanou, M. ; Junginger, H. E. ; Borchard, G. Preparation and characterization of 6-O-carboxymethyl-N-trimethyl chitosan derivative as a potential carrier for targeted polymeric gene and drug delivery. *STP Pharm. Sci.* **2003**, *13*, 63-67.

39. Park, I.K. ; Kim, T.H.; Kim, S.I.; Akaike, T.;. Cho, C. S. Chemical modification of chitosan for gene delivery. *J. Dispersion Sci. Technol.* **2003**, *24*, 489-498.

40. Davis. S.S. ; Biomedical applications of nanotechnology—implications for drug targeting and gene therapy. *Trends Biotechnol* **1997**,*15*, 217–25.

41. Chellat, F.; Grandjean-Laquerriere, A.; Le Naour, R. ; Fernandes, J.; Yahia, L'H. ; Guenounou, M.; Laurent-Maquin, D. Metalloproteinase and cytokine production by THP-1 macrophages following exposure to chitosan-DNA nanoparticles. *Biomaterials*, **2005**, *26*, 961-970.

42. Merodio, M.; Irache J.M.; Eclancher, F.; Mirshahi, M.; Villarroya, H. Distribution of albumin nanoparticles in animals induced with the experimental allergic encephalomyelitis. *J Drug Target* **2000**, *8*, 289–303.

43. Soma C.E.; Dubernet, C.; Barratt, G.; Benita, S.; Couvreur, P.; Investigation of the role of macrophages on the cytotoxicity of doxorubicin and doxorubicin-loaded nanoparticles on M5076 cells in vitro. *J. Control. Release* **2000**, *68*, 283–289.

44. Bender A.R.; von Briesen H.; Kreuter J.; Duncan I. B.; Rubsamen-Waigmann H. Efficiency of nanoparticles as a carrier system for antiviral agents in human immunodeficiency virus-infected human monocytes/macrophages in vitro. *Antimicrob Agents Chemother* **1996**, *40*, 1467–71.

45. Lobenberg, R.; Kreuter, J.; Macrophage targeting of azidothymidine: a promising strategy for AIDS therapy. *AIDS Res Hum Retroviruses* **1996**, *12*, 1709–1715.

46. Peppas, L. B.; Blanchett, J. O. Nanoparticle and targeted systems for cancer therapy. *Adv. Drug Deliv. Rev.* **2004**, *56*, 1649– 1659.

47. Sashiwa, H.; Yamamori, N.; Ichinose, Y.; Sunamoto, J.; Aiba, S.-i. Michael reaction of chitosan with various acryl reagents in water. *Bimacromolecules* **2003**, *4*, 1250-1254.

48. Park, J. H.; Cho, Y. W.; Chung, H.; Kwon, I. C.; Jeong, S. Y., Synthesis and characterization of sugar-bearing chitosan derivatives: aqueous solubility and biodegradability, *Biomacromolecules* **2003**, *4*, 1087–1091.

49. Sugimoto, M.; Morimoto, M.; Sashiwa H.; Saimoto, H.; Shigemasa, Y.; Preparation and characterization of water-soluble chitin and chitosan derivatives. *Carbohydr. Polym.* **1998**, *36*, 49–59.

50. Muslim, T.; Morimoto, M.; Saimoto, H.; Okamoto, Y.; Minami, S.; Shigemasa, Y.; Synthesis and bioactivities of poly(ethylene glycol)–chitosan hybrids. *Carbohydr. Polym.* **2001**, *46*, 323–330.

51. Morimoto, M.; Saimoto, H.; Shigemasa, Y. Control of functions of chitin and chitosan by chemical modification. *Trends Glycosci. Glycotechnol.* **2002**, *14*, 205-222.

52. Mao, H.Q.; Roy, K.; Troung-Le, V.L.; Janes K.A.; Lin K.Y.; Wang, Y.; August, J.T.; Leong, K.W. Chitosan-DNA nanoparticles as genecarriers: synthesis, characterization and transfection efficiency. *J Control Release* **2001**, 70, 399–421.

53. Park, I. K.; Ihm, J. E.; Park, Y.H.; Choi, Y. J.; Kim, S. I.; Kim, W. J.; Akaike, T.; Cho, C. S.; Galactosylated chitosan (GC)-graft-poly(vinyl pyrrolidone) (PVP) as hepatocyte-targeting DNA carrier Preparation and physicochemical characterization of GC-graft-PVP/DNA complex (1), *J. Controlled Release* **2003**, *86*, 349-359.

54. Leamon, C. P.; Weigl, D.; Hendren, R. W.; Folate copolymer-mediated transfection of cultured cells. *Bioconjugate Chem.* **1999**, *10,* 947-957.

55. van Steenis, J.H.; van Maarseveen, E. M.; Verbaan, F. J.; Verrijk, R., Crommelin, D.J.A; Storm, G.; Hennink, W. E., Preparation and characterization of folate-targeted pEG-coated pDMAEMA-based polyplexes, *J. Controlled Release* 2003, *87*, 167–176.

56. Chan, V.; Mao, H. Q.; Leong, K. W.; Chitosan-induced perturbation of dipalmitoyl-sn-glycero-3-phosphocholine membrane bilayer. *Langmuir*, 2001, *17*, 3749-3756.

57. Ishii, T.; Okahata, Y.; Sato, T.; Mechanism of cell transfection with plasmid/chitosan complexes. *Biochim. Biophys. Acta* 2001, *1514*, 51–64.

58. Lee, M.; Nah, J .W.; Kwon, Y.; Koh, J. J.; Ko, K. S.; Kim, S. W. Water-soluble and low molecular weight chitosan-based plasmid DNA delivery. *Pharm. Res.* 2001, *18*, 427-431.

59. Wolfert, M. A.; Seymour, L. W.; Characterization of vectors for gene therapy formed by self-assembly of DNA with synthetic block co-polymers, *Human Gene Ther.*, 1996, *3*, 269–73.

60. Akbuga, J.; Ozbas-Turan, E.; Erdogan, N.; Plasmid-DNA loaded chitosan microspheres for in vitro IL-2 expression. *Eur. J. Pharmaceutics Biopharmaceutics* 2004, *58*, 501–507.

61. Liu, W. G.; Zhang, X.; Sun, S. J.; Sun, G. J.; De Yao, K. N-alkylated chitosan as a potential nonviral vector for gene transfection. *Bioconjugate Chem.* 2003, *14*, 782-789.

62. Wu, G. Y.; Wu., C. H.; Receptor-mediated delivery of foreign genes to hepatocytes. *Adv. Drug Deliv. Rev.* 1998, *29*, 243–248.

63. Ashwell, G.; Harford, J.; Carbohydrate-specific receptors of the liver. *Ann. Rev. Biochem.*, 1982, *51*, 531–554.

64. Fallon, R. J.; Schwartz., A. L.; Asialoglycoprotein receptor phosphorylation and receptor-mediated endocytosis in hepatoma cells. Effect of phorbol esters. *J. Biol. Chem.* 1988, *63*, 13159–13166.

65. Knowles, B. B.; Howe, C. C.; Aden, D. P. Human hepatocellular carcinoma cell lines secrete the major plasma proteins and hepatitis B surface antigen, *Science* 1980, *209*, 497–499.

66. Ashwell, G.; Morell, A.; The role of surface carbohydrates in the hepatic recognition and transport of circulating glycoproteins, *Adv. Enzymol.* 1974, *41*, 99–128.

67. Ciechanover, A.; Schwartz, A. L.; Lodish., H. F.Sorting and recycling of cell surface receptors and endocytosed ligands: the asialoglycoprotein and transferrin receptors. *J. Cell Biochem.*, 1983, *23*, 107–130.

68. Murata , J-I; Ohya , Y.; Ouchi, T. Possibility of application of quaternary chitosan having pendant galactose residues as gene delivery tool. *Carbohydr. Polym.*, 1996, *29*, 69-74.

69. Park, Y. K.; Park, Y. H.; Shin, B. A.; Choi, E. S.; Park, Y. R.; Akaike, T.; Cho, C. S.; Galactosylated chitosan-graft-dextran as hepatocyte-targeting DNA carrier, *J. Controlled Release*, 2000, *69*, 97-108.

70. Park, I. K.; Park, Y. H.; Shin., B. A.; Choi, E. S.; Kim,Y. R.; Akaike, T.; Cho, C. S. Galactosylated chitosan-graft-dextran as hepatocyte-targeting DNA carrier. *J. Controlled Release* **2001**, *75*, 433.

71. Kim, T. H.; Park, I. K.; Nah, J. W.; Choi, Y. J.; Cho, C. S.; Galactosylated chitosan/DNA nanoparticles prepared using water-soluble chitosan as a gene carrier. *Biomaterials* **2004**, *25*, 3783-3792.

72. Park, I. K.; Kim, T. H.; Park,;Y. H.; Shin, B. A.; Choi, E. S.; Chowdhury, E. H.; Akaike, T.; Cho, C. S. Galactosylated chitosan-graft-poly(ethylene glycol) as hepatocyte-targeting DNA carrier. *J. Controlled Release* **2001**, *76*, 349-362.

73. Park, I. K.; Kim, T. H.; Kim, S. I.; Park, Y. H.; Kim, W. J; Akaike, T.; Cho, C. S. Visualization of transfection of hepatocytes by galactosylated chitosan-graft-poly(ethylene glycol)/DNA complexes by confocal laser scanning microscopy. *Int. J. Pharm.* **2003**, *257*, 103-110.

74. Gao, S. Y.; Chen, J. N.; Xu, X. R.; Ding, Z.; Yang, Y. H.; Hua, Z. C.; Zhang, J. F. Galactosylated low molecular weight chitosan as DNA carrier for hepatocyte-targeting. *Int. J. Pharm.* **2003**, *255*, 57-68.

75. Barabas, K.; Faulk., W. P. transferrin receptors associate with drug-resistance in cancer-cells. *Biochem. Biophys. Res. Commun.* **1993**, *197*, 702–708.

76. McPhail, D., Tetley, L.; Dufes C.; Uchegbu I. F., Liposomes encapsulating polymeric chitosan based vesicles — a vesicle in vesicle system for drug delivery, *Int. J. Pharm.* **2000**, *200*, 73-86.

77. Uchegbu, I. F.; Schatzlein, A.G.; Tetley, L.; Gray, A. I.; Sludden, J.; Siddique, S.; Mosha, E. Polymeric chitosan-based vesicles for drug delivery. *J. Pharm. Pharmacol*, **1998**, *50*, 453-458.

78. Dufes, C.; Schatzlein, A. G.; Tetley, L.; Gray, A. I.; Watson, D. G.; Olivier, J.-C.; Couet, W.; Uchegbu, I. F. Niosomes and polymeric chitosan based vesicles bearing transferrin and glucose ligands for drug targeting. *Pharm. Res.*, **2000**, *17*, 1251-1258.

79. Dufes, C.; Muller, J.-M.; Couet, W.; Olivier, J.-C.; Uchegbu, I. F.; Schatzlein, A. G. Anticancer drug delivery with transferrin targeted polymeric chitosan vesicles, *Pharm. Res.*, **2004**, *21*, 101-107.

80. Guo, W.; Lee , R. L. Receptor-targeted gene delivery via folate-conjugated polyethylenimine *AAPS Pharm. Sci.* **1999**, *1*, 1-7.

81. Benns, J. M.; Maheshwari, A.; Furgeson, D. Y.; Mahato, R. I.; Kim, S. W. Folate-PEG-folate-graft-polyethylenimine-based gene delivery, *J. Drug Target.* **2001**, *9*, 123-139.

82. Benns, J. M.; Mahato, R. I.; Kim, S. W.; Optimization of factors influencing the transfection efficiency of folate-PEG-folate-graft-polyethylenimine. *J. Control. Release* **2002**, *79*, 255-269.

83. Mansouri, S., Cui, Y., Winnik, F., Shi, Q., Benderdour, M., Lavigne, P., Beaumont, E. , Fernandes, J.C. Characterization of Folate-Chitosan-DNA nanoparticles for gene therapy. in press, *Biomaterials*, **2005**.

84. Cho, Y. W.; Kim J. D.; Park, K. Pollycation gene delivery systems: escape from endosomes to cytosol. *J. Pharm. Pharmacol.* **2003**, *55*, 721-734.

85. Boussif, O.; Lezoualch, F.; Zanta M. A.; Mergny, M. D.; Scherman, D.; Demeneix, B.; Behr, J.-P. A versatile vector for gene and oligonucleotide transfer into cells in culture and in vivo: polyethylenimine. *Proc. Natl. Acad. Sci. U.S.A.* **1995**, *92*, 7297-7301.

86. Kiang, T.; Bright, C.; Cheung, C. Y.; Stayton, P. S.; Hoffman, A. S.; Leong, K. W. Formulation of chitosan-DNA nanoparticles with poly(propyl acrylic acid) enhances gene expression. *J. Biomaterials Sci., Polym. Ed.* **2004**, *15*, 1405-1421.

87. Funhoff, A. M.; van Nostrum, C. F.; Koning, G. A. Endosomal escape of polymeric gene delivery complexes is not always enhanced by polymers buffering at low pH. *Biomacromolecules* **2004**, *5*, 32-39.

88. Kim, T.H.; Ihm, J.E.; Choi, Y.J.; Nah, J.W.; Cho, C.S. Efficient gene delivery by urocanic acid-modified chitosan. *J. Controlled Release* **2003**, *93*, 389-402.

89. Bieber, T.; Meissne,r W.; Kostin, S.; Niemann, A.; Elsasser, H.-P. Intracellular route and transcriptional competence ofpolyethylenimine-DNA complexes. *J. Controlled Rel.* **2002**, *82*, 441–454.

90. Godbey, W.T.; Wu, K.K.; Mikos, A.G. Tracking the intracellular path of poly(ethylenimine)/DNA complexes for gene delivery. *Proc. Natl. Acad. Sci. USA* **1999**, *96*,5177–5181

91. Godbey, W.T.; Barry, M.A.; Saggau, P.; Wu, K.K.; Mikos, A.G. Polyethylenimine-mediated transfection: a new paradigm for gene delivery. *J. Biomed. Mater. Res.* **2000**, *51*,321–328

92. Hébert, E. Improvement of exogenous DNA nuclear importation by nuclear localization signal-bearing vectors: a promising way for non-viral gene therapy? *Biology of the Cell* **2003**, *95*, 59–68.

93. Chan, C-K; Jans, D.A.; Using nuclear targeting signals to enhance non-viral gene transfer. *Immunol. Cell Biol.* **2002**, *80*,119–130.

94. Cartier, R.; Reszka, R. Utilization of synthetic peptides containing nuclear localization signals for nonviral gene transfer systems. *Gene Ther* **2002**, *9*,157–167.

95. Dean, D.A.; Dean, B.S.; Muller, S.; Smith, L.C. Sequence requirements for plasmid nuclear import. *Exp. Cell. Res.* **1999**, *253*, 713–722.

96. Vacik. J.; Dean, B.S.; Zimmer, W.E.; Dean, D.A. Cell-specific nuclear import of plasmid DNA. *Gene Ther.* **1999**, *6*:1006–1014.

97. Schaffer, D.V.; Fidelman, N.A.; Dan, N.; Lauffenburger, D.A. Vector unpacking as a potential barrier for receptor-mediated polyplex gene delivery. *Biotechnol* **2000**, *67*, 598–606.

98. Ziady, A.G.; Ferkol, T.; Dawson, D.V.; Perlmutter, D.H.; Davis, P.B. Chain length of the polylysine in receptor-targeted gene transfer complexes affects duration of reporter gene expression both in vitro and in vivo. *J Biol Chem* **1999**, *274*,4908–4916

Chapter 10

Chitosan as a Biomaterial for Preparation of Depot-Based Delivery Systems

J. Grant and C. Allen[*]

Department of Pharmaceutical Sciences, University of Toronto, Toronto, Ontario M5S 2S2, Canada

Chitosan is a natural polysaccharide that has become established as a material with great potential for use in biomedical applications. In addition, the recent success of many polymeric depot delivery systems has encouraged further research to identify new materials and develop new systems for use in regional therapy. Several strategies have been developed for preparation of stable chitosan-based depot systems for drug delivery. Chitosan can be physically or chemically crosslinked to prepare microspheres, films and gels. Chitosan has also been blended with a wide range of polymers to produce mostly two-phase systems, which have unique properties that are required for specific applications. These stable chitosan-based depot systems have been investigated for treatment of various diseases including cancer and bacterial infection. This review includes a summary of the favorable chemical, physical and biological properties of chitosan. In addition, the properties of an ideal depot system for use in drug delivery are outlined. Overall, it is hoped that the reader gains an appreciation for the use of chitosan-based systems for the regional or localized delivery of drugs.

Depot systems include films (wafers), gels and particulates (i.e. microspheres, microparticles) that provide a storage reservoir of drug, which is then released at a specific rate within the body. As outlined in Figure 1, depot systems are often administered via parenteral or local routes. The parenteral mode of administration refers to injection through a needle into the body at various sites and depths. The primary routes of parenteral delivery are intravenous, intramuscular and subcutaneous, although intradermal, intracardiac and intraspinal are also included. Depot systems such as microparticles or gels are commonly administered via a parenteral route. Systems such as films or wafers may be administered locally by surgical implantation at a specific site (1).

As shown in Figure 1, following administration the depot system may provide local and/or systemic delivery of the drug. Systemic delivery refers to a drug that is administered, and reaches the blood stream. By contrast, for local delivery the drug is directly delivered at the site of implantation of the depot system (1). Rational design of a depot system requires consideration of the desired mode of administration of the system and type of drug delivery required. In general, depot systems are designed to: 1) provide sustained drug release in order to extend the duration of treatment, 2) provide a therapeutic concentration of drug to a specific site and 3) improve patient compliance. These systems have been investigated for a wide variety of indications including psychological disorders, wound healing, cancer treatment, periodontal disease, and gene therapy (2-5). The drug delivery devices explored to date have been prepared from synthetic and/or natural polymer materials (6). Indeed, several depot systems based on polymer materials have received FDA approval for the treatment of various diseases (e.g. Atridox®, Decapeptyl®, Gliadel®, Lupron Depot®, Nutropin Depot®, Sandostatin LAR®, Trelstar Depot® and Zoladex®). These approved systems are mostly formed from the synthetic polyesters poly(d,l-lactide) (PLA) and poly(lactide-co-glycolide) (PLGA). Lupron Depot® was the first depot delivery system to be approved for clinical use and includes PLGA microspheres that provide sustained systemic delivery of Leuprolide acetate. This product was approved in 1989 and in 2001 its total annual sales were $833 million (7). Specifically, Lupron Depot® has been approved for treatment of fibroids, prostate cancer, endometriosis, and central precocious puberty. Currently, over 600 patents on drug delivery devices formed from PLGA and PLA have been filed and the annual sales for these products have reached approximately $2.5 billion (7). PLGA and PLA are known to be biocompatible, biodegradable materials that are stable under physiological conditions (8). However, they have also been associated with some limitations including foreign body responses that lead to capsid formation (9, 10). Capsid formation involves the encapsulation of the depot in collagenous tissue and thus may alter both the degradation and drug release profiles of the delivery system (11). Also, the degradation products of these synthetic polyesters may cause an increase in the acidity within the area of implantation

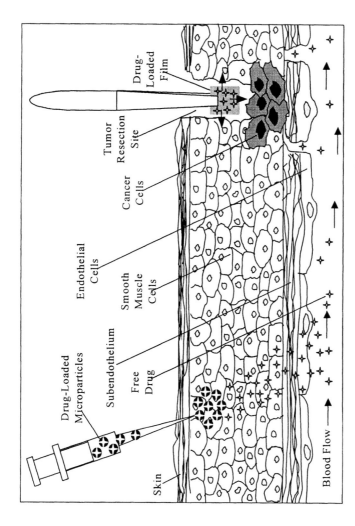

Figure 1. A schematic illustrating the concept of systemic and localized drug delivery. For systemic delivery, microparticles encapsulated with drug are injected subcutaneously; drug is released and able to reach the blood stream. For localized drug delivery, a film loaded with drug is directly placed in a tumor resection site using surgical forceps and drug is released locally.

resulting in local irritation and accelerated hydrolysis or degradation of some drugs (12). In this way, there is a need to develop and explore new materials for preparation of depot or implantable drug delivery systems. In addition, it is known that the degree of compatibility or interaction between a drug and the material employed to prepare the delivery system may influence many of the performance related parameters of the system including stability, drug loading efficiency and drug release profile. In this way, it is unlikely that any one system prepared from a specific material will universally serve as a universal delivery system for all drugs.

Recently, there has been increased interest in the use of natural polymers such as polysaccharides for biomedical applications. This class of polymers includes: alginates, amylose, carrageenans, cellulose, chitin, chitosan, dextran, glycogen, inulin, pectin, pullulan and starch. Cellulose, chitin and chitosan are the most naturally abundant polysaccharides in the world. Many of these biopolymers, especially chitosan, have been shown to have good biological and film-forming properties.

Chitosan was first described in the literature in 1811 and has been traditionally used in America for the treatment of machete gashes and in the Orient for the treatment of abrasions. Due to many of chitosan's favorable properties it has been explored for commercial use in nutrition, foods, cosmetics as well as environmental and biomedical applications (13, 14). The potential biomedical applications for chitosan, as summarized by Sanford and Skaugrud, includes: sutures, haemostatic, fungistatic, spermicidal, anti-carcinogen, anti-cholesterolimic, contact lenses, wound healing, eye bandages, dental-bioadhesives and orthopaedic materials (15, 16). The focus of the present review is the design and use of chitosan-based depot systems for sustained drug delivery. Special attention is given to these systems, as they have shown great potential for treatment of a range of diseases. It is our hope that the reader will gain an appreciation for the exploitable properties of chitosan, the interactions involved in preparing stable chitosan-based systems and the potential of localized drug delivery. Furthermore, this review may serve as a useful tool for preparation of novel chitosan-based depot systems for drug delivery.

Chitosan: Chemical, Physical, and Biological Properties

Chitosan is naturally present in crustacean shells, such as shrimp, crab and lobster, some insects, microorganisms and fungi but can also be derived from the deacetylation of chitin (17). Chitosan is one of nature's most versatile biomaterials as it can be prepared as a powder, film, fiber, gel, beads, paste or solution (13). Chitosan is a cationic copolymer that consists mainly of β-(1,4)-2-deoxy-2-amino-D-glucopyranose and partially of β-(1,4)-2-deoxy-2-acetamido-D-glucopyranose units. Its properties, such as molecular weight, degree of deacetylation, viscosity and purity can be made to vary over a wide range. For example, chitosan can be prepared to have a molecular weight of 50 to 2000 kDa, its degree of deacetylation can be varied from 40-98% and its viscosity, which is concentration and temperature dependent, can be increased to 2000 MPa (18). Chitosan has been more commonly explored for pharmaceutical applications rather than chitin as the former is more readily dissolved in various solvents. The solubility of chitosan is dependent on both the degree of deacetylation of the material and the pH of the medium. For example, 85% deacetylated chitosan is soluble in water at pH<6.5, whereas 40% deacetylated chitosan is soluble at pH<9 (18). Acetic acid has been commonly used to protonate chitosan's amine groups (pK=6.3), in addition to hydrochloric acid, lactic acid, formic acid and citric acid. When chitosan is dissolved in a 1% acetic acid solution, the amine groups become protonated ($NH3^+$) and associate with acetate counter ions (19). If the chitosan acetate salt is heated, it can be converted into chitin (20). From our results, chitosan flakes have a relatively low degree of crystallinity with three broad short-range orders at $2\theta= 10.5°$, $20.1°$ and $22.0°$ which is in agreement with previous reports (21, 22). When chitosan is dissolved in a 1% (v/v) acetic acid solution and dried, an amorphous film is produced with three frequent atomic distances of 9.9, 7.3 and 6.6 Å appeared with lower intensity at $2\theta= 9.0°$, $12.1°$ and $20.1°$, respectively (21). DSC analysis of a chitosan sample reveals two endothermic peaks at 100 °C and 290 °C that are owed to the evaporation of water and the onset of degradation, respectively (23, 24). However, these values may vary with the moisture content of chitosan. The glass transition temperature for chitosan cannot be observed without degrading the polymer due to the strong network of hydrogen bonds that exist within the biomaterial (25). The structure of the polymer can be controlled by the degree of deacetylation; at high degrees of deacetylation the chitosan chains are more flexible whereas the chains are rod-shaped at lower degrees of deacetylation (26). The degree of deacetylation can also affect the rate of degradation, which in turn can alter the drug release profile of a system formed from this material (27). Furthermore, the molecular weight, elongation at break, tensile strength and biological properties of chitosan are influenced by the degree of deacetylation (27-30).

In order to gain approval of this material for clinical use, a large number of studies have examined the biocompatibility, biodegradation and sterilization of chitosan. Biodegradation of polysaccharides involves two processes: erosion and degradation. Erosion can be further subclassified into bulk and surface erosion. Bulk erosion normally occurs when water can easily penetrate into the bulk, whereas for surface erosion, material is lost from the surface which is common for hydrophobic polymers (31). Biodegradable polymers degrade via bond cleavage into metabolites or products that are completely removed from the body (31). The mechanism of degradation for most natural polymers such as polysaccharides is via enzymatic hydrolysis of the polymer (27). *In vivo,* chitosan is degraded primarily by the hydrolytic enzyme, lysozyme, and the degradation products are glucosamine and N-acetylglucosamine residues (32). These degradation products can be incorporated into the glycosaminoglycan and glycoprotein metabolic pathways and/or excreted directly. The kinetics of degradation of chitosan has been found to be inversely proportional to the degree of crystallinity, which is primarily determined by the degree of deacetylation of the material. Therefore, systems formed from highly deacetylated chitosan have been shown to last up to months *in vivo*. Ikada and Tomihata studied the *in vivo* degradation of chitosan films implanted subcutaneously in rats (27). From their results, the stability of the films increased when formed from chitosan having higher degrees of deacetylation. Specifically, two weeks post-implantation a significant reduction in weight was observed for a pure chitin film, whereas no signs of biodegradation were shown for a film composed of pure chitosan (27).

Our thermogravimetric analysis of chitosan has revealed that the temperature at which the onset of degradation occurs is approximately 240 °C, which is in agreement with previous reports (33). The thermal stability of this material allows for safe steam sterilization which is typically carried out at temperatures of 120 °C, as observed by Rao and Sharma (34). However, it has been shown that exposure to high temperatures can change the physical properties of chitosan. For example, Ling et al obtained a lower solubility for chitosan films exposed to dry heat which may be attributed to increased interchain crosslinking (35). Another proposed mechanism is the conversion of the water-soluble chitosan acetate to the water-insoluble chitin (20). Safe sterilization methods for chitosan have been reviewed in detail by Khor and Lim (36). The successful sterilization of chitosan-based systems will further enhance their biocompatibility and potential for use in biomedical applications.

Clinical trials on chitosan-based biomaterials have not shown any inflammatory or allergic responses following implantation, injection, topical application or ingestion in the human body (28, 37). This has also been shown in animal models as chitosan systems have been demonstrated to evoke a minimal foreign body reaction in rat and mouse models; observed as a very mild inflammatory reaction and in most cases no major fibrous encapsulation (38). In a study by Khan et al, chitosan was dissolved in lactic acid or acetic acid and injected intra-cutaneously in New Zealand rabbits. From their results, no gross signs of toxicity were observed (39). Therefore, the physical and biological properties of chitosan have been established and can be manipulated to design systems that are suitable for specific applications.

The Use of Chitosan in Drug Delivery

Chitosan has been employed in the preparation of drug delivery systems such as implants, granules, beads, tablets, emulsions, pellets, gels, micelles, microparticles and nanoparticles. These drug delivery systems have been explored for treatment of several medical conditions including: antacid/antiulcer, hypocholesterol, orthopedic/periodontal, cancer, tissue engineering, wound healing and gene therapy. Previous reviews provide detailed discussion of many of the delivery systems that have been prepared from chitosan (17, 18, 36, 40-42). This section of the present review discusses the strategies that are commonly used to stabilize chitosan-based depot systems. In each case, examples are provided of current systems that have been successfully designed using these strategies.

The properties of films and gels formed from chitosan can be controlled by varying the processing parameters such as: type and pH of solvent used, concentration of chitosan as well as the temperature employed (43). However, films formed from chitosan alone are generally brittle under dry conditions and unstable in aqueous environments (44). From our studies, a film formed from chitosan alone has a Young's modulus of 3850 MPa and when placed in phosphate buffer solution (pH=7.4), it dissolves within 24 hours (21). Washes of NaOH solution can remove the acetyl groups and stabilize chitosan in aqueous environments; however, this treatment is considered harsh with economic and ecological consequences (45). Therefore, in order to form stable systems for drug delivery many groups have explored both the chemical and/or physical crosslinking of chitosan.

A crosslinking agent is described as a molecule with at least two reactive functional groups that form bridges between the polymer chains. Chemically crosslinked chitosan hydrogels are formed by irreversible covalent bonds with agents such as glyoxal, glutaraldehyde and epoxy compounds. For example, glutaraldehyde chemically crosslinks chitosan's amine groups to form covalent imine bonds (Schiff reaction) (45). These chemically crosslinked chitosan-based hydrogels have been used to prepare films, gels and microspheres (46-48). Yet, these systems are often non-biodegradable and the chemical crosslinking agents have been found to be toxic (49). A recent trend includes the use of natural materials as crosslinking agents as they have been shown to enhance the use of chitosan systems for wound healing and tissue engineering. For example, genipin, obtained from its parent compound geniposide which is extracted from the fruits of Gardenia jasminoides Ellis, has been used to crosslink chitosan in order to produce films, gels and microspheres for drug delivery (50-52). This common herbal medicinal agent or food dye has been shown to react with amino acids and proteins and is about 5000-10,000 times less toxic *in vitro* than glutaraldehyde (53). Mi et al, reported that genipin directly reacts with chitosan's amino groups by forming an intermediate aldehyde group via a ring opening reaction (54).

Recently, various groups have also blended chitosan with polymers or small molecules in order to produce stable films. These chitosan blends are stabilized by non-covalent or physical crosslinking that is achieved via electrostatic and/or hydrogen bonding interactions. Hydrophobic interactions have also been exploited in order to increase the stability of chitosan-based systems. For example, negatively charged molecules such as β-glycerol phosphate and tripolyphosphate form ionic interactions with chitosan's positively charged amine groups (55-57). In addition, small molecules such as glycerol have been used to physically crosslink chitosan chains by forming hydrogen bonds between glycerol and the amine group of chitosan (19). Several studies have also investigated the blending of chitosan with homopolymers. Generally, chitosan interacts with homopolymers by inter and intramolecular hydrogen bonding and/or ionic interactions as outlined in Table 1. From the literature, most chitosan blends (and blends in general) are two-phase systems due to the very small mixing entropy for the materials. These immiscible or partially miscible blends have unique properties that make them suitable for specific applications (Table 1).

A chitosan-based thermosensitive gel that has been studied extensively by Leroux's group has shown potential as a local delivery device. This system is formed from a solution of chitosan and glycerophosphate that gels at 37 °C in neutral pH (58). In this way, the chitosan is crosslinked with the glycerol-phosphate disodium salt and the interactions responsible for providing the sol/gel transition include: increased interchain hydrogen bonding, electrostatic interactions between chitosan and glycerophosphate and hydrophobic interactions between the chitosan chains (58). Interestingly, upon heating to 37 °C the glycerol moiety of glycerophosphate has been shown to promote interaction between the chitosan chains by the removal of water. Thus, although electrostatic and hydrogen bonding forces are operative, hydrophobic interactions play a major role in the gelation of the chitosan-glycerophosphate solutions.

Our studies have focused on characterizing biodegradable and biocompatible films formed from blends of chitosan and phospholipids. The specific combination of chitosan and egg phosphatidylcholine (ePC) produced films with a minimal degree of swelling and high stability for use in localized drug delivery (59). The possible interactions that stabilize the chitosan-lipid films are shown in Figure 2. The properties of the chitosan-ePC films are discussed in further detail below.

Physically crosslinked gels or films have also been prepared from chitosan-based graft copolymers. For example, D,L-lactic acid and poly(ethylene glycol) (PEG) have been successfully grafted onto chitosan's amine group to produce pH sensitive and thermosensitive hydrogels, respectively (60). The mechanism of gelation for the chitosan-g-PEG thermogel is similar to that for the chitosan-glycerophosphate blend: water molecules are removed from PEG and chitosan chains which then interact with increasing temperature. It should be noted that there is some degree of difficulty associated with synthesizing chitosan graft copolymers owing to the strong hydrogen bonding network that exists between the chitosan chains.

Design of the Ideal Depot System

There are numerous factors that should be considered when designing depot systems for drug delivery. First, the therapeutic dose and rate of delivery required for successful treatment of the disease must be known. The system may then be designed to deliver the required dose over the specified period of time. It is advantageous for the system to have a high loading capacity for the drug of interest in order to achieve a high drug to material ratio. It is also important to note that the dose considered to be therapeutically relevant when the drug is administered systemically may differ from that required when the drug is delivered locally at a sustained rate. In the same manner, the therapeutically relevant dose for sustained, local delivery may differ from that for bolus, local delivery. The drug must also remain physiologically active prior to and following release from the delivery system.

Secondly, the mode of administration must also be considered and in general may include implantation (e.g. film) or injection (e.g. microspheres, solution that gels in-situ). Thirdly, the conditions at the site of administration must be evaluated including volume and types of fluid present, pH of the environment as well as the level of enzymes at this site. Furthermore, films or gels used for drug delivery will often be subjected to physical deformation. Therefore, evaluation of the mechanical strength of an implant under relevant physiological conditions (pH = 7.4, temperature = 37°C) is necessary as it provides an indication of the physical deformations that the system may sustain *in vivo*. In addition, the softness of the medical device should also be measured in order to prevent damage to surrounding tissues. Other properties of biomedical devices that should be considered when designing a depot system include: biodegradability, biocompatibility, toxicity, bioadhesion, swelling behavior, stability, sterilization and patient compliance.

Recently, several studies have focused on establishing relationships between the composition and the performance related properties of the delivery system (28, 61). Research in this area will allow systems to be designed and optimized for specific applications. For example, studies on films formed from chitosan-gelatin blends demonstrated that the gelatin content was related to the crystallinity, water content, Young's modulus and the degree of adsorption of fibronectin to the film. The presence of chitosan in the films was shown to improve nerve cell affinity, when compared to films formed from gelatin alone (62). From our studies, we established relationships between the composition and properties of films formed from the partially miscible blend of chitosan and ePC (Table 2). Specifically, the amount of lipid within the chitosan-ePC films was related to the swelling behavior, stability, storage modulus, morphology and protein binding properties. In addition, for phase separated blends consisting of chitosan and PVA, the molecular weight of PVA was related to the swelling, mechanical properties and the amount of drug released from the blend (63).

Furthermore, studies have focused on understanding the relationships between the properties of the system and clinical performance. For example, Jones et al. have demonstrated that the rheological properties of biomaterials directly affect their clinical performance (64). In their study, the formulation compressibility and hardness were related to ease of product removal from a container, ease of application onto a substrate and product comfort within the oral cavity (65, 66). The elastic and viscous modulus of the implant was also

Table I. Interactions Between Chitosan and Homopolymers

Polymer Blended with Chitosan	Miscibility with Chitosan	Interactions	Application	References
Poly(lactic acid)	Immiscible	No interactions (FTIR)	Food packaging	(85)
Poly(caprolactone)	Miscible	Intermolecular hydrogen bonding		(86)
Cellulose	Immiscible	Intramolecular and intrastand hydrogen bonding	Wound dressing	(87)
Gelatin	Miscible	Hydrogen bonding	Nerve regeneration	(62)
Collagen	Miscible	Hydrogen bonding and ionic interactions		(88)
Poly(ethylene glycol)	Miscible below 50% (wt/wt) of chitosan	Intermolecular hydrogen bonds		(89, 90)
Poly(ethylene oxide)	Miscible	Hydrogen bonding		(63)
Poly(vinyl alcohol)	Imiscibile	Intermolecular interaction	Oral mucosal delivery Fibroblast adhesion	(63, 91)
Silk Fibroin	Miscible	Hydrogen bonds	Wound dressing and artifical skin Artifical muscle	(61, 92)
Poly(3-hydroxybutyrate) and Poly(3-hydroxybutyrate-co-3-hydroxyvalerate)	Miscible	Intermolecular hydrogen bonding		(93)
Polycarbonate (Liquid crystalline) (LCPC)	Immiscible	Repulsive interactions		(94)
LCPC/PVA	Miscible	Intermolecular interaction		(94)
Pectin		Intramolecular hydrogen bonding and ionic interactions	Film coating	(95)
Poly(4-vinyl-N-butyl) Pyridine (quaternized)	Partial miscibility	Intermolecular interaction		(96)

				(97)
Nylon-6	Immiscible			
Nylon-4	Partial miscibility	Hydrogen bonding	Catalyst for destruction of toxic chemical by hydrolysis	(97)
Poly(N-Vinyl-2-pyrrolidone)	Miscible	Hydrogen bonding	Oral Mucosal delivery Wound Healing	(63, 98-100)
Poly(N-methyl-N-vinylacetamide) Poly(N,N-dimethylacrylamide) Poly(2-methyl-2-oxazoline) Poly(2-ethyl-2-oxazoline)	Miscible	Hydrogen bonding		(98)

212

1. HYDROPHOBIC INTERACTIONS

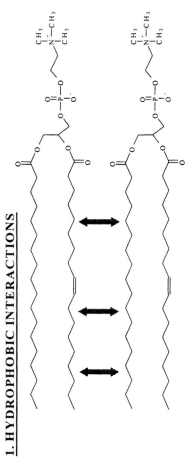

Hydrophobic interactions between lipid hydrocarbon chains.

2. HYDROGEN BONDING

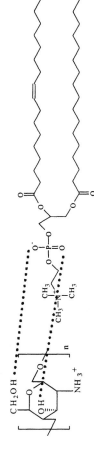

Hydrogen bonding between chitosan repeat unit and head group of lipid.

3. IONIC INTERACTIONS

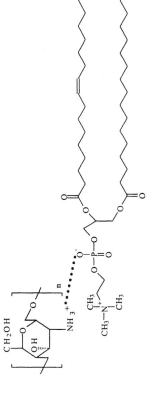

Ionic interactions between chitosan repeat unit and head group of lipid.

Figure 2. Examples of hydrogen bonding, hydrophobic and ionic interactions that may act to stabilize the chitosan-lipid drug delivery system.

Table II. Relationship between Composition and Properties for Chitosan : ePC Films

Composition Chitosan:ePC (wt/wt)	Swelling Ratio in 0.01M PBS at 37°C	%Weight Loss Following 1 Week in FBS at 37°C	Storage Modulus at 37°C (MPa)	Young's Modulus For Dry Films (MPa)	Young's Modulus For Wet Films (MPa)	Morphology (Size of Domains) (μm)	Protein Binding (following PBS wash) (μg/cm²)
1:0 (chitosan)	****	****	7200	3850	----	----	****
1:0.2	4.8	50	6000	2400	0.7	1-10	337
1:0.4	3.1	40	4300	2360	0.6	----	350
1:0.8	3.2	33	3500	1180	0.9	----	355
1:1.7	2.4	28	2350	1060	1.0	----	416
1:2.5	1.6	12	1300	350	1.4	20-30	434

Note: **** signifies that the chitosan films dissolve in solution preventing analysis. Also, ePC alone is unable to form a film. Adapted from (21, 59).

directly related to the drug release profile from the biomedical device. Furthermore, Needleman et al. found that the degree of swelling of a gel *in vitro* could be used as a reliable means to predict the duration of bioadhesion *in vivo* (67, 68).

Chitosan-Based Depot Systems for Regional Therapy

There are numerous diseases that are suited for treatment using localized delivery of therapeutic agent. The implantation of a polymer-based device containing an active agent can provide a high-dose to a specific area for a prolonged period of time. Localized delivery can also reduce or eliminate the harmful side effects that are commonly associated with the systemic administration of agents. In addition, localized delivery of clinically relevant concentrations of cell cycle specific drugs may result in improved efficacy due to the increased time of cell exposure or duration of treatment. Also, the complete biodegradation of implantable materials prevents the need for invasive surgery to remove the biomedical device. Other advantages of these implantable systems are reviewed by Langer (69). One of the most successful and extensively studied local delivery devices is Gliadel® which is composed of a biodegradable polyanhydride (i.e. poly [1,3-bis (carboxyphenoxy) propane-co-sebacic-acid]) matrix that contains the drug, carmustine (70). This controlled release polymeric system has been FDA approved for the treatment of malignant gliomas. Specifically, Gliadel® provides a high local concentration of carmustine at the tumor resection site while reducing the systemic toxicity of the drug and protecting the drug from hydrolytic degradation. A few other implant systems, mostly based on PLGA, have been FDA approved or are currently in clinical development for the treatment of other cancers.

Implant systems that have been studied for the local delivery of anti-cancer agents include: microspheres, pastes, films, discs or rods. For example, Burt's group developed a poly(caprolactone) paste as a delivery system for the anticancer agent, paclitaxel, for the treatment of prostate cancer (71). Paclitaxel, isolated from the bark of the pacific yew tree, prevents mitotic cell division by interacting with tubulin dimers, thus promoting polymerization and stabilization of microtubules (72, 73). Commercially, Bristol-Myers Squibb produces Taxol®, a formulation consisting of paclitaxel dissolved in Cremophor EL vehicle (30 mg paclitaxel in 5 ml of Cremophor EL:dehydrated ethanol (1:1 v/v)). However, Cremophor EL has been found to cause serious and sometimes fatal hypersensitivity (allergic) reactions in patients (74, 75). Therefore, many research groups have focused on designing novel formulations for this compound including systems for localized delivery. Paclitaxel is a lipophilic compound with limited water solubility ($< 0.1 \mu g/ml$), which makes it difficult to solubilize and achieve high drug loading in hydrogel systems. However, Chitosan-based systems have been designed for delivery of paclitaxel. For example, Panchanula and Dhanikula prepared chitosan-based films with high loading efficiencies for paclitaxel using two methods (76). In the first method, chitosan was mixed with glycerol (a plasticizer that physically crosslinks chitosan) and liposomes were incorporated into the matrix to solubilize paclitaxel. The second formulation included paclitaxel incorporated in films formed from poloxamer 407 and chitosan. However, the *in vitro* release of paclitaxel from the films containing liposomes was negligible and the poloxamer containing films showed a 10% burst release followed by no release for 144 hours. The release of paclitaxel from the films was then studied in the presence of lysozyme in order to simulate *in vivo* conditions, yet no significant difference in the release profile for the drug was found (76).

Our studies incorporated a therapeutically relevant concentration of paclitaxel within nanoparticles that were dispersed throughout a chitosan-lipid matrix (Figure 3, adapted from ref (59)). From our studies the system was found to provide a sustained release of paclitaxel over a four-month period in phosphate buffer solution (PBS, pH=7.4) containing lysozyme. The release was attributed to the three primary mechanisms for drug release from matrix systems: swelling, diffusion and erosion (77, 78). In addition, when the chitosan-lipid films were implanted in the peritoneal cavity of CD-1 mice the release of paclitaxel was found to proceed at a sustained rate (79). Post-mortem examination of the mice revealed no observable signs of inflammation, infection or injury two weeks following the intraperitoneal implantation of the films. Immunohistochemical analysis of the films revealed no signs of protein adherence or fibrous encapsulation. As a comparison, films formed from PLA or PCL were also implanted into the peritoneal cavities of CD-1 mice. Post-implantation, the PLA and PCL films were found to display significant fibrous encapsulation (79). In this way, the chitosan-based films may be more suitable than polyester-based systems for implantation into sites such as the peritoneal cavity.

The thermosensitive chitosan-glycerophosphate gel developed by Leroux's group was also found to provide sustained release of paclitaxel over a prolonged period in PBS containing 0.3% sodium dodecyl sulfate (SDS) at 37 °C under sink conditions (80). When the chitosan-glycerophosphate gel was injected intratumorally into a mouse tumor model (EMT-6), it was found to be as efficacious as four intravenous injections of Taxol® (80). Similarly, a study on paclitaxel loaded chitin and chitin-Pluronic F-108 microparticles were found to provide a significant reduction in tumor volume six days following a subcutaneous injection at the base of the tumor in the hind left flank region of the mice (81). Therefore, chitosan based formulations have shown potential for use in the localized treatment of cancer.

Regional biodegradable depot systems have also been investigated for use in treatment of infectious diseases. Buranapanitkit et al. studied the efficacy of a local biodegradable composite composed of hydroxyapatite and chitosan loaded with antibiotics to treat methicillin-resistant Staphylococcus aureus (MRSA) (82). From their results, the chitosan-hydroxyapatite implant showed a significant inhibitory effect over a three-month period against MRSA. A

217

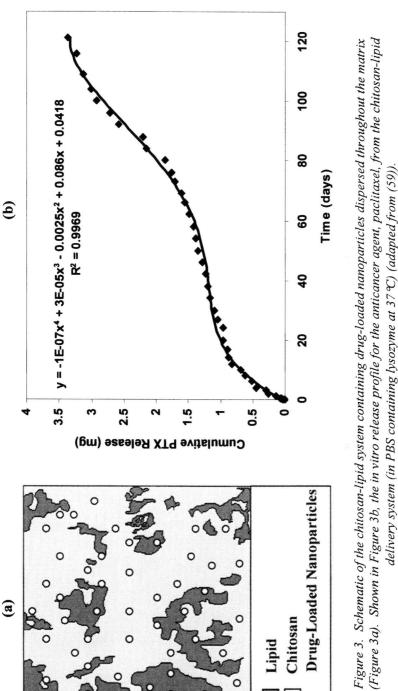

Figure 3. Schematic of the chitosan-lipid system containing drug-loaded nanoparticles dispersed throughout the matrix (Figure 3a). Shown in Figure 3b, the in vitro release profile for the anticancer agent, paclitaxel, from the chitosan-lipid delivery system (in PBS containing lysozyme at 37 °C) (adapted from (59)).

218

biodegradable chitosan bar was also found to provide sustained release of the antibiotic gentamicin when evaluated for treatment of bone infection in a rabbit model. The local delivery of gentamicin was found to inhibit osteomyelitis in the bone tissue surrounding the bar for approximately eight weeks following implantation (83). No side effects commonly associated with the systemic administration of gentamicin were observed over the course of treatment (83).

Finally, Perugini et al. combined chitosan with PLGA by an emulsification process and loaded ipriflavone, a poorly water soluble agent, to prepare a film for local treatment of periodontitis. The PLGA/chitosan film was shown in vitro to provide a sustained release of the agent for 20 days (84).

Conclusions

The success of polymer-based depot delivery systems such as Gliadel® and Lupron Depot® encourages the design of other polymeric systems. Yet, the number of materials that are approved for use in humans is limited. The natural abundance and favorable properties of chitosan have made this material attractive for fabrication of drug delivery systems. However, further studies are required in order to fully exploit chitosan as a biomaterial for use in drug delivery. Specifically, studies focused on gaining an understanding of the composition-property and property-performance relationships will allow for these systems to be designed to suit specific applications. Further pre-clinical evaluation of the chitosan-based systems in relevant animal models will also encourage the advancement of these systems for use in humans. The existing and emerging literature on this material is promising and therefore supports the progression of this material for preparation of depot systems for clinical use. Overall, chitosan is a promising and exciting material for design of advanced depot-based drug delivery systems.

Acknowledgements

The authors would like to thank Helen Lee for her assistance with the illustrations. The authors are also grateful to the Natural Science and Engineering Research Council of Canada (NSERC) for an operating grant and the Ontario Cancer Research Network (OCRN) for a translational research grant to C. Allen.

References

1. Allen, L.V., N.G. Popovich, and H.C. Ansel, *Ansel's pharmaceutical dosage forms and drug delivery systems*. 8th ed. 2005, Philadelphia: Lippincott Williams & Wilkins. xi, 738.

2. Messer, T., et al., *Risperidone microspheres - Renaissance of neuroleptic depot therapy.* Psychopharmakotherap, **2002**, 9,133-139.

3. Brem, H. and P. Gabikian, *Biodegradable polymer implants to treat brain tumors.* J Control Release, **2001**, 74,63-67.

4. Yusof, N.L.B.M., et al., *Flexible chitin films as potential wound-dressing materials: wound model studies.* J Biomed Mater Res A, **2003**, 66A,224-232.

5. Lacevic, A., E. Vranic, and I. Zulic, *Endodontic-periodontal locally delivered antibiotics.* Bosn J Basic Med Sci, **2004**, 4,73-8.

6. Hedley, M.L., *Formulations containing poly(lactide-co-glycolide) and plasmid DNA expression vectors.* Expert Opinion on Biological Therapy, **2003**, 3,903-910.

7. Chauba, M., *Polylactides/Glycolides - Excipients for Injectable Drug.* Drug Delivery Technology, **2002**, 2,34-36.

8. Grayson, A.C.R., et al., *Differential degradation rates in vivo and in vitro of biocompatible poly(lactic acid) and poly(glycolic acid) homo- and co-polymers for a polymeric drug-delivery microchip.* Journal of Biomaterials Science-Polymer Edition, **2004**, 15,1281-1304.

9. Hickey, T., et al., *In vivo evaluation of a dexamethasone/PLGA microsphere system designed to suppress the inflammatory tissue response to implantable medical devices.* Journal of Biomedical Materials Research, **2002**, 61,180-187.

10. Fulzele, S.V., P.M. Satturwar, and A.K. Dorle, *Study of the biodegradation and in vivo biocompatibility of novel biomaterials.* Eur J Pharm Sci, **2003**, 20,53-61.

11. Ratner, B.D. and S.J. Bryant, *Biomaterials: Where we have been and where we are going.* Annual Review of Biomedical Engineering, **2004**, 6,41-75.

12. Agrawal, C.M. and K.A. Athanasiou, *Technique to control pH in vicinity of biodegrading PLA-PGA implants.* J Biomed Mater Res, **1997**, 38,105-14.

13. Skj*k-Br*k, G., T. Anthonsen, and P.A. Sandford, *Chitin and chitosan : sources, chemistry, biochemistry, physical properties, and applications.* 1989, London ; New York: Elsevier Applied Science. xxii, 835.

14. Li, Q., et al., *Applications and Properties of Chitosan.* J Bioact Compat Pol, **1992**, 7,370-397.

15. Skaugrud, O., *Chitosan-New biopolymer for cosmetics and drugs.* Drug Cosmetic Ind., **1991**, 148,24-29.

16. Sandford, P.A., *Chitosan and Alginate - New Forms of Commercial Interest.* Amer Chem Soc Div Polym Chem, **1990**, 31,628.
17. Singla, A.K. and M. Chawla, *Chitosan: some pharmaceutical and biological aspects - an update.* J Pharm Pharmacol, **2001**, 53,1047-1067.
18. Illum, L., *Chitosan and its use as a pharmaceutical excipient.* Pharmaceut Res, **1998**, 15,1326-1331.
19. Brown, C.D., et al., *Release of PEGylated granulocyte-macrophage colony-stimulating factor from chitosan/glycerol films.* J Control Release, **2001**, 72,35-46.
20. Toffey, A., et al., *Chitin derivatives .1. Kinetics of the heat-induced conversion of chitosan to chitin.* J Appl Polym Sci, **1996**, 60,75-85.
21. Grant, J., Tomba, J.P., Lee, H., Allen, C.J., *Relationship Between Composition and Properties for Stable Chitosan Films Containing Lipid Microdomains.* J Appl Polym Sci, **2005**, Submitted.
22. Jaworska, M., et al., *Influence of chitosan characteristics on polymer properties. I: Crystallographic properties.* Polym Int, **2003**, 52,198-205.
23. Screenivasan, K., *Thermal stability studies of some chitosan-metal ion complexes using differential scanning calorimetry.* Polym Degrad Stabil, **1996**, 52,85-87.
24. Tirkistani, F.A.A., *Thermal analysis of chitosan modified by cyclic oxygenated compounds.* Polym Degrad Stabil, **1998**, 61,161-164.
25. Dufresne, A., et al., *Morphology, phase continuity and mechanical behaviour of polyamide 6 chitosan blends.* Polymer, **1999**, 40,1657-1666.
26. Errington, N., et al., *Hydrodynamic Characterization of Chitosans Varying in Degree of Acetylation.* Int J Biol Macromol, **1993**, 15,113-117.
27. Tomihata, K. and Y. Ikada, *In vitro and in vivo degradation of films of chitin and its deacetylated derivatives.* Biomaterials, **1997**, 18,567-575.
28. Chatelet, C., O. Damour, and A. Domard, *Influence of the degree of acetylation on some biological properties of chitosan films.* Biomaterials, **2001**, 22,261-268.
29. Varum, K.M., et al., *In vitro degradation rates of partially N-acetylated chitosans in human serum.* Carbohyd Res, **1997**, 299,99-101.
30. Blair, H.S., et al., *Chitosan and Modified Chitosan Membranes .1. Preparation and Characterization.* J Appl Polym Sci, **1987**, 33,641-656.
31. Katti, D.S., et al., *Toxicity, biodegradation and elimination of polyanhydrides.* Adv Drug Deliver Rev, **2002**, 54,933-961.
32. Hirano, S., H. Tsuchida, and N. Nagao, *N-Acetylation in Chitosan and the Rate of Its Enzymic-Hydrolysis.* Biomaterials, **1989**, 10,574-576.
33. Liu, Y.H., et al., *Graft copolymerizaztion of methyl acrylate onto chitosan initiated by potassium diperiodatocuprate (III).* J Appl Polym Sci, **2003**, 89,2283-2289.
34. Rao, S.B. and C.P. Sharma, *Sterilization of chitosan: implications.* J Biomater Appl, **1995**, 10,136-43.

35. Lim, L.Y., E. Khor, and C.E. Ling, *Effects of dry heat and saturated steam on the physical properties of chitosan.* J Biomed Mater Res, **1999**, 48,111-116.

36. Khor, E. and L.Y. Lim, *Implantable applications of chitin and chitosan.* Biomaterials, **2003**, 24,2339-2349.

37. Muzzarelli, R.A.A., *Chitin and the human body. In: First international conference of the European Chitin Society.* Advances in Chitin Science, **1995**, 448-461.

38. Suh, J.K.F. and H.W.T. Matthew, *Application of chitosan-based polysaccharide biomaterials in cartilage tissue engineering: a review.* Biomaterials, **2000**, 21,2589-2598.

39. Khan, T.A., K.K. Peh, and H.S. Ch'ng, *Mechanical, bioadhesive strength and biological evaluations of Chitosan films for wound dressing.* J Pharm Pharm Sci, **2000**, 3,303-311.

40. Felt, O., P. Buri, and R. Gurny, *Chitosan: A unique polysaccharide for drug delivery.* Drug Dev Ind Pharm, **1998**, 24,979-993.

41. Genta, I., et al., *Microparticulate drug delivery systems.* Exs, **1999**, 87,305-13.

42. *Chitosan per os: from dietary supplement to drug carrier*; Muzzarelli, R.A.; Atec, Grottammare: Ancona, Italy, 2000.

43. Arvanitoyannis, I.S., A. Nakayama, and S. Aiba, *Chitosan and gelatin based edible films: state diagrams, mechanical and permeation properties.* Carbohyd Polym, **1998**, 37,371-382.

44. Nunthanid, J., et al., *Physical properties and molecular behavior of chitosan films.* Drug Development and Industrial Pharmacy, **2001**, 27,143-157.

45. Kumar, M.N.V.R., et al., *Chitosan chemistry and pharmaceutical perspectives.* Chem Rev, **2004**, 104,6017-6084.

46. Kumbar, S.G., K.S. Soppimath, and T.M. Aminabhavi, *Synthesis and characterization of polyacrylamide-grafted chitosan hydrogel microspheres for the controlled release of indomethacin.* J Appl Polym Sci, **2003**, 87,1525-1536.

47. Oztop, H.N., D. Saraydin, and S. Cetinus, *pH-sensitive chitosan films for baker's yeast immobilization.* Appl Biochem Biotechnol, **2002**, 101,239-49.

48. Chen, L., Z. Tian, and Y. Du, *Synthesis and pH sensitivity of carboxymethyl chitosan-based polyampholyte hydrogels for protein carrier matrices.* Biomaterials, **2004**, 25,3725-32.

49. Nishi, C., N. Nakajima, and Y. Ikada, *In-Vitro Evaluation of Cytotoxicity of Diepoxy Compounds Used for Biomaterial Modification.* J Biomed Mater Res, **1995**, 29,829-834.

50. Jin, J., M. Song, and D.J. Hourston, *Novel chitosan-based films cross-linked by genipin with improved physical properties.* Biomacromolecules, **2004**, 5,162-8.

51. Mi, F.L., et al., *In vivo biocompatibility and degradability of a novel injectable-chitosan-based implant.* Biomaterials, **2002**, 23,181-91.
52. Mwale, F., et al., *Biological evaluation of chitosan salts cross-linked to genipin as a cell scaffold for disk tissue engineering.* Tissue Eng, **2005**, 11,130-140.
53. Sung, H.W., et al., *In vitro surface characterization of a biological patch fixed with a naturally occurring crosslinking agent.* Biomaterials, **2000**, 21,1353-1362.
54. Mi, F.L., H.W. Sung, and S.S. Shyu, *Synthesis and characterization of a novel chitosan-based network prepared using naturally occurring crosslinker.* J Polym Sci Pol Chem, **2000**, 38,2804-2814.
55. Molinaro, G., et al., *Biocompatibility of thermosensitive chitosan-based hydrogels: an in vivo experimental approach to injectable biomaterials.* Biomaterials, **2002**, 23,2717-2722.
56. Zhang, H., et al., *Monodisperse chitosan nanoparticles for mucosal drug delivery.* Biomacromolecules, **2004**, 5,2461-2468.
57. Shu, X. and K.J. Zhu, *A novel approach to prepare tripolyphosphate/chitosan complex beads for controlled release drug delivery.* Int J Pharm, **2000**, 201,51-58.
58. Chenite, A., et al., *Novel injectable neutral solutions of chitosan form biodegradable gels in situ.* Biomaterials, **2000**, 21,2155-2161.
59. Grant, J., Blicker, M., Piquette-Miller, M., Allen, C.J., *Hybrid Films from Blends of Chitosan and Egg Phosphatidylcholine for Localized Delivery of Paclitaxel.* J Pharm Sci, **2005**, In Press.
60. Qu, X., A. Wirsen, and A.C. Albertsson, *Synthesis and characterization of pH-sensitive hydrogels based on chitosan and D,L-lactic acid.* J Appl Polym Sci, **1999**, 74,3193-3202.
61. Kweon, H., et al., *Physical properties of silk fibroin/chitosan blend films.* J Appl Polym Sci, **2001**, 80,928-934.
62. Cheng, M.Y., et al., *Study on physical properties and nerve cell affinity of composite films from chitosan and gelatin solutions.* Biomaterials, **2003**, 24,2871-2880.
63. Khoo, C.G., et al., *Oral gingival delivery systems from chitosan blends with hydrophilic polymers.* Eur J Pharm Biopharm, **2003**, 55,47-56.
64. Jones, D.S., *Dynamic mechanical analysis of polymeric systems of pharmaceutical and biomedical significance.* Int J Pharm, **1999**, 179,167-178.
65. Jones, D.S., A.D. Woolfson, and A.F. Brown, *Textural analysis and flow rheometry of novel, bioadhesive antimicrobial oral gels.* Pharmaceut Res, **1997**, 14,450-457.
66. Jones, D.S., Woolfson, A.D., Brown, A.F., *Textural, viscoelastic and mucoadhesive properties of gels composed of celluloser polymers.* Int. J. Pharm., **1997**, 151,223-233.

67. Needleman, I.G., G.P. Martin, and F.C. Smales, *Characterisation of bioadhesives for periodontal and oral mucosal drug delivery.* J Clin Periodontol, **1998**, 25,74-82.

68. Needleman, I.G., F.C. Smales, and G.P. Martin, *An investigation of bioadhesion for periodontal and oral mucosal drug delivery.* J Clin Periodontol, **1997**, 24,394-400.

69. Langer, R., *Implantable controlled release systems.* Pharmacol Ther, **1983**, 21,35-51.

70. Sampath, P. and H. Brem, *Implantable Slow-Release Chemotherapeutic Polymers for the Treatment of Malignant Brain Tumors.* Cancer Control, **1998**, 5,130-137.

71. Jackson, J.K., et al., *The suppression of human prostate tumor growth in mice by the intratumoral injection of a slow-release polymeric paste formulation of paclitaxel.* Cancer Res, **2000**, 60,4146-51.

72. Hui, A., et al., *Paclitaxel selectively induces mitotic arrest and apoptosis in proliferating bovine synoviocytes.* Arthritis Rheum, **1997**, 40,1073-84.

73. Miller, M.L. and I. Ojima, *Chemistry and chemical biology of taxane anticancer agents.* Chem Rec, **2001**, 1,195-211.

74. Friedland, D., G. Gorman, and J. Treat, *Hypersensitivity reactions from taxol and etoposide.* J Natl Cancer Inst, **1993**, 85,2036.

75. Spencer, C.M. and D. Faulds, *Paclitaxel. A review of its pharmacodynamic and pharmacokinetic properties and therapeutic potential in the treatment of cancer.* Drugs, **1994**, 48,794-847.

76. Dhanikula, A.B. and R. Panchagnula, *Development and characterization of biodegradable chitosan films for local delivery of Paclitaxel.* AAPS J, **2004**, 6,e27.

77. Peppas, N.A., *Analysis of Fickian and non-Fickian drug release from polymers.* Pharm Acta Helv, **1985**, 60,110-1.

78. Peppas, L.B., *Polymer in controlled drug delivery.* Med. Plast. Biomater., **1997**, 4,34-44.

79. Ho, E., Vassileva,V., Allen,C., Piquette-Miller,M., *In vitro and in vivo characterization of a novel biocompatible polymer–lipid implant system for the sustained delivery of paclitaxel.* J. Control. Rel., **2005**, In Press.

80. Ruel-Gariepy, E., et al., *A thermosensitive chitosan-based hydrogel for the local delivery of paclitaxel.* Eur J Pharm Biopharm, **2004**, 57,53-63.

81. Nsereko, S. and M. Amiji, *Localized delivery of paclitaxel in solid tumors from biodegradable chitin microparticle formulations.* Biomaterials, **2002**, 23,2723-31.

82. Buranapanitkit, B., et al., *The efficacy of a hydroxyapatite composite as a biodegradable antibiotic delivery system.* Clin Orthop Relat Res, **2004**, 244-52.

83. Aimin, C., et al., *Antibiotic loaded chitosan bar. An in vitro, in vivo study of a possible treatment for osteomyelitis.* Clin Orthop Relat Res, **1999**, 239-47.

84. Perugini, P., et al., *Periodontal delivery of ipriflavone: new chitosan/PLGA film delivery system for a lipophilic drug.* Int J Pharm, **2003**, 252,1-9.

85. Suyatma, N.E., et al., *Mechanical and barrier properties of biodegradable films made from chitosan and poly (lactic acid) blends.* J Polym Environ, **2004**, 12,1-6.

86. Honma, T., T. Senda, and Y. Inoue, *Thermal properties and crystallization behaviour of blends of poly(epsilon-caprolactone) with chitin and chitosan.* Polym Int, **2003**, 52,1839-1846.

87. Wu, Y.B., et al., *Preparation and characterization on mechanical and antibacterial properties of chitsoan/cellulose blends.* Carbohyd Polym, **2004**, 57,435-440.

88. Sionkowska, A., et al., *Molecular interactions in collagen and chitosan blends.* Biomaterials, **2004**, 25,795-801.

89. Zhao, W.W., et al., *The Compatibility and Morphology of Chitosan-Poly(Ethylene-Oxide) Blends.* J Macromol Sci Phys, **1995**, B34,231-237.

90. Zhang, M., et al., *Properties and biocompatibility of chitosan films modified by blending with PEG.* Biomaterials, **2002**, 23,2641-8.

91. Chuang, W.Y., et al., *Properties of the poly(vinyl alcohol)/chitosan blend and its effect on the culture of fibroblast in vitro.* Biomaterials, **1999**, 20,1479-1487.

92. Chen, X., et al., *pH sensitivity and ion sensitivity of hydrogels based on complex-forming chitosan/silk fibroin interpenetrating polymer network.* J Appl Polym Sci, **1997**, 65,2257-2262.

93. Cheung, M.K., K.P.Y. Wan, and P.H. Yu, *Miscibility and morphology of chiral semicrystalline poly-(R)-(3-hydroxybutyrate)/chitosan and poly-(R)-(3-hydroxybutyrate-co-3-hydroxyvalerate)/chitosan blends studied with DSC, H-1 T-1 and T-1 rho CRAMPS.* J Appl Polym Sci, **2002**, 86,1253-1258.

94. Sato, M., et al., *Miscibility study of polymer blends composed of semirigid thermotropic liquid crystalline polycarbonate, poly(vinyl alcohol), and chitosan.* J Appl Polym Sci, **2004**, 93,1616-1622.

95. Macleod, G.S., J.H. Collett, and J.T. Fell, *The potential use of mixed films of pectin, chitosan and HPMC for bimodal drug release.* J Control Release, **1999**, 58,303-310.

96. Liu, C.H. and C.B. Xiao, *Characterization of films from chitosan and quaternized poly(4-vinyl-N-butyl) pyridine solutions.* J Appl Polym Sci, **2004**, 92,559-566.

97. Ratto, J.A., C.C. Chen, and R.B. Blumstein, *Phase behavior study of chitosan polyamide blends.* J Appl Polym Sci, **1996**, 59,1451-1461.

98. Fang, L. and S.H. Goh, *Miscible chitosan/tertiary amide polymer blends.* J Appl Polym Sci, **2000**, 76,1785-1790.

99. Cao, S.G., Y.Q. Shi, and G.W. Chen, *Blend of chitosan acetate salt with .poly(N-vinyl-2-pyrrolidone): Interaction between chain-chain.* Polym Bull, **1998**, 41,553-559.
100. Risbud, M., A. Hardikar, and R. Bhonde, *Growth modulation of fibroblasts by chitosan-polyvinyl pyrrolidone hydrogel: Implications for wound management?* J Bioscience, **2000**, 25,25-31.

Chapter 11

Characterization of Glycol Chitosan: A Potential Material for Use in Biomedical and Pharmaceutical Applications

A Comparison of Fractionation Techniques

Darryl K. Knight, Stephen N. Shapka, and Brian G. Amsden[*]

Department of Chemical Engineering, Queen's University, Kingston, Ontario K7L 3N6, Canada
*Corresponding author: amsden@chee.queensu.ca

Glycol chitosan, a water soluble chitosan derivative being proposed as a material for pharmaceutical and biomedical engineering applications, was modified to further promote its use *in vivo*. Initial characterization of the glycol chitosan with ^{1}H NMR spectroscopy illustrated the presence of both secondary and tertiary amine groups. Fractionation of glycol chitosan with nitrous acid resulted in a significant reduction in the number average molecular weight, specifically, from 210 to approximately 7-8 kDa. However, the structural integrity of the glycol chitosan was lost following fractionation, as the secondary amine groups were converted to N-nitrosamines, which are potentially carcinogenic. An increase in the pH of the reaction limited their formation, but not entirely; therefore, a second approach to reducing the molecular weight was sought. The free radical degradation, initiated with potassium persulfate, was not as effective at reducing the molecular weight, which ranged from 17 to 20 kDa post fractionation, but did retain the structural integrity of the glycol chitosan.

Introduction

Chitosan is a cationic polysaccharide derived from the abundant biopolymer chitin, which is obtained from the exoskeleton of crustaceans. (*1*) Chitosan, shown in Figure 1, (*2,3*) is the term given to the family of polysaccharides consisting of N-acetyl D-glucosamine and D-glucosamine residues coupled through a β(1→4) (glycosidic) (*4*) linkage where the degree of N-acetylation is less than 50%. (*5-7*)

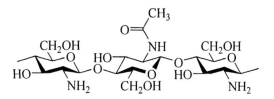

Figure 1. Structure of chitosan.

Chitosan is being examined extensively in the biomedical field for several applications including wound dressings, (*8*) scaffolds for tissue engineering including the development of artificial skin, and as drug delivery vehicles, which have been reviewed elsewhere. (*9*) Chitosan is garnering much attention because of its cytocompatible nature as seen from *in vitro* studies with human endothelial cells, (*10,11*) fibroblasts (*11-13*) and keratinocytes. (*12-14*) Chitosan can bind directly with cell membranes, particularly those of red blood cells, inducing the formation of a coagulum of erythrocytes (*15*) while accelerating the wound healing process. (*8,14,16,17*) Although chitosan has demonstrated biocompatibility in various applications thus far, it suffers from very limited water solubility. The achievement of a chitosan-based derivative soluble at physiologic conditions would promote its use in the pharmaceutical and biomedical engineering fields.

Glycol chitosan, a water soluble derivative whose proposed structure is shown in Figure 2, (*18-20*) owes its water solubility to the incorporation of the hydrophilic glycol group. More importantly, the free amine groups along the backbone would allow for future modification or interaction with the host cells. Although very little characterization has been conducted to elucidate glycol chitosan's actual structure, it is being proposed as a suitable material for various pharmaceutical and biomedical applications.

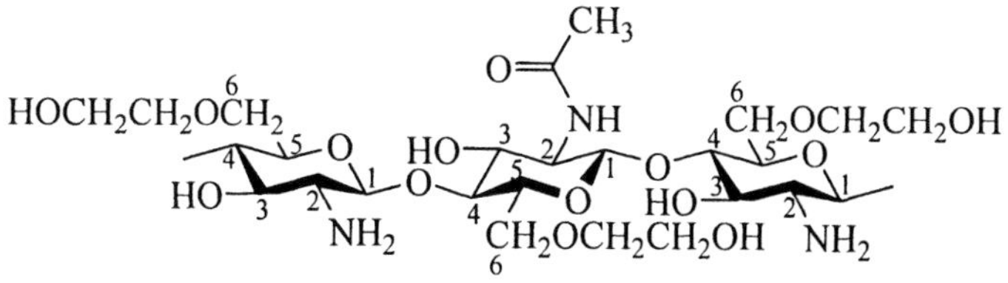

Figure 2. Currently accepted structure of glycol chitosan.

The focus of this study was to fully characterize glycol chitosan in terms of its molecular weight and degree of acetylation, as well as its degree of glycolation prior to and following its reduction in molecular weight. The development of a procedure to consistently realize a low molecular weight glycol chitosan would be desirable to facilitate its potential application *in vivo* through an improvement in its bioeliminability and a reduction in its viscosity. Specifically, two approaches for the fractionation of glycol chitosan were studied. Cleavage of the glycosidic bond along the glycol chitosan backbone was achieved through diazotization of the primary amine via reaction with nitrous acid. (2) The second approach involved free radical degradation of the polysaccharide with the thermal dissociation initiator, potassium persulfate. (21)

Materials and Methods

Glycol chitosan, potassium nitrite, potassium persulfate, sodium acetate, sodium borohydride and sodium hydroxide were all obtained from Sigma-Aldrich Canada Ltd. Deuterium oxide was purchased from Cambridge Isotope Laboratories, Inc. Acetic and hydrochloric acids were obtained from Fisher Scientific Limited and all reagents were used as received. Type I was obtained from a Millipore Milli-Q Plus Ultra-Pure Water System.

Purification of High Molecular Weight Glycol Chitosan

Glycol chitosan (1 g) was dissolved in water (75 mL) and filtered under vacuum to remove insoluble impurities. The filtrate was then dialyzed with molecular weight cutoff 50 kDa dialysis tubing against water for 8 hours. Both the membranes and media were replaced at 4 hours. The purified high molecular weight glycol chitosan was frozen at –20°C for 8 hours and lyophilized for 48 hours under a reduced pressure of approximately 300 µbar.

Fractionation of Glycol Chitosan via Nitrous Acid

The fractionation of glycol chitosan with nitrous acid was adapted from a procedure proposed by Allan and Peyron, (2) which is illustrated in Scheme 1. (22) Purified glycol chitosan (~ 0.8 g, 3.8 x 10^{-3} mmol) was dissolved in water (75 mL) and under magnetic stirring, 1 M hydrochloric acid was added followed immediately by 1 M potassium nitrite (1 mL, 1 mmol) to give solutions of pH 1.6, 2.9 and 5.1. Upon 4 hours of fractionation, sodium borohydride (100 mg, 2.64 mmol) was added and allowed to react for an additional 30 minutes. The pH of the low molecular weight glycol chitosan solution was neutralized and concentrated on a rotary evaporator at 35°C and then dialyzed with molecular weight cutoff 1 kDa dialysis tubing against water for 4 hours. The membranes and media were changed at 2 hours. The purified low molecular weight glycol chitosan solution was again neutralized prior to freezing at –20°C for 8 hours. The sample was lyophilized for 48 hours under a reduced pressure of approximately 300 μbar.

Fractionation of Glycol Chitosan via Potassium Persulfate

The fractionation of glycol chitosan with potassium persulfate was adapted from a procedure proposed by Hsu et al. (21) and is illustrated in Scheme 2. Purified glycol chitosan (~ 0.8 g, 3.8 x 10^{-3} mmol) was dissolved in a 2% (v/v) hydrochloric acid solution (75 mL), followed immediately by the addition of potassium persulfate (135 mg, 0.5 mmol) at 70°C. Upon 2, 4 or 8 hours of fractionation, sodium borohydride (100 mg, 2.64 mmol) was added and allowed to react for an additional 30 minutes at room temperature. The pH of the low molecular weight glycol chitosan solution was neutralized with 1 M sodium hydroxide and concentrated on a rotary evaporator at 35°C and then dialyzed with molecular weight cutoff 1 kDa dialysis tubing against water for 4 hours. The membranes and media were again changed at 2 hours. The purified low molecular weight glycol chitosan solution was quenched under a low flowrate of air for 24 hours. The solution was again dialyzed with molecular weight cutoff 1 kDa dialysis tubing against water for 2 hours followed by its subsequent neutralization. The resulting purified fractionated glycol chitosan solution was frozen at –20°C for 8 hours and lyophilized for 48 hours under a reduced pressure of approximately 300 μbar.

Nuclear Magnetic Resonance (NMR) Spectroscopy

NMR spectra were conducted with a Bruker Avance-600 Ultrashield spectrometer equipped with a 5 mm TBI S3 probe with Z gradient and variable temperature capability. Samples were prepared at 20 mg·mL^{-1} in deuterium oxide, preheated at 60°C for 6 hours, then adjusted to pH > 10 with 1 M sodium hydroxide (30 μL) prior to running. Samples were allowed to stand for 10

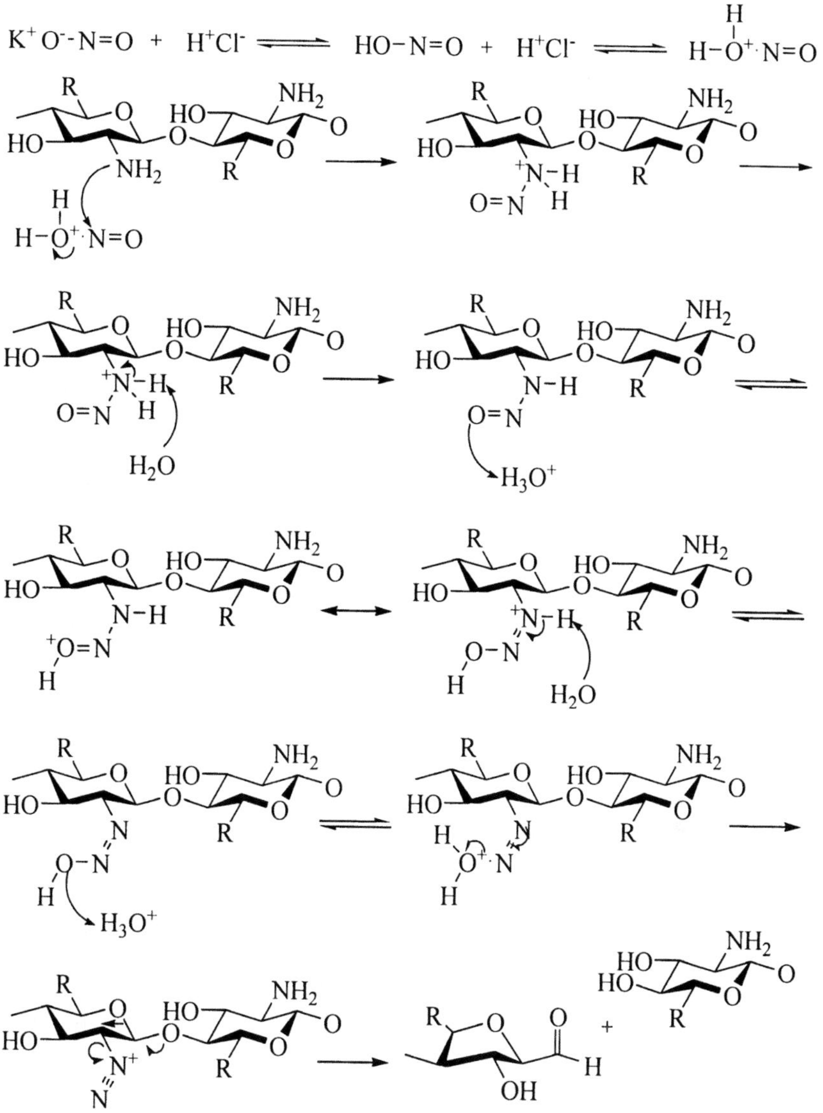

Scheme 1. Proposed fractionation of glycol chitosan via diazotization of the primary amine with the nitrous acidium ion.

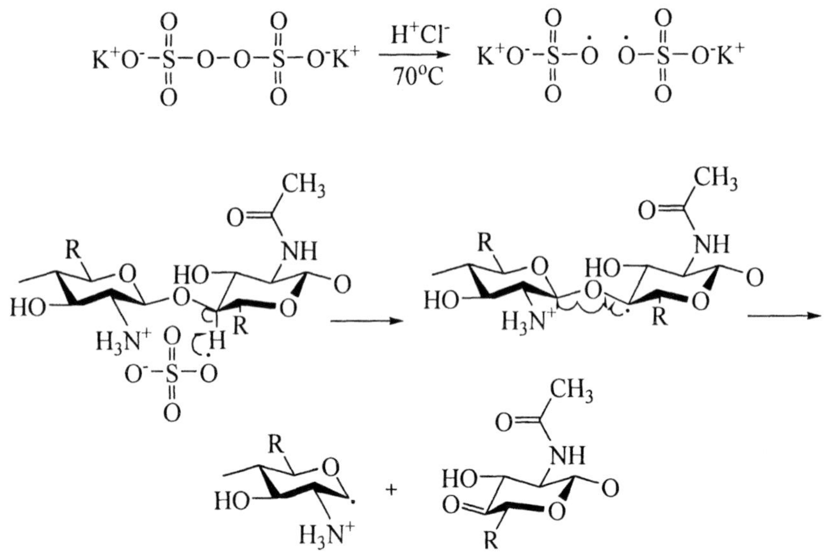

Scheme 2. Proposed reaction for the fractionation of glycol chitosan with potassium persulfate at 70°C. (21)

minutes at 90°C within the spectrometer prior to shimming to ensure a homogenous sample temperature. All chemical shifts were referenced to the HOD peak.

Gel Permeation Chromatography (GPC) with Light Scattering

GPC with light scattering data were obtained with a Waters 1525 Binary HPLC pump and a Precision Detectors Enterprise[MDP] PD2100 Series equipped with refractive index and light scattering detectors with angles of 15 and 90°. GPC was achieved using Waters Ultrahydrogel 2000, 250 and 120 columns connected in series. Samples, dissolved in a 0.3 M sodium acetate / 0.2 M acetic acid (pH 4.8) eluant, were filtered (0.45 µm) and injected (100 µL) with a Waters 717plus Autosampler onto the column at either 0.6 mL·min⁻¹ (Ultrahydrogel 2000) or 0.8 mL·min⁻¹ (Ultrahydrogel 250 and 120) and 30°C at a concentration of around 30 mg·mL⁻¹. All data were obtained and processed in Precision Detectors' Precision Aquire32 and Discovery32 software programs using an absolute refractive index of 1.3255 mL/g determined on a Wyatt Optilab rEX and a refractive index increment (dn/dc) value of 0.08 mL/g. (*19*)

Results and Discussion

The initial glycol chitosan was obtained from Sigma-Aldrich; however, very little information about its properties was provided. Unfortunately, a manner to determine the degree of acetylation of glycol chitosan has not yet been published; however, many research groups have used ^{1}H NMR spectroscopy to calculate the degree of acetylation of chitosan, (5,23-27) which may be extrapolated to glycol chitosan. Assignment of the major peaks in the ^{1}H NMR spectrum of the purified glycol chitosan, illustrated in Figure 3, can be attributed to the protons as numbered in Figure 2.

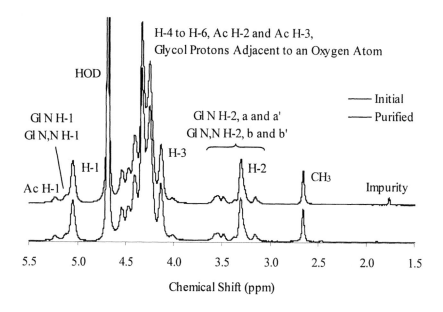

Figure 3. Offset ^{1}H NMR spectra of the initial (above) and purified (bottom) glycol chitosan.

The peak at 2.66 ppm arises due to the methyl protons (CH_3) of the acetyl group. The peak at 3.30 ppm arises from the proton at position 2 (H-2) in the primary amine residue, as it is on a carbon adjacent to a nitrogen atom; whereas, the protons at positions 3, 4, 5 and 6 (H-3 through H-6) are all on carbons adjacent to an oxygen atom and therefore give rise to the overlapped peaks between existing 4.13 and 4.54 ppm. The proton at position 1 (H-1) is on a carbon adjacent to two oxygen atoms and occurs furthest downfield at 5.05 ppm.

The chemical shifts of the peaks arising from the protons at positions 1, 2 and 3 all shift downfield in the acetylated residue (denoted as Ac H-1 to Ac H-3), where the proton at position 2 shifts the furthest downfield because of its relative proximity to the acetyl group. The peaks due to Ac H-2 and Ac H-3 are, in all likelihood, overlapped by the other peaks between 4.13 and 4.54 ppm. In the case of Ac H-1, the peak can be seen at 5.23 ppm.

Upon examination of the ^{1}H NMR spectra, it becomes apparent that the amine groups must also be glycolated to account for the small peaks seen between 3.1 and 3.6 ppm. The peaks at 3.11, 3.15, 3.37, 3.48, 3.54 and 3.57 ppm likely arise due to protons on a carbon adjacent to a nitrogen atom, as was the case with H-2. Given that at least six peaks can be seen, it is expected that both secondary and tertiary amines exist, indicating mono- and disubstitution. Knowing that some amine groups are glycolated, it would therefore be expected that these groups would also have an impact on the protons at positions 1, 2 and 3 giving rise to additional peaks, including the peak at 5.13 ppm, which can be attributed to the proton at position 1 of an N-glycolated residue. Based solely on the ^{1}H NMR spectrum of the purified glycol chitosan, it is not known which peaks are contributed by the secondary or tertiary amines.

With this knowledge, a new structure of glycol chitosan, shown in Figure 4, which illustrates these residues, is being proposed in this study. The protons of each methylene (CH_2) group, as part of the glycol groups, are diastereotopic due to the stereogenic centres along the chitosan backbone. As such, protons denoted as a and a' in the secondary amine case are magnetically different and should give rise to separate peaks in the ^{1}H NMR spectrum, thus producing three small peaks within the region of 3.1 to 3.6 ppm, of similar integration to each other – the proton at position 2, a and a'. In the case of the tertiary amine, where two glycol groups are present, the protons of each methylene group are still magnetically different; however, the protons between each glycol group should be similar; therefore, the peaks attributed to b and b' should be twice that of the proton at position 2.

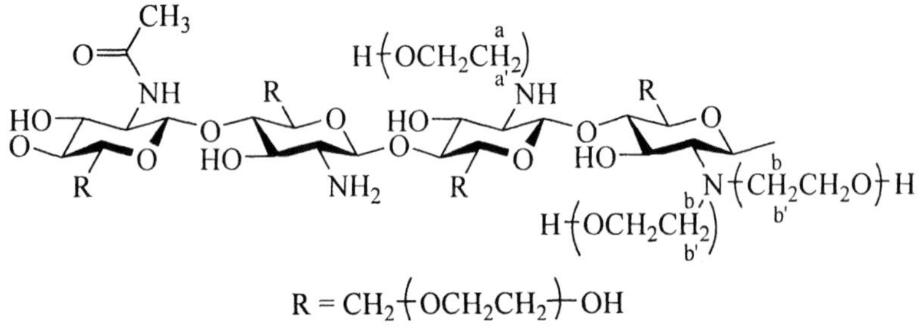

Figure 4. Proposed glycol chitosan structure illustrating the different residues.

With the knowledge of the possible residues along the glycol chitosan backbone, the degree of acetylation can be determined. Because the peaks between 5.05 and 5.23 ppm arise from the proton at position 1 for each substituent at position 2, the sum of all of these integrations will give an indication of the average number of residues per chain. As such, a ratio to the methyl protons of the acetyl group will give the degree of acetylation as illustrated in Equation 1:

$$\text{Degree of Acetylation} = \frac{I_{CH_3}/3}{I_{H-1\,\text{Total}}} \qquad (1)$$

where I represents the integration of the peak as indicated by the subscript. Using this equation, the degree of acetylation was calculated to be 12% in the purified glycol chitosan. The remaining 88% would be the total contribution of the primary, secondary and tertiary amine groups; however, their individual contributions are not yet known, as their corresponding peaks can not be fully resolved in the purified glycol chitosan ^{1}H NMR spectrum.

In addition to calculating the degree of acetylation, the number average, M_n, and weight average, M_w, molecular weight of the initial and purified glycol chitosan were also determined and are shown in Table I along with their corresponding polydispersities (PI). The increase in the number average molecular weight from 171 to 210 kDa, resulting from the removal of the low molecular weight components, would normally result in a smaller polydispersity index; however, at an initial value of 1.1, very little improvement will be observed.

Table I. Molecular weights of the initial and purified glycol chitosan

Sample	M_n, kDa	M_w, kDa	PI
Initial Glycol Chitosan	171	195	1.1
Purified Glycol Chitosan	210	232	1.1

NOTE: Average of three runs

Fractionation of Glycol Chitosan with Nitrous Acid

Upon characterization of the purified glycol chitosan, a method to achieve a fractionated glycol chitosan of reproducible molecular weight was desired. The first approach used to reduce the molecular weight was diazotization of the primary amine through its reaction with nitrous acid proposed by Allan and Peyron (2) and shown in Scheme 1. (22) Nitrous acid forms a number of nitrosating agents in aqueous solution, which are dependent on the pH as well as

236

the anion present. (*28*) Although in this study, the anion was not changed, the pH of the solution was altered to determine whether a change in pH would impact the fractionation of glycol chitosan. Offset ¹H NMR spectra of the purified glycol chitosan along with three fractionated samples conducted at pH 1.6, 2.9 and 5.1 are shown in Figure 5.

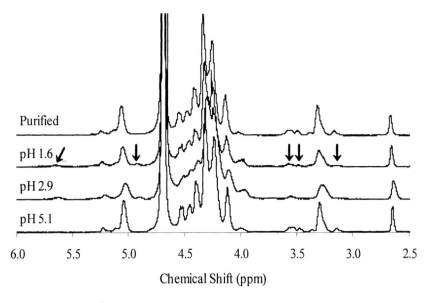

Figure 5. Offset ¹H NMR spectra of purified and fractionated glycol chitosan with nitrous acid at varying pHs.

Upon examination of the ¹H NMR spectra, it is apparent that the reaction pH does impact the structural integrity of the glycol chitosan. Firstly, band broadening can be observed on the methyl protons of the acetyl group, as well as at the protons at carbon positions 1 and 2 in the trials at pH 1.6 and 2.9. Also of note in these two trials is the reduction in the intensity of the peaks attributed to the secondary and tertiary amine groups between 3.1 and 3.6 ppm, indicating that the electronic environment of these protons has been modified. Finally, peaks at 4.90 and 5.63 ppm also appear following fractionation with nitrous acid at pH of 1.6 and 2.9 and to a much lesser extent in the pH 5.1 trial. Unlike the two low pH trials, the structural integrity appears intact in the pH 5.1 trial, as it is similar to the ¹H NMR spectrum of the purified glycol chitosan.

Allan and Peyron (*2*) proposed the nitrous acidium ion as the nitrosating agent from kinetic data conducted with chitosan in aqueous hydrochloric acid (50.0 – 125 mM) solutions, where a pH below 3 would be achieved. Some authors (*28,29*) have noted that nitrosation of amide groups will occur below pH 3, and that a ten fold increase is observed for each reduction in pH from 3 to 1. Nitrosation of an amide results in the formation of an N-nitrosamide, which is unstable at physiologic pH and decomposes to a diazohydroxyl and finally a diazonium cation, as seen in Scheme 3, (*28*) which is the same intermediate as in the primary amine case resulting in the cleavage of the glycosidic bond. (*22*)

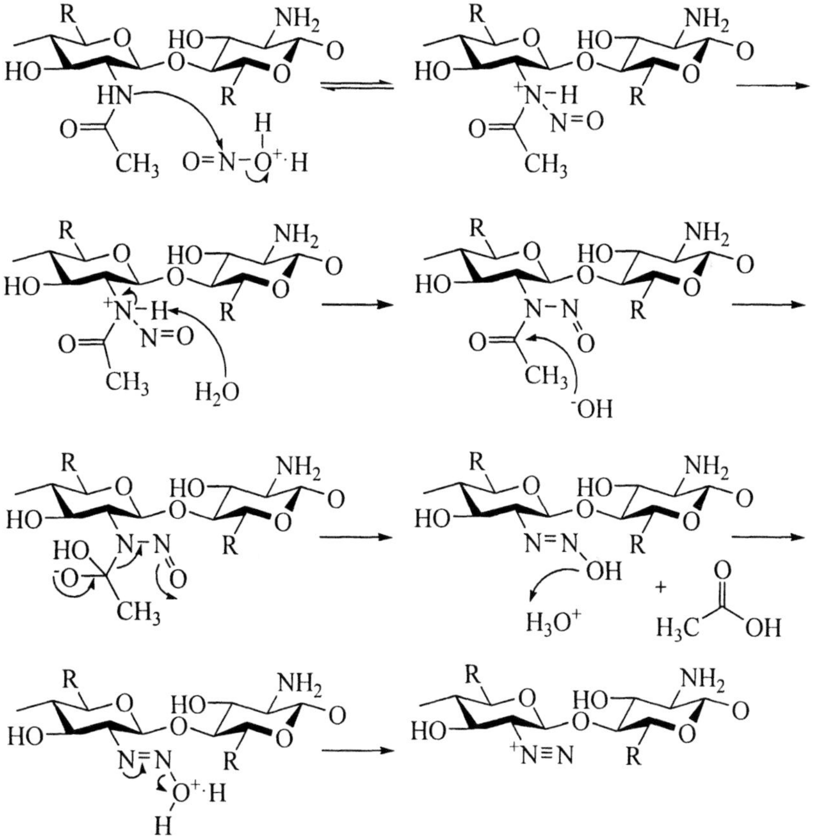

Scheme 3. Nitrosation of the amide as part of the acetyl group and its subsequent decomposition to the diazonium cation. (40)

The large change in electronic effects seen in the ^{1}H NMR spectra of the pH 1.6 and 2.9 trials would not be expected to arise from the formation of an N-nitrosamide, as the fractionated glycol chitosan is neutralized to terminate the reaction, which would promote the formation of the diazonium cation; however, a reduction in the degree of acetylation may therefore be observed.

Although N-nitrosamides are unstable at physiologic conditions, N-nitrosamines formed from the nitrosation of secondary amines are quite stable (*28*) and are formed under milder conditions. (*29*) The formation of N-nitrosamines, shown in Scheme 4, (*22*) is not desirable, as most N-nitroso compounds studied have demonstrated some carcinogenicity. (*28*)

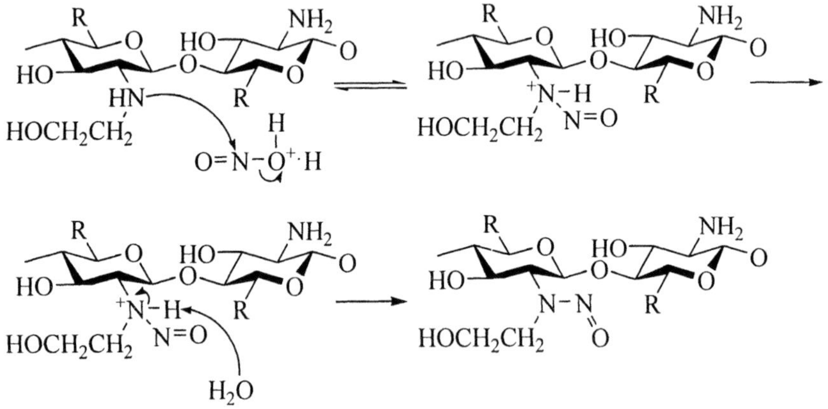

Scheme 4. Nitrosation of the secondary amine resulting in an N-nitrosamine.

Because N-nitrosamines form readily from the reaction of secondary amines with nitrosating agents, it might be expected that the reduction in the intensity of the peaks between 3.1 and 3.6 ppm might be due to their conversion to the N-nitrosamine, which would give rise to peaks further downfield. The formation of the N-nitroso derivatives likely accounts for the peaks around 4.90 and 5.63 ppm. Specifically, the peak seen at 4.90 ppm may be the proton at carbon position 3 (denoted as N-nsa H-3) in the N-nitrosamine case, while the peak observed at 5.63 ppm would be the corresponding proton at carbon position 1 (N-nsa H-1) in the same residue. Although nitrosation of tertiary amines is possible, yielding N-nitrosamines resulting from the cleavage of a glycol group, (*29,30*) their occurrence would only be significant at elevated temperatures. (*29-32*) As such, it is assumed their formation does not occur under the current conditions.

Based on the above logic, the degrees of N-Nitrosamination can be determined by taking a similar relationship to that presented in Equation 1. Specifically, the degree of N-nitrosamination following the fractionation of glycol chitosan with nitrous acid can be approximated as:

$$\text{Degree of N-Nitrosamination} = \frac{I_{\text{N-nsa H-1}}}{I_{\text{H-1 Total}}} \quad (2)$$

where $I_{\text{H-1 Total}}$ now includes the contribution of the N-nitrosamine. The degree of N-nitrosamination was calculated to be 13, 14 and 5% in the pH 1.6, 2.9 and 5.1 trials respectively. This indicates that fractionating the glycol chitosan with nitrous acid at higher pH does inhibit the formation of the N-nitrosamines. The degree of acetylation was unaffected in all three trials indicating that N-nitrosamides were not formed. Confirmation that nitrous acid resulted in fractionation of the glycol chitosan backbone was seen in the GPC data presented in Table II. The lowest molecular weight was achieved with the pH 2.9 trial; however, significant fractionation was achieved in all three cases as the M_n was determined to be 7.8, 6.6 and 8.7 kDa in the pH trials of 1.6, 2.9 and 5.1 respectively.

Table II. Molecular weights of the fractionated glycol chitosan with nitrous acid at varying pHs

Sample	M_n, kDa	M_w, kDa	PI
Fractionated GC, pH 1.6	7.8	11.4	1.5
Fractionated GC, pH 2.9	6.8	9.3	1.4
Fractionated GC, pH 5.1	8.7	12.8	1.5

NOTE: Average of three runs; GC = Glycol chitosan

The lower molecular weight in the pH 2.9 trial compared to the pH 1.6 trial may be attributed to a greater degree of protonation in the pH 1.6 trial, restricting the reaction between the nitrous acidium ion and an unionized primary amine group and hence the formation of the diazonium cation. In the case of the pH 5.1 trial, where a higher proportion of the primary amine groups would be unionized, the nitrosating agent may not be the same. The study conducted by Allan and Peyron (2) demonstrated that the nitrosating species was the nitrous acidium ion for pHs below 3. At a pH of 5.1, the nitrosating agent may be nitrosyl chloride or even nitrous anhydride, which are not as effective nitrosating agents, (28) resulting in a higher molecular weight.

Fractionated Glycol Chitosan via Potassium Persulfate

Although fractionation of glycol chitosan has been achieved with nitrous acid, the formation of potentially carcinogenic compounds necessitated the evaluation of a secondary approach. Free radical degradation of the glycol

240

chitosan backbone was achieved with the thermal dissociation initiator, potassium persulfate, as proposed by Hsu et al. (*21*) and is shown in Scheme 2. Although these authors observed a saturation effect in the reduction of the molecular weight of chitosan after 1 hour, fractionation of glycol chitosan in this study was conducted up to 8 hours, as the half-life of potassium persulfate is 20.9 hours. (*21*) The effect of the potassium persulfate on the structural integrity of the glycol chitosan was assessed through ^{1}H NMR spectroscopy. Offset spectra of purified and fractionated glycol chitosan with potassium persulfate at 2, 4 and 8 hours are shown in Figure 6. Unlike the nitrous acid fractionated glycol chitosan, the ^{1}H NMR spectra overlap quite nicely, indicating that the integrity of the glycol chitosan backbone is retained following potassium persulfate fractionation.

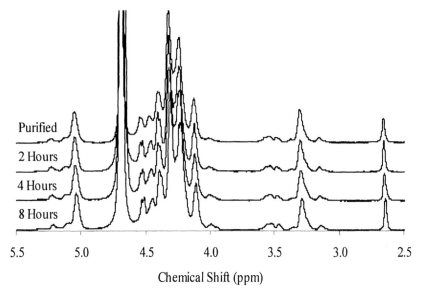

Figure 6. Offset ^{1}H NMR of purified and fractionated glycol chitosan with potassium persulfate at varying times.

The effectiveness of potassium persulfate as a fractionating agent was assessed through the GPC data shown in Table III. There is a further reduction in the molecular weight with longer reaction times; however, the molecular weights are not as low as the nitrous acid approach and may be due to the lack of specificity of the free radical degradation. This would also explain the higher polydispersity indices then what was observed in the nitrous acid trials.

Table III. Molecular weights of the fractionated glycol chitosan with potassium persulfate at varying times

Sample	M_n, kDa	M_w, kDa	PI
Fractionated GC, 2 Hours	20.6	35.6	1.7
Fractionated GC, 4 Hours	18.5	32.6	1.8
Fractionated GC, 8 Hours	17.2	32.3	1.9

NOTE: Average of three runs; GC = Glycol chitosan

Conclusions

Chitosan is already being examined in the biomedical field for several applications, but the achievement of a chitosan-based derivative soluble at physiologic conditions would further promote its use in the pharmaceutical and biomedical engineering fields. The water soluble derivative, glycol chitosan, was fractionated by two separate methods, reducing its viscosity, which would facilitate its use *in vivo*. The reduction in molecular weight would also enhance its rate of bioelimination. Initial characterization of the purified glycol chitosan demonstrated that some of the amine groups were glycolated indicating the presence of both secondary and tertiary amines. Fractionation of glycol chitosan with nitrous acid at pHs of 1.6, 2.9 and 5.1 all resulted in a reduction in molecular weight, specifically, from 210 to approximately 7-8 kDa. Unfortunately, the structural integrity of the glycol chitosan was lost as the secondary amine groups were converted to N-nitrosamines, which are potentially carcinogenic. Fractionation with the pH 5.1 trial limited their formation, but not completely; therefore, a second approach to reducing the molecular weight was examined. The free radical degradation with the thermal dissociation initiator, potassium persulfate, was found to be not as effective at reducing the molecular weight, which ranged from 17 to 20 kDa at reaction times between 2 and 8 hours, but did retain the structural integrity of the glycol chitosan.

References

1. Muzzarelli, R. A. A. *Chitin;* Pergamon of Canada Ltd.: Toronto, Canada, 1977; pp 5-44.
2. Allan, G. G.; Peyron, M. *Carbohyd. Res.* **1995,** *277,* 257.
3. Chellat, F.; Tabrizian, M.; Dumitriu, S.; Chornet, E.; Rivard, C.-H.; Yahia, L. *J. Biomed. Mater. Res. Part B: Appl. Biomater.* **2000,** *53,* 592.
4. Nordtveit, R. J.; Vårum, K. M.; Smidsrød, O. *Carbohyd. Polym.* **1994,** *23,* 253.
5. Rinaudo, M.; Milas, M.; Le Dung, P. *Int. J. Biol. Macromol.* **1993,** *15,* 281.
6. le Dung, P.; Milas, M.; Rinaudo, M.; Desbrières, J. *Carbohyd. Polym.* **1994,** *24,* 209.

7. Brugnerotto, J.; Desbrières, J.; Heux, L.; Mazeau, K.; Rinaudo, M. *Macromol. Symp.* **2001**, *168*, 1.
8. Azad, A. K.; Sermsintham, N.; Chandrkrachang, S.; Stevens, W. F. *J. Biomed. Mater. Res. Part B: Appl. Biomater.* **2004**, *69B*, 216.
9. Khor, E.; Lim, L. Y. *Biomaterials* **2003**, *24*, 2339.
10. Malette, W. G.; Quigley Jr., H. J.; Adickes, E. D. In *Chitin in Nature and Technology;* Muzzarelli, R.; Jeuniaux, C.; Gooday, G. W., Eds.; Plenum Press: New York, 1986; pp 435-442.
11. Ono, K.; Saito, Y.; Yura, H.; Ishikawa, K.; Kurita, A.; Akaike, T.; Ishihara, M. *J. Biomed. Mater. Res.* **2000**, *49*, 289.
12. Chatelet, C.; Damour, O.; Domard, A. *Biomaterials* **2001**, *22*, 261.
13. Howling, G. I.; Dettmar, P. W.; Goddard, P. A.; Hampson, F. C.; Dornish, M.; Wood, E. J. *Biomaterials* **2001**, *22*, 2959.
14. Denuzière, A.; Ferrier, D.; Damour, O.; Domard, A. *Biomaterials* **1998**, *19*, 1275.
15. Rao, S. B.; Sharma, C. P. *J. Biomed. Mater. Res.* **1997**, *34*, 21.
16. Obara, K.; Ishihara, M.; Ishizuka, T.; Fujita, M.; Ozeki, Y.; Maehara, T.; Saito, Y.; Yura, H.; Matsui, T.; Hattori, H.; Kikuchi, M.; Kurita, A. *Biomaterials* **2003**, *24*, 3437.
17. Cho, Y.-W.; Cho, Y.-N.; Chung, S.-H.; Yoo, G.; Ko, S.-W. *Biomaterials* **1999**, *20*, 2139.
18. Carreño-Gómez, B.; Duncan, R. *Int. J. Pharm.* **1997**, *148*, 231.
19. Wang, W.; McConaghy, A. M.; Tetley, L.; Uchegbu, I. F. *Langmuir* **2001**, *17*, 631.
20. Kwon, S.; Park, J. H.; Chung, H.; Kwon, I. C.; Jeong, S. Y. *Langmuir* **2003**, *19*, 10188.
21. Hsu, S.-C.; Don, T.-M.; Chiu, W.-Y. *Polym. Degrad. Stabil.* **2002**, *75*, 73.
22. Wade Jr., L. G. *Organic Chemistry;* Prentice-Hall Canada Inc.: Toronto, Canada, 1987; pp 910-971.
23. Hirai, A.; Odani, H.; Nakajima, A. *Polym. Bull.* **1991**, *26*, 87.
24. Vårum, K. M.; Anthonsen, M. W.; Grasdalen, H.; Smidsrød, O. *Carbohyd. Res.* **1991**, *211*, 17.
25. Shigemasa, Y.; Matsuura, H.; Sashiwa, H.; Saimoto, H. *Int. J. Biol. Macromol.* **1996**, *18*, 237.
26. Kubota, N.; Tatsumoto, N.; Sano, T.; Toya, K. *Carbohyd. Res.* **2000**, *324*, 268.
27. Lavertu, M.; Xia, Z.; Serreqi, A. N.; Berrada, M.; Rodrigues, A.; Wang, D.; Buschmann, M. D.; Gupta, A. *J. Pharmaceut. Biomed.* **2003**, *32*, 1149.
28. Digenis, G. A.; Issidorides, C. H. *Bioorg. Chem.* **1979**, *8*, 97.
29. Williams, D. L. H. *Adv. Phys. Org. Chem.* **1983**, *19*, 381.
30. Lijinsky, W.; Keefer, L.; Conrad, E.; Van de Bogart, R. *J. Natl. Cancer I.* **1972**, *49*, 1239.
31. Loeppky, R. N.; Tomasik, W. *J. Org. Chem.* **1983**, *48*, 2751.
32. Smith, P. A. S.; Loeppky, R. N. *J. Am. Chem. Soc.* **1967**, *89*, 1147.

Chapter 12

Chitosan: A Natural Polycation with Multiple Applications

Eve Ruel-Gariépy and Jean-Christophe Leroux

Faculty of Pharmacy, University of Montréal, C.P. 6128 Succ. Centre-ville, Montréal, Québec H3C 3J7, Canada

Chitosan is one of the most abundant polysaccharide in nature. It is a versatile polymer used in various fields like water treatment, food conservation, and drug delivery. Lately, a novel chitosan-based thermosensitive solution has been developed. This system, made of chitosan and β–glycerophosphate, is liquid at or below room temperature, and forms monolithic gels at body temperature. *In vitro* release studies showed that this solution can sustain the delivery of a compound over a few hours up to several days, depending on its water solubility and molecular weight. Other experiments revealed the ability of this preparation to dramatically improve cartilage repair in a safe manner.

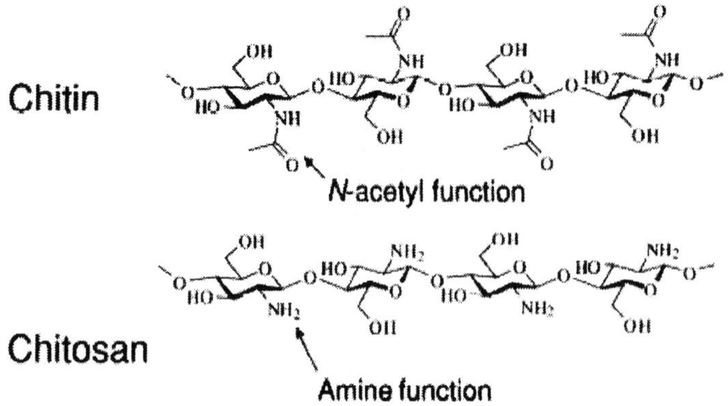

Figure 1. Structural formula of chitin and chitosan.

The term chitosan usually refers to a family of polymers derived from chitin, which is one of the most abundant polysaccharide in nature, second only to cellulose. Chitin is the major structural component of the exoskeleton of invertebrates and the cell walls of fungi. Chitosan is the partially deacetylated form of chitin. It is a linear copolymer composed of glucosamine and *N*-acetylglucosamine units linked in a β(1→4) manner (Figure 1). Chitosan production from crustacean shells begins with protein and calcium salts removal. Afterwards, the chitin is deacetylated by use of concentrated NaOH. This process causes hydrolysis of the aminoacetyl groups. The deacetylation degree (DD) of the polymer chain ranges from 70 to 95%, according to the method used. For most applications, pure chitosan is needed. Therefore, the crude product is dissolved in an acidic solution and filtered. Microfiltration is used to remove insoluble compounds while ultrafiltration removes low molecular weight compounds. The precipitate is then washed and dried. For pharmaceutical applications in particular, the endotoxin content is reduced to a level acceptable for use of the polymer in humans. Figure 2 illustrates the different steps involved in the production of Protasan™ chitosan from Pronova (1).

The characteristics of chitosan are mainly influenced by its composition and molecular weight. The different batches are first characterized by their DD, which is the ratio between the glucosamine and *N*-acetylglucosamine units. This value can be determined by titration, nuclear magnetic resonance, infrared or ultraviolet spectroscopies. It is one of the most important intrinsic properties of chitosan. It also determines the number of free amine groups in the polymer chain. The polymer chains are equally characterized by their molecular weight (MW). It usually ranges from 10,000 to 1,000,000. The most common methods for MW determination are calculations based on intrinsic viscosity and light scattering measurements.

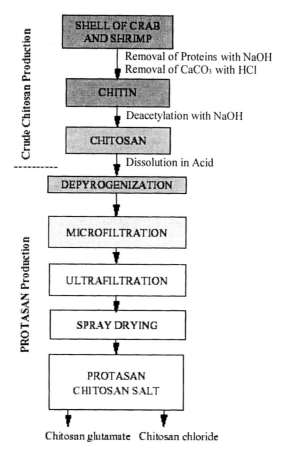

Figure 2. Chitosan production. Reprinted with permission from ref. (1).

Chitosan is insoluble in most organic solvents. It dissolves in acidic aqueous solutions but is insoluble at pH > 6. Some commercial chitosans contain a stochiometric amount of an appropriate acid as a counterion to allow dissolution in water. The solubility of chitosan depends, to a certain extent, on its molecular weight and DD. The apparent pKa value of the amino group of the glucosamine moiety is 6.5. Therefore, under mildly acidic conditions, the polymer is positively-charged and can interact with negative surfaces, such as cell membranes. Chitosan has both reactive amino and hydroxyl groups, which can be used to chemically alter its properties under mild reaction conditions. Chitosan also possesses multiple biological properties.

Chitosan is a versatile polymer with applications in various fields. Herein will be presented some examples in different areas of research. The aim of the present document is to give an overview of the different possibilities of chitosan rather than to review every possible application.

Industrial Applications of Chitosan

Chelation

The binding capacity of chitosan has proved useful in a wide variety of applications, particularly in wastewater management. The polymer is normally used as a flocculation agent. Chitosan represents an interesting alternative to commonly used compounds (e.g. alum and lime) for these applications since it is more environment-friendly (biodegradable and less toxic). For instance, pretreatment of cork processing wastewaters with chitosan has been utilized to enhance the performance of the consecutive filtrations (2). Chitosan was also efficiently employed to remove kaolinite, an agent found in the effluents of the clay processing industry (3). The flocs produced were coarse and settled rapidly. The total time required for flocculation and settling was less than 1 h. Investigators from Thailand developed a chitosan derivative which is an attractive candidate for the removal of dyes from textile wastewater (4). Decolorization reached 90% within 10 min. Moreover, the chitosan derivative could be regenerated and reused more than 10 times. Additionally, chitosan effectively reduced turbidity due to silt and could be used in the preparation of potable water from river water (5). Likewise, chitosan reduced the algal content of water used for decorative or recreational purposes (6). These examples show the variety of compounds that can be bound to chitosan. It is a very interesting chelating agent once the proper working conditions are determined (*e.g.* pH, concentration...).

Paper Treatment

Recently, the role of chitosan in improving the ageing resistance of paper was assessed (7). Samples were dipped in chitosan solutions, submitted to accelerated ageing (100°C for 36, 72, 108, 144 h) and their mechanical properties evaluated. Chitosan treatment improved the strength and durability of the paper. Further research in this area could allow the development of novel, more resistant types of paper.

Food Processing

Utilization of chitosan in the food industry has received considerable attention in recent years (8, 9). In this field, it is often used as a preservative owing to its antimicrobial properties. Indeed, chitosan shows antibacterial and antifungal activity against several microorganisms. The antibacterial activity may be either bactericidal or bacteriostatic, depending on strains and chitosan characteristics (9, 10). The exact mechanism of this activity is still unknown, but different mechanisms have been proposed. For example, chitosan appears to bind to the outer membrane of gram-negative bacteria, explaining the loss of the barrier function (11). The antifungal effect of chitosan allows control of numerous post-harvest fungal diseases found in fruits and vegetables. Some investigators used chitosan to develop a novel preservation system for fresh pork sausages (12). This system allowed reduction of the sulphite concentrations required. Chitosan is also employed as a film or coating in food storage applications due to its ecofriendly and biodegradable nature. Chitosan films were shown to extent the shelf-life of mangoes up to 18 days compared to 9 days for the controls (13). Likewise, chitosan coating of fresh fillets of Atlantic cod (*Gadus morhua*) and herring (*Clupea harengus*) significantly reduced moisture loss, lipid oxidation, and microbial growth (14). Romanazzi *et al.* (15) showed the effectiveness of pre- and post harvest treatments of table grapes with chitosan to control *Botrytis cinerea*. Applications of chitosan in this field are varied. The limiting factor for a given purpose being the determination of the appropriate chitosan DD, MW and concentration required.

Biomedical Applications

Chitosan is currently receiving a great deal of interest for medical and pharmaceutical applications because of its biocompatible and biodegradable nature. Still, investigators continue to evaluate the biocompatibility of their proposed system as the term chitosan refers to group of molecules exhibiting variable properties. Chitosan in diverse forms (particles, films, implants) has been studied following oral, subcutaneous, intramuscular, intraperitoneal or intrabony administration (16-20). Hidaka *et al.* (17) showed that chitosan membranes implanted over rat calvaria elicited an inflammatory reaction whose intensity was inversely related to the DD. VandeVord *et al.* (18) obtained similar results in mice with their implant made of 92% deacetylated chitosan. Macroscopic inspection of the implantation site revealed no pathological inflammatory response. As for the biodegradability of chitosan-based biomaterials, studies show that it increases as the DD decreases (21).

Blood Coagulation

Chitosan and some of its derivatives significantly alter blood coagulation, it is an effective hemostatic agent. It enhances platelet adhesion and aggregation, one of the first steps involved in the early wound healing process (22, 23). It also reduces the blood coagulation time (23). On the other hand, some sulfated derivatives exhibit noticeable anticoagulant activity (24). One sulfated chitosan showed anticoagulant potency similar to heparin (25). Partially N-acylated chitosan derivatives showing good antithrombogenic properties could be used to coat medical instruments in order to increase their blood compatibility (26).

Wound healing

Chitosan is an interesting polymer for wound healing applications. Since it can stop bleeding, it was initially incorporated in wound dressings. Later, it was found that it could also accelerate wound healing. The potential wound healing mechanism of chitosan *in vivo* and *in vitro* has been reviewed by Ueno *et al.* (27). The polymer is usually applied topically as a film (28) or an ointment (29). A chitosan derivative forming a hydrogel when exposed to ultra-violet light was recently proposed as a biological adhesive for soft tissues (30-33). The hydrogel covered a wound effectively, firmly adhered two pieces of skin together and accelerated wound closure and healing. In order to increase treatment efficacy, various growth factors can be added to the formulation. For instance, Mi *et al.* (34) developed a bilayer chitosan wound dressing containing silver sulfadiazine for the control of wound infection. *In vivo* antibacterial tests confirmed that this dressing was effective for long-term inhibition of the growth of *Pseudomonas aeruginosa* and *Staphylococcus aureus* at an infected wound site. Mizuno *et al.* (35) prepared chitosan films containing basic fibroblast growth factor to accelerate wound healing. Likewise, Ishihara *et al.* (32) incorporated various growth factors and heparin in a chitosan hydrogel.

Tissue reconstruction

Tissue repair is a biomedical application for which chitosan has been extensively investigated. Park *et al.* (36) demonstrated the beneficial periodontal tissue regenerative effect of a chitosan collagen sponge. It has also been used to prepare a human reconstructed skin *in vitro* (37-39). This new type of skin substitute could be of great help in the treatment of burnt patients. The transplantation of this reconstructed skin on nude mice induced revascularization, promoted the rapid remodeling of an extracellular matrix nearly similar to normal dermis and encouraged nerve growth. Moreover, chitosan is a key component of various biomaterials proposed for bone reconstruction (40). It has been shown that osteoblasts can proliferate in porous chitosan matrices and promote bone regeneration (41).

Effect on serum lipids

The ability of chitosan to alter serum lipids concentrations has been demonstrated in some studies. In one of them, chitosan has been administered to female volunteers with confirmed mild-to-moderate hypercholesterolemia to investigate its effectiveness in reducing serum cholesterol without concomitant diet therapy (42). Eight weeks of chitosan therapy produced a statistically significant reduction in total cholesterol and LDL cholesterol. Triglycerides levels did not change significantly and HDL cholesterol was maintained within normal limits. The hypocholesterolemic effect of chitosan in animals and humans is well established but the mechanism remains undetermined. Increased fecal excretion of bile acids is one possibility. Chitosan is a weak anion exchanger, and consequently binds bile acids. This has been demonstrated in several studies. A reduction in cholesterol absorption could also explain the cholesterol lowering effect of chitosan. Most *in vivo* studies support this hypothesis (43).

Anti-tumor activity

Many investigators have reported antitumoral effects of chitosan. Nishimura *et al.* (44) showed that some chitosans have a potency to activate peritoneal macrophages and suppress the growth of tumor cells in mice. Likewise, the intraperitoneal injection or oral administration of a low MW chitosan, prepared from enzymatic hydrolysis, inhibited the growth of sarcoma 180 tumor cells in mice (45). Moreover, some oligosaccharides exhibited a growth-inhibitory effect against Meth-A solid tumor transplanted into mice (46, 47).

Chitosan-Based Drug delivery

In recent years, a great deal of research involving chitosan has been directed towards the development of chitosan-based drug delivery systems. Many reviews have dealt with this subject (48-55). Some properties of chitosan make this polymer very attractive for this purpose. For example, chitosan can increase mucosal absorption by interacting with the tight junctions and provoking their opening as well as through mucoadhesion by extending the residence time of drug carriers at the absorption site (56-64).

Chitosan has been used to design oral (57, 65, 66), nasal (60, 67-73), ocular (58), intra-articular (74) and intra-tumoral (75) formulations. Tozaki *et al.* (66) demonstrated that 5-ASA (5-aminosalicylic acid) can be specifically delivered to the colon using chitosan capsules. These could be useful carriers of anti-inflammatory drugs in ulcerative colitis and Chrohn's disease. Illum *et al.* (69, 70) developed novel nasal morphine formulations based on chitosan. Pilot studies in cancer patients have shown the efficacy of this approach as a means of improving the treatment of breakthrough pain. An intra-nasal chitosan-based vaccine against diphtheria has also been developed (72). Significant systemic humoral immune responses were found after nasal administration of diphtheria toxoid associated chitosan microparticles to mice. A recent study in healthy volunteers revealed that a single nasal immunization was well tolerated and boosted antitoxin neutralizing activity (73). The neutralizing activity was equivalent to that induced by standard intramuscular vaccine. It has also been shown that the intratumoral administration of a specific photosensitizer, indocyanine green, in a chitosan gel increased tumor response to laser treatment (75).

Chitosan has recently emerged as a nonviral gene delivery system. The cationic nature of the polymer in solution allows interaction with the negatively-charged DNA molecule. Chitosan presents the advantage of being less toxic than most cationic polymers currently used in this field of research. Köping-Höggård *et al.* (76) reported the formation of stable colloidal polyplexes with DNA and the protection of DNA from serum degradation. Kumar *et al.* (77) and Iqbal *et al.* (78) used chitosan to develop an intranasal vaccine against respiratory syncytial virus (RSV) infection. This vaccine contained chitosan-DNA nanospheres. In mice, its administration resulted in a significant reduction of viral titers and viral antigen load after acute RSV infection. Moreover, it resulted in significant induction of antibodies production.

Lately, Chenite *et al.* (79, 80) developed a novel chitosan-based thermosensitive solution. This system is made of chitosan and a polyol counterionic monohead salt (*e.g.* β-glycerophosphate, GP). The latter is used to neutralize the chitosan solution. It increases pH to physiologically acceptable levels (6.8 to 7.2) without making the polymer precipitate. Moreover, the resulting solution acquires thermo-responsive properties. Chitosan/GP solutions are liquid at or below room temperature, and form monolithic gels at body temperature (Figure 3).

Our group evaluated the chitosan/GP solution as a drug delivery system (81-83). The physico-chemical characterization of the solution revealed that the gelation rate was dependent on the temperature and on the chitosan DD. Stability studies showed that solutions made with 84% deacetylated chitosan could be stored for 3 months at 4°C without apparent change in viscosity. The porous

Figure 3. The chitosan-based thermosensitive solution at room temperature (left) and at 37°C (right). Reprinted with permission from ref. (81)

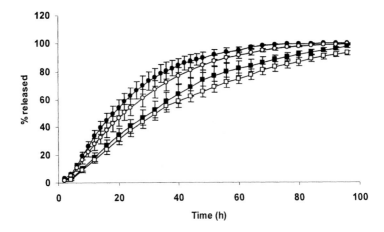

Figure 4. Release profiles of FITC-dextran of different molecular weights from a C/GP solution in PBS at pH 7.4 and 37±0.1°C. 12,000 (closed circles), 19,500 (open circles), 38,260 (closed squares), 148,000 (open squares). In each experiment, the loading was 3%. Each point represents the mean value ± SEM (n=5). Reprinted with permission from ref. (82)

structure of the gel formed at 37°C and its weight loss profile over time were not influenced by the DD of the chitosan used. Erosion of the polymeric matrix occurred mostly during the first 4 hours. *In vitro* release experiments revealed that the system could deliver macromolecules (MW 12,000-148,000) over a period of several hours to a few days (Figure 4) (82).

However, with low MW hydrophilic compounds (<1000 g/mol), the release was generally completed within 24 h. The relatively fast release was attributed to the high water content of the hydrogel (>90%) and, thus, to the presence of very large pores. To achieve sustained delivery that would be independent of the drug's molecular weight, hydrophilic compounds were first loaded into liposomes, which were then mixed with the thermosensitive solution (83). The addition of liposomes to the thermosensitive solution did not change the overall gelling behavior of the thermosensitive system. This approach substantially slowed the release rate (Figure 5). Furthermore, the release profile of the incorporated compound could be controlled by adjusting liposome characteristics, such as size and composition (Figures 5-6).

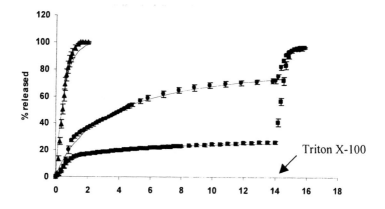

Figure 5. In vitro cumulative release of carboxyfluorescein (CF), free (triangles) or encapsulated in 100-nm (circles) or 280-nm (squares) egg phosphatidylcholine(EPC)/cholesterol(Chol) liposomes from the C/GP gel in PBS at 37°C. Each point represents the mean value ± SEM (n=5). Reprinted with permission from ref. (83)

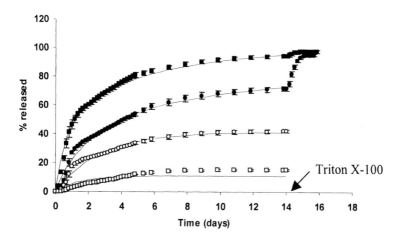

Figure 6. In vitro release kinetics of CF, encapsulated in EPC (squares) or EPC/Chol (circles) 100 nm liposomes from the C/GP gel in PBS at 37°C. The total amount (closed symbols) and the liposome-associated fraction (open symbols) of CF released are shown in the figure. Each point represents the mean value ± SEM (n=5). Reprinted with permission from ref. (83)

Possible benefits of the thermosensitive solution for the sustained delivery of paclitaxel, a hydrophobic anticancer drug, were also investigated (81). The *in vitro* release profiles demonstrated controlled delivery over 1 month. The efficacy of the thermosensitive hydrogel in delivering paclitaxel intratumorally was evaluated using Balb/c mice bearing an EMT-6 tumor. These experiments showed that one intratumoral injection of the new formulation was as efficacious as 4 intravenous injections of the commercially-available Taxol® formulation in inhibiting the growth of EMT-6 cancer cells, and proved to be less toxic (Figure 7).

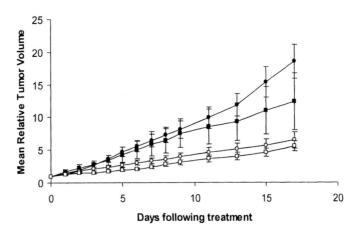

Figure 7. In vivo antitumor effect in Balb/c mice bearing an EMT-6 tumor. Saline 0.2 mL/day intravenous x 4 days (closed circles, n=8), Taxol® 10 mg/kg/day intravenous x 4 days (open circles, n=8), chitosan/GP solution 10 μL intratumoral (closed squares, n = 8) and paclitaxel/chitosan/GP solution (64 mg/mL paclitaxel) 10 μL (40 mg/kg paclitaxel) intratumoral (open squares, n = 8). Reprinted with permission from ref. (81).

Berrada *et al.* (84) proposed to use the thermosensitive system for the sustained intra-tumoral release of camptothecin, another anti-cancer drug exhibiting low solubility. The effectiveness of treatment was evaluated using a RIF-1 mouse tumor model. Animals treated with the thermosensitive formulation containing camptothecin had significantly longer tumor growth delays compared to untreated animals as well as to animals treated systemically with camptothecin by intra-peritoneal injection with no evidence of toxicity in terms of loss of body weight.

The chitosan/GP solution has also been evaluated as a tool for cartilage repair (85, 86). The authors first showed that primary calf chondrocytes could proliferate in solidified chitosan/GP solutions both *in vitro* and *in vivo* (85). Later, a hybrid implant, made of the chitosan/GP solution and whole blood, was developed to improve cartilage healing (86). This new system, named CarGel®, was investigated in rabbits, sheep, and horses. These studies collectively confirmed that CarGel®-treatment of marrow-stimulated defects has the ability to dramatically improve cartilage repair over control in a safe manner. Recently, human studies were undertaken with this new product.

Other experiments revealed that the biocompatibility of the thermosensitive chitosan/GP solutions was mainly influenced by the chitosan used and the injection site. Their injection in the rat hind paw elicited an inflammatory reaction which was inversely related to the chitosan's DD (87) and substantially more important than the one observed after a transdermal injection in the rat dorsal region (88).

Because the chitosan/GP solutions are essentially designed to be injected, sterilization becomes necessary. Owing to their thermosensitive behavior, chitosan/GP solutions cannot be simply sterilized by autoclave; it is preferable to sterilize separately the chitosan and GP solution. Two approaches were assessed for the chitosan solution: steam sterilization and γ-irradiation. Steam sterilization induced MW reduction, which led to a decrease in dynamic viscosity, gelling rate, and mechanical strength of the chitosan/GP hydrogels but the thermogelling character remained clearly observable even after a 60 min autoclave cycle (88). Gamma-irradiation strongly decreased the chitosan solution viscosity and the thermogelling properties were dramatically altered (89). Thus, the steam sterilization process may be considered when high mechanical performance is not critical or essential for implant function, such as in the case of controlled drug delivery whereas γ-irradiation would not be an appropriate method to sterilize chitosan solutions.

In conclusion, chitosan constitutes a promising tool for pharmaceutical and bioengineering applications. This polymer is biocompatible, biodegradable and very versatile.

Achowledgements

Financial support from the Natural Sciences and Engineering Research Council of Canada, Biosyntech Inc. (Laval, Qc, Canada) and the Canada Research Chair program is acknowledged.

References

(1) Skaugrud, O.; Hagen, A.; Borgersen, B.; Dornish, M. ; Pronova Biomedicals, 1998.

(2) Minhalma, M.; De Pinho, M. N. *Envion. Sci. Technol.* **2001**, *35*, 4916-4921.

(3) Divakaran, R.; Pillai, V. N. S. *Wat. Res.* **2001**, *35*, 3904-3908.

(4) Trung, T. S.; Ng, C. H.; Stevens, W. F. *Biotechnol. Lett.* **2003**, *25*, 1185-1190.

(5) Divakaran, R.; Pillai, V. N. S. *Wat. Res.* **2002**, *36*, 2414-2418.

(6) Divakaran, R.; Pillai, V. N. S. *J. Appl. Phycol.* **2002**, *14*, 419-422.

(7) Basta, A. H. *Restaurator* **2003**, *24*, 106-117.

(8) Shahidi, F.; Arachchi, J. K. V.; Jeon, Y. J. *Trends Food Sci. Technol.* **1999**, *10*, 37-51.

(9) Muzzarelli, C.; Muzzarelli, R. A. A. *Agro Food Ind. Hi Tec.* **2003**, *14*, 39-42.

(10) No, H. K.; Park, N. Y.; Lee, S. H.; Meyers, S. P. *Int. J. Food Microbiol.* **2002**, *74*, 65-72.

(11) Helander, I. M.; Nurmiaho-Lassila, E. L.; Ahvenainen, R.; Rhoades, J.; Roller, S. *Int. J. Food Microbiol.* **2001**, *71*, 235-244.

(12) Roller, S.; Sagoo, S.; Board, R.; O'Mahony, T.; Caplice, E.; Fitzgerald, G.; Fogden, M.; Owen, M.; Fletcher, H. *Meat Sci.* **2002**, *62*, 165-177.

(13) Srinivasa, P. C.; Baskaran, R.; Ramesh, M. N.; Harish Prashanth, K. V.; Tharanathan, R. N. *Eur. Food Res. Technol.* **2002**, *215*, 504-508.

(14) Jeon, Y. J.; Kamil, J. Y. V. A.; Shahidi, F. *J. Agric. Food Chem.* **2002**, *50*, 5167-5178.

(15) Romanazzi, G.; Nigro, F.; Ippolito, A.; Di Venere, D.; Salerno, M. *J. Food Sci.* **2002**, *67*, 1862-1867.

(16) Tanaka, Y.; Tanioka, S.; Tanaka, M.; Tanigawa, T.; Kitamura, Y.; Minami, S.; Okamoto, Y.; Miyashita, M.; Nanno, M. *Biomaterials* **1997**, *18*, 591-595.

(17) Hidaka, Y.; Ito, M.; Mori, K.; Yagasaki, H.; Kafrawy, A. H. *J. Biomed. Mater. Res.* **1999**, *46*, 418-423.

(18) VandeVord, P. J.; Matthew, H. W. T.; DeSilva, S. P.; Mayton, L.; Wu, B.; Wooley, P. H. *J. Biomed. Mater. Res.* **2002**, *59*, 585-590.

(19) Wang, X.; Ma, J.; Wang, Y.; He, B. *Biomaterials* **2002**, *23*, 4167-4176.

(20) Mi, F. L.; Tan, Y. C.; Liang, H. F.; Sung, H. W. *Biomaterials* **2002**, *23*, 181-191.

(21) Dubé, A. Mémoire de maîtrise, Université de Montréal, Montréal, 2003.

(22) Chou, T. C.; Fu, E.; Wu, C. J.; Yeh, J. H. *Biochem. Biophys. Res. Commun.* **2003**, *302*, 480-483.

(23) Okamoto, Y.; Yano, R.; Miyatake, K.; Tomohiro, I.; Shigemasa, Y.; Minami, S. *Carbohydr. Polym.* **2003**, *53*, 337-342.

(24) Huang, R.; Du, Y.; Yang, J.; Fan, L. *Carbohydrate Res.* **2003**, *338*, 483-489.

(25) Vongchan, P.; Sajomsang, W.; Subyen, D.; Kongtawelert, P. *Carbohydr. Res.* **2002**, *337*, 1239-1242.

(26) Lee, K. Y.; Ha, W. S.; Park, W. H. *Biomaterials* **1995**, *16*, 1211-1216.

(27) Ueno, H.; Mori, T.; Fujinaga, T. *Adv. Drug Deliv. Rev.* **2001**, *52*, 105-115.

(28) Khan, T. A.; Peh, K. K. *J. Pharm. Pharmaceut. Sci.* **2003**, *6*, 20-26.

(29) Kweon, D. K.; Song, S. B.; Park, Y. Y. *Biomaterials* **2003**, *24*, 1595-1601.

(30) Ono, K.; Saito, Y.; Yura, H.; Ishikawa, K.; Kurita, A.; Akaike, T.; Ishihara, M. *J. Biomed. Mater. Res.* **2000**, *49*, 289-295.

(31) Ishihara, M.; Nakanishi, K.; Ono, K.; Sato, M.; Kikuchi, M.; Saito, Y.; Yura, H.; Matsui, T.; Hattori, H.; Uenoyama, M.; Kurita, A. *Biomaterials* **2002**, *23*, 833-840.

(32) Ishihara, M.; Obara, K.; Ishizuka, T.; Fujita, M.; Sato, M.; Masuoka, K.; Saito, Y.; Yura, H.; Matsui, T.; Hattori, H.; Kikuchi, M.; Kurita, A. *J. Biomed. Mater. Res.* **2002**, *64A*, 551-559.

(33) Obara, K.; Ishihara, M.; Ishizuka, T.; Fujita, M.; Ozeki, Y.; Maehara, T.; Saito, Y.; Yura, H.; Matsui, T.; Hattori, H.; Kikuchi, M.; Kurita, A. *Biomaterials* **2003**, *24*, 3437-3444.

(34) Mi, F. L.; Wu, Y. B.; Shyu, S. S.; Schoung, J. Y.; Huang, Y. B.; Tsai, Y. H.; Hao, J. Y. *J. Biomed. Mater. Res.* **2002**, *59*, 438-449.

(35) Mizuno, K.; Yamamura, K.; Yano, K.; Osada, T.; Saeki, S.; Yakimoto, N.; Sakurai, T.; Nimura, Y. *J. Biomed. Mater. Res.* **2003**, *64A*, 177-181.

(36) Park, J. S.; Choi, S. H.; Moon, I. S.; Cho, K. S.; Chai, J. K.; Kim, C. K. *J. Clin. Periodontol.* **2003**, *30*, 443-453.

(37) Li, H.; Berthod, F.; Xu, W.; Damour, O.; Germain, L.; Auger, F. A. *J. Surg. Res.* **1997**, *73*, 143-148.

(38) Berthod, F.; Germain, L.; Li, H.; Xu, W.; Damour, O.; Auger, F. A. *Matrix Biol.* **2001**, *20*, 463-473.

(39) Gingras, M.; Paradis, I.; Berthod, F. *Biomaterials* **2003**, *24*, 1653-1661.

(40) Hu, Q.; Li, B.; Wang, M.; Shen, J. *Biomaterials* **2004**, *25*, 779-785.

(41) Lee, J. Y.; Nam, S. H.; Im, S. Y.; Park, Y. J.; Lee, Y. M.; Seol, Y. J.; Chung, C. P.; Lee, S. J. *J. Controlled Release* **2002**, *78*, 187-197.

(42) Bokura, H.; Kobayashi, S. *Eur. J. Clin. Nutr.* **2003**, *57*, 721-725.

(43) Gallaher, D. D. *Agro Food Ind. Hi Tec.* **2003**, *14*, 32-35.

(44) Nishimura, K.; Nishimura, S.; Nishi, N.; Saiki, I.; Tokura, S.; Azuma, I. *Vaccine* **1984**, *2*, 93-98.

(45) Qin, C.; Du, Y.; Xiao, L.; Li, Z.; Gao, X. *Int. J. Biol. Macromol.* **2003**, *31*, 111-117.

(46) Tokoro, A.; Tatewaki, N.; Suzuki, K.; Mikami, T.; Suzuki, S.; Suzuki, M. *Chem. Pharm. Bull.* **1988**, *36*, 784-790.

(47) Seljelid, R. *Biosci. Rep.* **1986**, *6*, 845-851.

(48) Kas, H. S. *J. Microencapsulation* **1997**, *14*, 689-711.

(49) Illum, L. *Pharm. Res.* **1998**, *15*, 1326-1331.

(50) Dodane, V.; Vilivalam, V. D. *PSTT* **1998**, *1*, 246-253.

(51) Felt, O.; Buri, P.; Gurny, R. *Drug. Dev. Ind. Pharm.* **1998**, *24*, 1-15.

(52) Singla, A. K.; Chawla, M. *J. Pharm. Pharmacol.* **2001**, *53*, 1047-1067.

(53) Khor, E.; Lim, L. Y. *Biomaterials* **2003**, *24*, 2339-2349.

(54) Berger, J.; Reist, M.; Mayer, J. M.; Felt, O.; Peppas, N. A.; Gurny, R. *Eur. J. Pharm. Biopharm.* **2004**, *57*, 19-34.

(55) Berger, J.; Reist, M.; Mayer, J. M.; Felt, O.; Gurny, R. *Eur. J. Pharm. Biopharm.* **2004**, *57*, 35-52.

(56) Lehr, C. M.; Bouwstra, J. A.; Schacht, E. H.; Junginger, H. E. *Int. J. Pharm.* **1992**, *78*, 43-48.

(57) Takeuchi, H.; Yamamoto, H.; Niwa, T.; Hino, T.; Kawashima, Y. *Pharm. Res.* **1996**, *13*, 896-901.

(58) Felt, O.; Furrer, P.; Mayer, J. M.; Plazonnet, B.; Buri, P.; Gurny, R. *Int. J. Pharm.* **1999**, *180*, 185-193.

(59) Pan, Y.; Li, Y. J.; Zhao, H. Y.; Zheng, J. M.; Xu, H.; Wei, G.; Hao, J. S.; Cui, F. D. *Int. J. Pharm.* **2002**, *249*, 139-147.

(60) Fernandez-Urrusuno, R.; Calvo, P.; Remunan-Lopez, C.; Vila-Jato, J. L.; Alonso, M. J. *Pharm. Res.* **1999**, *16*, 1576-1581.

(61) Thanou, M.; Verhoef, J. C.; Junginger, H. E. *Adv. Drug Deliv. Rev.* **2001**, *52*, 117-126.

(62) Thanou, M.; Verhoef, J. C.; Junginger, H. E. *Adv. Drug Deliv. Rev.* **2001**, *50*, S91-S101.

(63) Hamman, J. H.; Stander, M.; Kotzé, A. F. *Int. J. Pharm.* **2002**, *232*, 235-242.

(64) Sinswat, P.; Tengamnuay, P. *Int. J. Pharm.* **2003**, *257*, 15-22.

(65) Bernkop-Schnurch, A. *Int. J. Pharm.* **2000**, *194*, 1-13.

(66) Tozaki, H.; Odoriba, T.; Okada, N.; Fujita, T.; Terabe, A.; Suzuki, T.; Okabe, S.; Muranishi, S.; Yamamoto, A. *J. Controlled Release* **2002**, *82*, 51-61.

(67) Illum, L.; Davis, S. S.; Pawula, M.; Fisher, A. N.; Barrett, D. A.; Farraj, N. F.; Shaw, P. N. *Biopharm. Drug. Disp* **1996**, *17*, 717-724.

(68) Dyer, A. M.; Hinchcliffe, M.; Watts, P.; Castile, J.; Jabbal-Gill, I.; Nankervis, R.; Smith, A.; Illum, L. *Pharm. Res.* **2002**, *19*, 998-1008.

(69) Illum, L.; Watts, P.; Fisher, A. N.; Hinchcliffe, M.; Norbury, H.; Jabbal-Gill, I.; Nankervis, R.; Davis, S. S. *J. Pharmacol. Exp. Ther.* **2002**, *301*, 391-400.

(70) Pavis, H.; Wilcock, A.; Edgecombe, J.; Carr, D.; Manderson, C.; Church, A.; Fisher, A. *J. Pain Symptom Manag.* **2002**, *24*, 598-602.

(71) Illum, L. *J. Controlled Release* **2003**, *87*, 187-198.

(72) van der Lubben, I.; Kersten, G.; Fretz, M. M.; Beuvery, C.; Verhoef, J. C.; Junginger, H. E. *Vaccine* **2003**, *21*, 1400-1408.

(73) Mills, K. H. G.; Cosgrove, C.; McNeela, E. A.; Sexton, A.; Giemza, R.; Jabbal-Gill, I.; Church, A.; Lin, W.; Illum, L.; Podda, A.; Rappuoli, R.; Pizza, M.; Griffin, G. E.; Lewis, D. J. M. *Infect. Immun.* **2003**, *71*, 726-732.

(74) Thakkar, H.; Sharma, R. K.; Mishra, A. K.; Chuttani, K.; Murthy, R. S. R. *J. Pharm. Pharmacol.* **2004**, *56*, 1091-1099.

(75) Chen, W. R.; Adams, R. L.; Carubelli, R.; Nordquist, R. E. *Cancer Lett.* **1997,** *115,* 25-30.

(76) Koping-Hoggard, M.; Tubulekas, I.; Guan, H.; Edwards, K.; Nilsson, M.; Varum, K. M.; Artursson, P. *Gene Ther.* **2001,** *8,* 1108-1121.

(77) Kumar, M.; Behera, A. K.; Lockey, R. F.; Zhang, J.; Bhullar, G.; Perez de la Cruz, C.; Chen, L. C.; Leong, K. W.; Huang, S. K.; Mohapatra, S. S. *Human Gene Ther.* **2002,** *13,* 1415-1425.

(78) Iqbal, M.; Lin, W.; Jabbal-Gill, I.; Davis, S. S.; Steward, M. W.; Illum, L. *Vaccine* **2003,** *21,* 1478-1485.

(79) Chenite, A.; Chaput, C.; Wang, D.; Combes, C.; Buschmann, M. D.; Hoemann, C. D.; Leroux, J. C.; Atkinson, B. L.; Binette, F.; Selmani, A. *Biomaterials* **2000,** *21,* 2155-2161.

(80) Chenite, A.; Buschmann, M.; Wang, D.; Chaput, C.; Kandani, N. *Carbohydr. Polymers* **2001,** *46,* 39-47.

(81) Ruel-Gariépy, E.; Shive, M.; Bichara, A.; Berrada, M.; Le Garrec, D.; Chenite, A.; Leroux, J. C. *Eur. J. Pharm. Biopharm.* , *in press.*

(82) Ruel-Gariépy, E.; Chenite, A.; Chaput, C.; Guirguis, S.; Leroux, J. C. *Int. J. Pharm.* **2000,** *203,* 89-98.

(83) Ruel-Gariépy, E.; Leclair, G.; Hildgen, P.; Gupta, A.; Leroux, J. C. *J. Controlled Release* **2002,** *82,* 373-383.

(84) Berrada, M.; Serreqi, A.; Dabbarh, F.; Owusu, A.; Gupta, A.; Lehnert, S. *Biomaterials* **2005,** *26,* 2115-2120.

(85) Hoemann, C. D.; Sun, J.; Légaré, A.; McKee, M. D.; Ranger, P.; Buschmann, M. D. , San Francisco, California, February 25-28, 2001 2001; 626.

(86) Hoemann, C. D.; Hurtig, M.; Sun, J.; McWade, D.; Rossomacha, E.; Shive, M. S.; McKee, M. D.; Buschmann, M. D. , Pittsburgh PA, March 17-20 2003.

(87) Molinaro, G.; Leroux, J. C.; Damas, J.; Adam, A. *Biomaterials* **2002,** *23,* 2717-2722.

(88) Jarry, C.; Chaput, C.; Chenite, A.; Renaud, M. A.; Buschmann, M.; Leroux, J. C. *J. Biomed. Mater. Res. (Appl. Biomater.)* **2001,** *58,* 127-135.

(89) Jarry, C.; Leroux, J. C.; Haeck, J.; Chaput, C. *Chem. Pharm. Bull.* **2002,** *50,* 1335-1340.

(90) Heymann, E. *Trans. Faraday Soc.* **1935,** *31,* 846-864.

(91) Sarkar, N. *J. Appl. Polym. Sci.* **1979,** *24,* 1073-1087.

(92) Carlsson, A.; Karlström, G.; Lindman, B.; Stenberg, O. *Colloid Polym. Sci.* **1988,** *266,* 1031-1036.

(93) Carlsson, A.; Karlström, G.; Lindman, B. *Colloids Surf.* **1990,** *47,* 147-165.

(94) Shirakawa, M.; Yamatoya, K.; Nishinari, K. *Food Hydrocolloids* **1998,** *12,* 25-28.

Polysaccharides

Chapter 13

Properties of Levan and Potential Medical Uses

J. Combie

Montana Polysaccharides Corporation, 1910–107 Lavington, Rock Hill, SC 29732

Levan is an unusual polysaccharide that does not swell in water and that has an uncommonly low intrinsic viscosity. Animal studies have shown levan can lower blood cholesterol and a derivative of levan will increase calcium absorption. As a strong adhesive and a water soluble film former, levan has the potential to make a temporary coating or bandage.

The term "levan" was introduced over 100 years ago to describe the exopolysaccharide produced by a *Bacillus* when grown on sucrose *(1)*. While the glucose portion of the substrate is used as a microbial energy source, the fructose units are linked together to build levan, a natural polymer of fructose. β-D-fructo-furanosyl residues are connected by β-2,6 linkages. Branching is accomplished through occasional β-2,1 bonds. The degree of branching varies with the organism used in production but has been reported as high as 20%. Several dozen bacteria are known to produce levan, including species of *Acetobacter, Aerobacter, Azotobacter, Bacillus, Corynebacterium, Erwinia, Gluconobacter, Mycobacterium, Pseudomonas, Streptococcus,* and *Zymomonas (1,2)*. The molecular weights of microbial levans are usually greater than 0.5 million and occasionally as high as 40 million *(3)*. Levans made by plants are much smaller with the molecular weight generally under 10,000.

Properties of Levan

Adhesive

One interesting property of levan is the adhesive strength. Although sugar based materials are often sticky, the adhesive strength of levan is significantly greater than that of most other natural polymers. For example, when polysaccharides were applied to ten sets of bare aluminum adherends, cured for 10 days at 35 °C and then tested for tensile strength, levan had an average tensile strength of 991 psi. Under the same conditions, dextran had only half the tensile strength at 479 psi. Polysaccharides commonly used for thickening such as guar gum and xanthan gum, had even lower adhesive strengths at 63 and 33 psi respectively. Entanglement of the branches extending from the surface of levan spheres contributes to the cohesive strength of levan. It should be noted that all materials tested were diluted only with water. No formulation was done to enhance adhesive strength or other properties such as flexibility, fatigue resistance and shrinkage *(4)*.

Levan is water soluble but does not swell in water. It has potential use in bonding of tablets when dissolution is desired shortly after ingestion. If a more gradual breakup of the tablet is desired, a more water resistant fructan would be useful. Indeed, there is another fructan with low water solubility. This fructan, inulin, is chemically identical to levan, but bonding through the 2 and 1 carbons (as opposed to the 2 and 6 carbons of levan) results in a largely water insoluble compound. However, the adhesive strength of inulin is only about one-tenth that of levan.

Although there are numerous methods for decreasing the water solubility of a material, the moieties responsible for the adhesive properties of levan are also responsible for the interaction with water. Numerous attempts were required to solve the dilemma. Ultimately, a cross-linking procedure was found to be the most successful. The formulation is being optimized and will be published in the near future.

Like most polysaccharides, levan is very resistant to solvents. Coupons bonded with levan were soaked in *d*-limonene, methylethylketone or toluene for 48 hrs. The solvent-soaked levan bond retained full adhesive strength *(4)*. This property is particularly useful when bonding materials that may be exposed to solvents during production or cleaning.

Spherical Shape

Levan is one of the few polysaccharides in which the carbohydrate ring is in the furanose form. With greater flexibility than the more common pyranose of most polysaccharides, the 5-membered ring permits repeated folding *(5,6)*. The end result is a densely packed spherical structure. In the case of the present levan, approximately 10,000 fructans are joined and crowded into the small sphere. The sphere diameter ranges from 50 to 200 nm in diameter.

Membrane Protection

In plants, fructans serve as carbohydrate storage compounds. Evidence suggests they may also provide plant cells with enhanced drought tolerance and freeze resistance. Studies to elucidate the mechanism of these properties have begun to reveal the interaction between membrane components and the fructans. Vereyken showed that both an intermediate sized levan (DP 125) and a low molecular weight inulin protected the membrane barrier more effectively than dextran during dehydration-rehydration cycles. It appears that levan is inserted in the headgroup region between lipid layers. Experiments were done *in vitro* using unilamellar vesicles of 1-palmitoyl-2-oleoyl-*sn*-glycero-3-phosphocholine *(7,8)*. Similar studies have not been done on high molecular weight levans.

Low Intrinsic Viscosity

Despite a high molecular weight, levan has an exceptionally low intrinsic viscosity, resulting from the compact spherical shape. Levan produced in this laboratory by an unidentified species of *Bacillus* has a measured intrinsic viscosity of 0.14 dl/gm. Compare this with the intrinsic viscosity of dextran at about 1 dl/gm or that of polysaccharides typically used as thickeners which are frequently over 10 dl/gm *(5)*. The low viscosity facilitates application of levan as an easily spread paint or as an aerosol, not subject to clogging of the nozzle. When dry, a levan coating is hard, although brittle as reflected by the glass transition temperature of 123 °C.

Film Formation

Levan can be readily formed into a film. Not allowing the film to completely dry by adding a small amount of plasticizer will keep the film flexible. Mixing levan with another polysaccharide, curdlan, will also result in a flexible film. Perhaps more unexpected, was the result of mixing levan with the clay, montmorillonite. One part of montmorillonite was mixed with either 2 or 5 parts of levan. A flexible film was formed. The surprise was that the film was water resistant. These properties suggest potential application as a flexible coating or bandage. For a natural material, levan is quite heat stable with a melting point of 225 °C. Although autoclave moisture will interfere with pre-formed bonds, once excess moisture is removed, the levan regains its adhesive strength.

Medical Applications of Levan

Calcium Absorption

Ingestion of certain sugar alcohols, oligosaccharides and polysaccharides is known to enhance calcium absorption *(9)*. Cyclic disaccharide derivatives of levan or inulin, difructose anhydrides (DFA) have been the subject of recent studies in Tomita's laboratory. Four difructose anhydrides have been identified. One of them, DFA IV, can be made by growing certain species of *Arthrobacter* and *Pseudomonas* on levan. DFA IV has about half the sweetness of sucrose and a melting point of 177-178 °C, sufficient stability to permit use in many food applications. It is not digested or absorbed from the intestine of the rat. Rats fed DFA IV absorbed significantly more calcium than control animals *(10)*. The absorption was mainly in the small intestine and it was suggested these DFAs have potential in preventing osteoporosis *(11)*.

Lowering Cholesterol

Perhaps the most interesting health-related property of levan is its ability to lower cholesterol levels. Several drugs are currently marketed for lowering blood cholesterol but for some people, there is the possibility of adverse side effects. Water soluble dietary fibers are often used as antihyperlipidemics but the high viscosity of these vegetable gums makes them difficult to ingest a sufficient amount. Levan overcomes both of these problems. Studies indicate levan is a safe material *(12,13,14)* and the low viscosity simplifies formulation for easy consumption.

Ishihara was among the first to establish the value of high molecular weight levan as a hypocholesterolemic agent. He found that high molecular weight levan or a partial hydrolysate of levan could effectively lower blood cholesterol and aorta lipid deposits in rabbits. Triglycerides and the amount of adipose tissue were lowered significantly in rats on diets including levan or levan hydrolysates *(13)*. These studies were later confirmed by Iizuka and colleagues *(15)*. Taking these findings to the next step, Ishihara also claimed levan or levan derivatives were an effective antiobesity agent. The low viscosity of the solutions facilitated consumption of sufficient levan to be effective (preferably 100 mg levan or hydrolysate per kg of body weight). Animal studies showed no acute or chronic toxicity *(13)*.

Work in Yamaguchi's lab used a high molecular weight (ca. 2,000,000) levan in a systematic testing done in rats. The animals were fed diets which included either 1% or 5% levan. Blood cholesterol fell 17% or 41% respectively. Neither triacylglycerol nor glucose was affected by the dietary levan. Total sterol excreted in the feces of levan-fed rats was approximately double that excreted by control animals. *In vitro* testing indicated levan was not fermented by the selected bifidobacteria. It is possible that the mechanism by which the cholesterol lowering effect is accomplished may differ between high molecular weight levan and the low molecular weight fructooligosaccharides. One possible mechanism is that the levan binds or entraps sterols in the intestine, interfering with their reabsorption (16).

Additional Applications

Levan has long been known to be an antitumor agent *(12, 17, 18)*. Multiple mechanisms have been attributed to this activity. The host immune response is modulated, there is a direct inhibitory effect on tumor cells and levan augments the activity of other antitumor compounds *(17,19)*. Administration of fructans has been shown to reduce the incidence of carcinogen-induced pre-cancerous lesions in rats *(20)*.

Additional findings related to the immunomodulatory effect of levan, include the fact that levan can delay the rejection of skin grafts *(21)* and reduce the number of macrophages attaching to subcutaneously implanted foreign bodies *(22)*. Also, levan has been shown to decrease the accumulation of polymorphonuclear leucocytes in an experimentally induced inflammatory lesion *(23)*. In actively sensitized animals, levan markedly reduced the incidence and severity of allergic encephalomyelitis in guinea pigs *(24)*.

Oligosaccharides have been found useful as prebiotics, metabolized by the beneficial bifidobacteria and lactobacilli in the large intestine *(25)*. Few studies have examined the value of high molecular weight levan. However, it is known that large levans can be fermented by these beneficial bacteria but not by the undesirable *Clostridium perfringens* and *E. coli (26)*.

Like dextran, levan can be used to create two phase liquid systems, of potential value in purification of biological materials *(5)*.

Garegg et. al. tested high molecular weight levan derivatives and found a number of potential applications. For example, in an assay for inhibition of smooth muscle cell proliferation, the activity of the levan sulfate was one log greater than for the commercial heparin used for comparison. They also found levan sulfate effective in reducing virus growth in an *in vitro* test. Phosphated levan caused water and certain solvents to gel. Suggested uses for this gelled form of levan were in pharmaceuticals and as a fat substitute (27).

Conclusions

Levan is a spherical polymer of fructose with branches extending from the surface. Entanglement of these branches contributes to the cohesive strength. Unlike polysaccharides used as thickeners, levan does not swell in water and has an intrinsic viscosity of only 0.14 dl/gm. Like some other polysaccharide-based compounds, levan can enhance calcium absorption and host immune responses. The mechanism by which levan effectively lowers blood cholesterol in animals has not been fully elucidated. Additional preliminary findings suggest a variety of medical applications for levan in the future.

Acknowledgement

The author gratefully acknowledges the support of the Strategic Environmental Research and Development Program (SERDP) and the U.S. Environmental Protection Agency for funding work reported here.

References

1. Han, Y. *Adv. Appl. Microbiol.* **1990,** *35,* 171-194.
2. Rhee, S-K.; Song, K-B.; Kim, C-H.; Park, B-S.; Jang, E-K.; Jang, K-H. In *Polysaccharides from Prokaryotes;* Vandamme, E.; De Baets, S.; Steinbuchel, A., Ed.; *Biopolymers Volume 5, Polysaccharides I*; Wiley-VCH: WeinHeim, Germany, **2002;** pp 351-377.
3. Mays, T.; Dally, E. U.S. Patent 4,769,254, **1988.**
4. Combie, J.; Steel, A.; Sweitzer, R. *Clean Techn. Environ. Policy.* **2004,** 6, 258-262.

5. Kasapis, S.; Morris, E.; Gross, G.; Rudolph, K. *Carbohydr. Polym.* **1994,** 23, 55-64.
6. Marchessault, R.; Bleha, T.; Deslendes, Y.; Revol, J-F. *Can. J. Chem.* **1980,** 58, 2415-2417.
7. Vereyken, I.; van Kuik, J.; Evers, T.; Rijken, P.; de Kruijff, B. *Biophys. J.* **2003,** 84, 3147-3154.
8. Vereyken, I.; Chupin, V.; Islamov, A.; Kuklin, A.; Hincha, D.; de Kruijff, B. *Biophys. J.* **2003,** 85, 3058-3065.
9. Mineo, H.; Hara, H.; Kikuchi, H.; Sakurai, H.; Tomita, F. *J. Nutr.* **2001,** 131, 3243-3246.
10. Saito, K.; Tomita, F. *Biosci. Biotechnol. Biochem.*, **2000,** 64, 1321-1327.
11. Saito, K.; Kondo, K.; Kojima, I.; Yokota, A.; Tomita, F. *Appl. Environ. Microbiol.*, **2000,** 66, 2252-256.
12. Calazans, G.; Lopes, C.; Lima, R.; de Franca, F. *Biotechnol. Lett.*, **1997,** 19, 19-21.
13. Ishihara, K. U. S. Patent 5,527,784, **1996.**
14. Rolant, F.; Herscovici, B.; Wolman, M. *Biochem. Exp. Biol.*, **1977,** 13, 187-191.
15. Iizuka, M.; Minamiura, N.; Ogura, T. In *Glycoenzymes.* Ohnishi, M., Ed.; Japan Scientific Societies Press: Tokyo, Japan, **2000;** pp 241-258.
16. Yamamoto, Y.; Takahashi, Y.; Kawano, M.; Iizuka, M.; Matsumoto, T.; Sacki, S.; Yamaguchi, H. *J. Nutr. Biochem.* **1999,** 10, 13-18.
17. Leibovici, J.; Stark, Y.; Eldar, T.; Brudner, G.; Wolman, M. *Recent Results Cancer Res.* **1980,** 75, 173-179.
18. Yoo, S-H.; Yoon, E.; Cha, J.; Lee, H. *Int. J. Biolog. Macromol.* **2004,** 34, 37-41.
19. Leibovici, J.; Stark, Y.; Wolman, M. *J. Exp. Pathol.* **1983,** 64, 239-244.
20. Rowland, I. In *Fructan 2004;* Arrieta, J., Ed.; Elfos Scientiae: Havana, Cuba, **2004;** p 120.
21. Leibovici, J.; Bleiberg, I.; Wolman, M. *Proc. Soc. Exp. Biol. Med.* **1975,** 149, 348-350.
22. Papadimitriou, J.; Robertson, T.; Wolman, M.; Walters, M. *Pathology.* **1978,** 10, 235-241.
23. Sedgwick, A.; Rutman, A.; Sin, Y.; Mackay, A.; Willoughby, D. *Br. J. Exp. Pathol.* **1984,** 65, 215-222.
24. Berman, Z.; Leibovici, J.; Wolman. *Isr. J. Med. Sci.* **1976,** 12, 1294-1297.
25. Ritsema, T.; Smeekens, S. *Curr. Opin.* **2003,** 6, 223-230.
26. Kang, S.; Park, S.; Lee, J.; T. *J. Korean Soc. Food Sci. Nutr.* **2000,** 29, 35-40.
27. Garegg, P.; Roberts, E. Patent WO9803184, **1998.**

Chapter 14

Iron–Polysaccharide Composites for Pharmaceutical Applications

Kirill I. Shingel and Robert H. Marchessault[*]

Chemistry Department, McGill University, 3420 University Street, Montreal, Québec H3A 2A7, Canada

Polysaccharide-stabilized iron complexes are an important class of biomaterials with numerous applications, including iron supplement for treating anemia, contrast agents for magnetic resonance imaging, and magnetic separation of cells and protein. An original procedure for the "*in situ*" synthesis of ferrites is proposed and applied successfully to a series of carbohydrate matrices, differing in chemical structure and molecular organization, i.e. semi-crystalline, cross-linked, composites, etc. The method implies introduction of the iron salt into microvoids of polysaccharide platforms, followed by oxidation of iron hydroxide under alkaline conditions. The particles synthesized via "*in situ*" method are characterized by high magnetic responsiveness and represent an excellent support for covalent attachment of proteins, which make them suitable for the magnetic separation of pharmaceutically important biomolecules.

271

Biological metallocomplexes play a key role in the vitally important processes of respiration, photosynthesis, energy transfer, molecular recognition, and even magnetotacticity of cells. Synthesis of the man-made composites that would mimic natural metallocomplexes both structurally and functionally was an intriguing challenge prompting researchers to study the possibility of combining minerals with organic matter, via the so-called "mineralization" procedure. It was rapidly recognized that the physicochemical characteristics of the organic shell forming an interface between a liquid biological milieu and insoluble metal particles is an important parameter determining biocompatibility of the complex and bioavailability of the incorporated mineral. Therefore, the choice of suitable organic molecules for most *in vivo* applications is mostly limited to natural compounds. Polysaccharides were the first natural compounds tested as a carrier for iron. As will be shown below, it was not historically accidental.

In this contribution we describe general approaches used for the formation of the iron-polysaccharide particles using various carbohydrate structures. Due to unique physicochemical properties these composites find numerous applications in the pharmaceutical field, including iron supplement for treating anemia, contrast agents for magnetic resonance imaging (MRI), and magnetic separation of cells and protein. Polysaccharide matrices appeared to be particularly desirable for *in vitro* and *in vivo* use thanks to the relatively high biocompatibility and inability to interact non-specifically with proteins.

Colloid Iron-Dextran Complexes

The first synthesis of iron-containing biomimetic compounds was the preparation of the iron-dextran complexes by London and Twigg in the early 1950's (*1*), soon after clinical approval of dextran for intravenous injections. Approval of dextran solutions for clinical use was itself an important milestone heralding a new era of therapeutic agents based on natural polymers. A vast body of experimental evidence accumulated in the late 1950's confirmed the high therapeutic efficiency of the dextran preparation, rendering this product a standard formulation of the United States and European pharmacopeias. This breakthrough in utilization of the bacterial polysaccharide as a biomaterial has spurred investigations into the chemical transformation of native dextran, leading to the development of the therapeutically active iron-dextran colloids (*1*).

The procedure patented in 1956 included thermal treatment of dextran in alkali solution, followed by neutralization in the presence of ferric chloride. In 1958, applicability of the method was extended by using ionic derivatives of dextran (carboxymethyl dextran) in the formation of the iron-containing composites (*2*). A synthetic iron-dextran complex appeared to possess properties similar to those of ferritin including the ability to preserve iron in a soluble, nontoxic form *in vivo* (*3*).

Perhaps surprisingly from the viewpoint of the today's regulatory concerns, the therapeutic value of iron-dextran complexes for treating anemia was recognized well before the exact molecular structure was elucidated. The first attempt to describe molecular structures of the iron-dextran complexes was performed by Cox et al. in 1972 (*4*). The study hypothesized a core-shell structure, where the iron-rich core is shielded by the dextran layer. This structure was criticized in subsequent studies (*5*), but only recently was finally refined by London (*6*), the author of the invention, almost 40 years after innovation.

Early investigations of ferritin and iron-dextran composites by means of extended X-ray absorption technique revealed that the molecular organization of iron-dextran is very similar to that found for ferritin (*7*). This similarity was not commonly accepted. Towe demonstrated that the ferritin core is formed by the crystalline ferrihydrite ($5Fe_2O_3 \times 5H_2O$), whereas alkali treatment of the iron salt in the iron-dextran complexes generates predominantly akaganeite β-FeOOH (*8*). This difference, however, could not be distinguished by means of Mössbauer spectroscopy, since both ferritin and iron-dextran structures contain paramagnetic and magnetic components related to the hydrous ferric oxide particles (*9*).

Magnetic properties of the iron-dextran particles attracted considerable practical interest. Colloidal iron-dextran complexes have been successfully used as a contrast agent for MRI (*10-12*) and tested as a reagent for labeling and magnetic separation of cells (*13*).

"*In situ*" Synthesis of Superparamagnetic Polysaccharide Particles

In the meantime, research progress in understanding of magnetic materials for the information technologies was ongoing. Preparation of magnetic paper for the secure print purposes also came into the focus of researchers, and several types of magnetic cellulosics were synthesized and described (*14*).

It was shown that cellulose is an excellent support for the "*in situ*" synthesis of ferrites. According to the procedure, carbohydrate-stabilized $Fe(OH)_2$ is converted into magnetic species by means of air (*15*) or hydrogen peroxide (*16, 17*) oxidation. The method of hydrogen peroxide oxidation was then applied for the synthesis of superparamagnetic particles in cross-linked starch material (*18*). Magnetic polysaccharide materials prepared by "*in situ*" synthesis were reported to be superparamagnetic: they were easily magnetized in a magnetic field but did not retain magnetization once removed from the field.

Superparamagnetic polysaccharide particles (SPMP) appeared to be particularly interesting for the magnetic separation of pharmaceutically important proteins (*19*). Soluble structures, in contrast, seems not to be suitable for magnetic separation of biomolecules, since the final product of purification could be contaminated by the dextran chains released from the relatively labile dextran shell.

"In situ" synthesis of ferrites

Our approach to the synthesis of superparamagnetic polysaccharide support includes *"in situ"* synthesis of ferrites in the microvoids of polysaccharide particles, either cross-linked or crystalline, followed by functionalization of the support. Table I summarizes polysaccharide structures that were found chemically acceptable as iron carriers. Functional groups are either naturally occurring or created by chemical transformations.

Table I. Polysaccharide Structures Used for *"In situ"* Synthesis of Superparamagnetic particles (*19*)

Name	Description	Functional groups
Cross-linked starch	Cross-linked high amylose starch particles ((~ 80 μm)	Hydroxyl (-OH) Carboxyl (-COOH) Aldehyde (-CHO)
Sephadex®	Cross-linked dextran beads with particle size 20-40 μm	Hydroxyl (-OH) Carboxyl (-COOH) Aldehyde (-CHO)
Cellulose fibers (CLD)	Cellulose fibers of wood pulp	Carboxymethyl ($-CH_2COOH$) Hydroxyl (-OH)
Avicel®	Microcrystalline cellulose with particle size 90-180 μm	Hydroxyl (-OH) Carboxyl (-COOH) Aldehyde (-CHO)
Chitin crystallites	Rod-like crystallites (~ 0.1-0.2 μm long) of chitin obtained via acid hydrolysis	Amino(acetyl) Hydroxyl (-OH) Amino ($-NH_2$)

Initially, the ferrite synthesis comprised incubation of the water-insoluble polysaccharides in the solution of Fe^{2+} salt, followed by isolation and purification of the iron-containing carbohydrate matrix. In the next steps, iron salt is precipitated in an alkaline solution, and subsequently oxidized into a magnetic form (*16, 18*) in the presence of air or hydrogen peroxide.

When hydrogen peroxide was used as an oxidant, several precipitation-oxidation cycles were needed in order to obtain high magnetic response of the material, and each cycle required addition of fresh Fe^{2+} solution. Magnetic responsiveness of the particles was estimated by means of extraction sample magnetometry (Fig. 1). In these experiments, magnetic material was vibrated in a magnetic field with varied intensity. The data of magnetization as a function of applied field were recalculated into the values of iron content using the standard value of saturation magnetization for maghemite (74 J/T/kg).

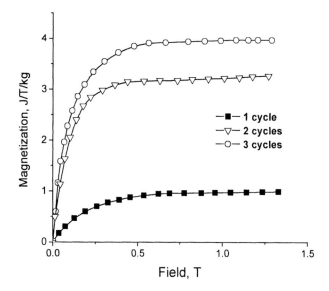

Figure 1. Magnetization curves of the starch SPMP obtained after several cycles of hydrogen peroxide oxidation. The saturation magnetization increases with increasing number of cycles.

However, comparative analysis of iron content determination by means of absorption spectroscopy (AAS) and magnetite-sensitive extraction sample magnetometry (ESM) revealed that a large proportion of iron was not magnetically responsive (Table II).

In addition to the VSM and AAS techniques, FTIR-spectroscopy was used to characterize molecular organization in the iron-polysaccharide systems. Spectral manifestations of molecular interactions in the iron-starch composites was monitored by an increase of the band at 3400 cm^{-1} due to stretching vibrations of the OH-bond and appearance of a few new bands at about 700 cm^{-1} and 430 cm^{-1} from Fe-O-Fe vibrations.

Table II. Characteristics of Superparamagnetic Particles Synthesized in The Cross-linked Starch via Peroxide Oxidation

Sample	Starting material (g)	Product (g)	Yield (%)	Oxidation cycles (No.)	Iron content (%)	
					ESM	AAS
Starch SPMP-1	1.00	0.32	32.0	3	36.1	33.1
Starch SPMP-2	3.50	1.62	46.3	3	10.3	28.8
Starch SPMP-3	2.00	0.43	21.5	3	13.8	23.3

The extra-absorbance of the peroxide-oxidized SPMP at 3400 cm^{-1} is conditioned by the combined contributions from carbohydrate hydroxyl groups and the OH-moieties of hydrated forms of iron oxides FeOOH (20). The shift of this band towards smaller wavenumbers is interpreted as being induced by strong interactions between iron ions and hydroxyl groups of starch due to hydrogen bonds (Fig. 2). Stabilization of the iron hydroxide by hydrogen bonds seems to be typical for iron-carbohydrate structures, since similar results were reported for polynuclear dextran and pullulan iron colloids (6, 21). A proposed structure of the FeOOH-starch composites synthesized via "*in situ*" method is shown in the Fig. 2.

Structural similarities between soluble iron-carbohydrate structures and "*in situ*" synthesized particles make it possible to assume an equivalent bioavailability of iron oxide in these formulations. In preparing pharmaceutical forms iron-loaded starch particles can be compressed into the tablets and administered as an excipient for oral delivery of exogenous iron.

An effect of cross-linking of polysaccharide shell on the stability and bioavailability of iron-containing composites was studied by Palmacci et al. (22). The results showed that cross-linking prevents heat-induced dissociation of the polysaccharide from the surface of iron oxide, thus extending plasma half-life of the complexes. Orally administered iron-starch particles are inevitably subjected to degradation under the action of amylolytic enzymes in the gastrointestinal tract, which may provide sustained delivery of iron *in vivo*.

Polysaccharide magnetic carriers

From the perspective of magnetic carrier preparation, the method of peroxide oxidation of iron hydroxide was found to be technically unacceptable. The procedure generates superparamagnetic particles with the yield of 20 – 45 % from the initial mass of starch (Table II). It appeared that the loss of material upon oxidation occurs due to degradation of the polysaccharide matrix by free radicals produced from hydrogen peroxide in the presence of iron ions (Fenton reaction).

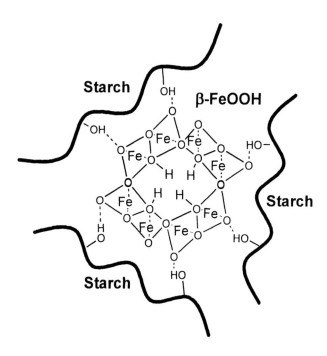

Figure 2. Proposed structure of the FeOOH-starch complexes. The structure is based on the results of FTIR-spectroscopy and molecular formula determination reported by Buchwald and Post for akaganeite (23).

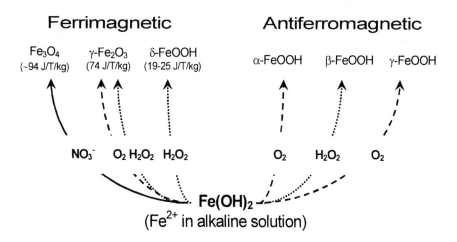

Figure 3. Chemical transformations of iron hydroxide Fe(OH)₂ under the action of oxidants

Therefore, in order to avoid contamination of the ferrites with non-magnetic material and improve the yield of the final product, different approaches to the conversion of iron hydroxide into magnetically active form were considered (Fig. 3).

As can be seen from the scheme in Fig. 3, neither hydrogen peroxide nor air oxidation converts $Fe(OH)_2$ selectively into magnetic form and, apart from γ-Fe_2O_3, produces non-magnetic material such as α-FeOOH and/or β-FeOOH. When present in the polysaccharide matrix, the latter species increase the total content of iron, but do not contribute to the magnetic responsiveness of the SPMP. This is the reason why the results of iron determination by means of VSM do not correlate with AAS data (Table II).

In contrast to hydrogen peroxide oxidation, the treatment of iron hydroxide with NO_3 ions was reported to produces only Fe_3O_4 (22), the desired magnetic material with maximal magnetization (Fig. 3). This reaction was used to prepare a series of SPMPs, physicochemical characteristics of which are summarized in Table III.

Table III. Characteristics of Superparamagnetic Particles Synthesized via Nitrate Oxidation of Iron Ions in Alkaline Conditions

Sample	Starting material (g)	Product (g)	Yield (%)	Oxidation cycles (No.)	Iron content (%)	
					ESM	AAS
Starch SPMP-1	5.00	3.91	78.2	1	6.80	6.20
Starch SPMP-2	5.00	4.86	97.2	1	10.1	11.4
Sephadex SPMP	2.00	1.86	93.0	1	14.5	9.02
Avicel SPMP-1	2.00	1.96	98.0	1	8.00	6.08
Avicel SPMP-2	5.00	5.30	106.0	1	n/d	5.70
Cellulose SPMP	0.75	0.75	100.0	1	31.0	24.5
Chitin SPMP	3.00	3.10	103.0	1	n/d	18.1

NOTE: n/d – values are not determined

As can be seen from the comparatively high yields of the SPMP preparation, the procedure of nitrate-mediated oxidation helps to avoid degradation of the polysaccharide shell, rendering resultant particles magnetically responsive in a single precipitation-oxidation cycle. The results of AAS are in good agreement with the data derived from ESM measurements, thus confirming that iron impregnated into polysaccharide matrix is entirely in the form of magnetite Fe_3O_4.

Typical appearance of the SPMP upon transmission electron microscopy (TEM) examination is shown in Fig. 4. All samples were found to have very similar morphology, albeit some differences in particle size were observed. Iron oxide crystals appeared in TEM as black dots occupying either the whole or partial surface of the polysaccharide support, depending on the avalailability of microvoids for iron salt penetration.

The mechanism for the formation of magnetite was proposed by Sugimoto et al. (24). According to the model, an oxidation of the alkali-precipitated iron hydroxide gel with nitrate ions generates small magnetite particles with regular crystalline structure. Being dispersed in the iron hydroxide gel, these small particles start to act as nucleation sites for the growing crystals, leading to formation of large aggregates of Fe_3O_4 (24).

The space constraints imposed by cross-linked or crystalline polysaccharide structures preclude aggregation of iron crystals and control the size of the "in situ" produced ferrites. The inhibition of the ferrite crystals growth has been recently achieved in the iron-surfactant systems (25). Organic layers of the metal-surfactant complex recovered from aqueous solution were subjected to controlled thermal degradation, which permitted obtention of relatively uniform nanoparticles of maghemite with sizes ranging from 8 nm to 11 nm (25).

As a particle of magnetite is reduced in size it becomes too small to accommodate domains and appears to be a single-domain structure. Further decrease below the single domain value leads to appearance of the superparamagnetic system, i.e. the particle exhibits a relatively large magnetization in the applied field, but no remanence. Owing to this property SPMPs remain in suspension for a long time to ensure efficient binding with target compounds and can be easily resuspended after magnetic migration.

Analysis of the size distribution of ferrite particles was performed using the theoretical approach developed previously for the magnetic cellulose material (17). According to the method, magnetization curves (Fig. 1) are fitted to the discrete distribution of Langevin functions:

$$M = M_s \sum_i a_i \mathrm{L}(\mu_i \mathrm{H}/kT)$$

where M_s corresponds to the global saturation magnetization, a_i is the proportion of particle i having magnetic moment μ_i, and L is the Langevin function. The volume of the particles is then estimated from the magnetic moment distribution, as described in Ref. (17). The size of ferrites calculated by this method varies in the range of 30–150 nm, which is comparable with the values observed by TEM (Fig. 4).

Functionalization of the SPMPs

Magnetic separation of individual proteins from a complex biological liquids relies on the selectivity of interactions between a target compound and a ligand attached to the magnetic carrier. Attachment of the ligand is usually achieved by creating covalent linkages between biomolecule and SPMP that is

Figure 4. Transmission electron micrograph of the SPMP synthesized in the cross-linked starch by using potassium nitrate as an oxidant for iron

initially modified by introduction of a number of reactive groups. In view of the fact that most of the ligands are proteins that can irreversibly lose their native conformation required for affinity interactions, the reactions between ligand and SPMP are carried out under mild "physiological" conditions.

An example of functionalization of the polysaccharide structure is the conversion of surface hydroxyl groups of SPMP into carboxylic groups by means of TEMPO-catalyzed hypochlorite oxidation. The method is known to produce ionic derivatives with a high degree of substitution predominantly at the C-6 position of a glucopyranose ring (26, 27). The degree of carboxylation is easily controlled by the amount of hypochlorite added to the reaction media (Fig. 5). The newly formed carboxylic groups readily react with free amino groups of proteins in the presence of carbodiimide. The protein-binding capacity of the carboxylated SPMP was found to increase with increasing degree of carboxylation (Fig. 6).

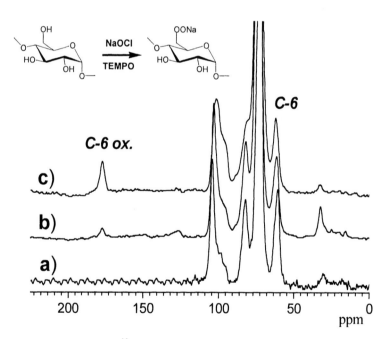

Figure 5. Solid state ^{13}C NMR spectra of cross-linked starch (a) and its carboxylated derivatives with carboxylation degree 10% (b) and 35 % (c).

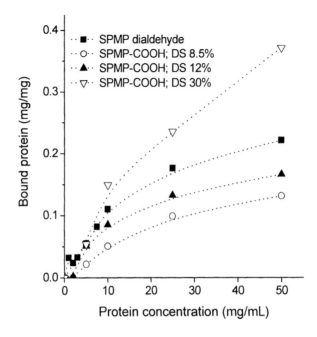

Figure 6. Covalent attachment of bovine serum albumin to the modified starch SPMPs as a function of protein concentration.

Periodate oxidation of the glucopyranose units generates dialdehyde structures that subsequently interact with amino-groups of proteins to give stable amide linkages. For both carboxylated and dialdehyde-modified starch SPMPs the concentration of the linked ligand increases with increasing concentration of the biomolecule in the solution (Fig. 6).

To confirm formation of the covalent linkages between protein molecules and polysaccharide moieties, FTIR-spectra of the conjugates were recorded and analyzed. In the spectrum of carboxylated starch FTIR the bands at 1730 cm^{-1} and 1625 cm^{-1} ascribed to the carboxylic groups were detected. The FTIR-spectra of the starch-protein conjugates showed an appearance of the Amide I band at about 1640 cm^{-1} and the band at 1307 cm^{-1} attributed to the vibrations of the –C–N– bond. Amide II and Amide III bands were also detected, thus confirming formation of covalent linkages between protein molecule and SPMP.

Another approach for covalent attachment of proteins was developed by introducing reactive aldehyde-terminated spacer arms decorating outer surface of the polysaccharide SPMPs. Amino-containing particles such as chitin or chitosan-bearing materials were found to be particularly suitable for this modification. Chitosan-bearing SPMP were synthesized using carboxylated starch SPMP as a starting magnetic support (Fig. 7). In the synthesis, carboxylated SPMPs are first incubated in the chitosan solution at acidic pH, and SPMP-chitosan salt is then precipitated by adjusting the pH to pH 10.0.

Precipitated SPMP-chitosan composites are separated magnetically and washed several times in phosphate buffer at neutral pH. Finally, amide linkages between carboxylic groups and amino-groups of chitosan are produced in the presence of carbodiimide.

The concentration of COOH-groups in the carboxylated starch particles is by far less than the concentration of the amino-groups available in the chitosan structure and therefore, the resultant material possesses cationic properties due to residual amino groups from chitosan (Fig. 7).

The product of the reaction between chitosan and carboxylated starch SPMP was studied by means of FTIR-spectroscopy in the carbohydrate fingerprint region 1200-800 cm^{-1}. (Fig. 8).

In this spectral range, chitosan is characterized by an intense band at 1075 cm^{-1} (Fig. 8a), whereas the starch moiety of the SPMP shows absorbance at 1024 cm^{-1} (Fig. 8b). As it is expected, the spectra of starch SPMP-chitosan composites comprise the bands from both parent polymers, and the relative intensity of these bands appear dependent on the weight ratio of chitosan and SPMP taken for the synthesis (Fig. 8).

In binding with proteins, amino groups at the surface of the SPMP are first reacted with dialdehyde that then condenses with amino-groups of protein ligand in the reaction of reductive amination (Fig. 9).

The conjugates of SPMP and protein molecules are characterized by high stability, which permits them to be used for multiple separations of the target biomolecules exerting affinity towards selected ligand.

Conclusion

The general approaches for the synthesis of the polysaccharide-based iron complexes are discussed in this contribution to demonstrate great versatility of the mineralization procedure. The original procedure for the "*in situ*" synthesis of superparamagnetic particles with numerous applications in the pharmaceutical fields is proposed and applied successfully to the series of carbohydrate matrices, differing in chemical properties and molecular organization, i.e. crystalline, cross-linked, self-assembled etc. The method implies introduction of the iron salt into microvoids of polysaccharide matrixes with subsequent oxidation of iron hydroxide under mild alkaline conditions. Different oxidation techniques were explored for the preparation of the SPMPs with respect to the desired biomedical application. For example, the particles synthesized via nitrate-mediated oxidation of iron ions are characterized by high magnetic responsiveness and represent an excellent support for covalent attachment of proteins, which make them exceptionally suitable for the magnetic separation of pharmaceutically important biomolecules, as described in the first chapter of this book.

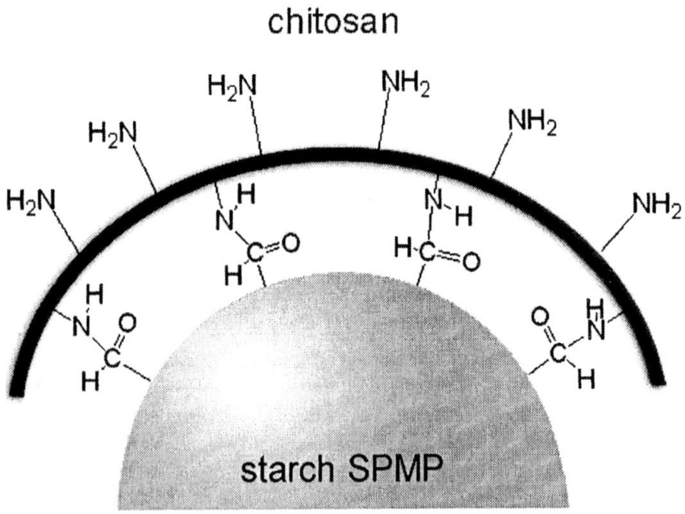

Figure 7. Schematic representation of the amino-containing SPMPs synthesized by grafting chitosan macromolecules onto the surface of starch-based superparamagnetic particles

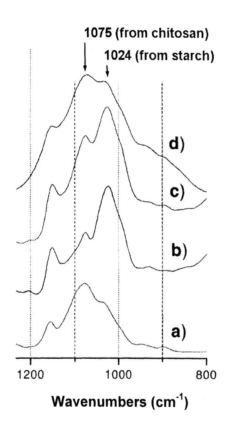

Figure 8. FTIR-spectra of chitosan (a), carboxylated starch SPMP (b) and their conjugates prepared with weight ratio of chitosan to SPMP 0.3:1.0 (c) and 1.0:1.0 (d).

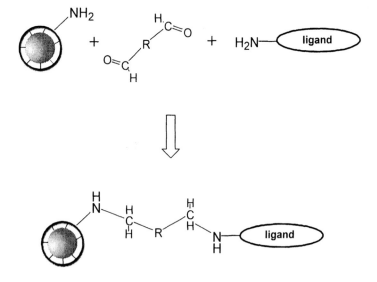

Figure 9. Covalent attachment of protein ligand to the amino-containing polysaccharide SPMP through spacer arm.

References

1. London, E., Twigg, G.D. Brit. Patent 748,024, 1956.
2. Novak, L.J. Brit. Patent 891,769, 1958.
3. Munro, H.N., Linder, M.C. *Physiol. Rev.* **1978**, *58*, 317.
4. Cox, J.S.G., Kennedy, G.R., King, J., Marshall, P.R., Rutherford, D., J. *Pharm. Pharmacol.* **1972**, *24*, 513.
5. Coe, E.M., Bowen, L.H., Wang, Z., Sayers, D.E., Bareman, R.D. *J. Inorg. Biochem.* **1995**, *58*, 269.
6. London, E. *J. Pharm. Sci.* **2004**, *93*, 1838.
7. Theil, E.C., Sayers, D.E., Brown, M.A. *J. Biol. Chem.* **1979**, *254*, 8132.
8. Towe, K. *J. Biol. Chem.* **1981**, *256*, 9377.
9. Oshtrakh, M.I. *Faraday Discuss.* **2004**, *126*, 119.
10. White, D.L., Aicher, K.P., Tzika, A.A., Kucharczyk, J., Engelstad B.L., Moseley, M.E. *Magn. Reson. Med.* **1992**, *24*, 14.
11. Magin, R.L., Bacic, G., Niesman, M.R., Alameda, J.C.Jr, Wright, S.M., Swartz, H.M. *Magn. Reson. Med.* **1992**, *24*, 14.
12. Palmacci, S., Josephson, L. US Patent 5,262,176, 1993.
13. Molday, R.S., MacKenzie, D. *J. Immunol. Methods* **1982**, *52*, 353.
14. Marchessault, R.H., Rioux, P., Raymond, L. Polymer **1991**, *33*, 4025.
15. Sourty, E., Ryan, D.H., Marchessault, R.H. *Chem. Mater.* **1998**, *10*, 1755.
16. Raymond, L., Revol, J.-F., Marchessault, R.H. *Polymer* **1995**, *36*, 5035.
17. Sourty, E., Ryan, D.H., Marchessault, R.H. *Cellulose* **1998**, *5*, 5.
18. Veiga, V., Ryan, D.H., Sourty, E., Llanes, F., Marchessault, R.H. *Carbohydr. Polym.* **2000**, *42*, 353.
19. Marchessault, R.H., Shingel, K.I., Vinson, R.K., Coquoz, D.G. US Patent Application 2004/0146855, 2003.
20. Ruan, H.D., Frost, J.T., Kloprogge, J.T. *Spectrochim. Acta* **2001**, *57*, 2575.
21. Ilic, L., Ristic, S., Cakic, M., Goran, N., Stankovic, S. World Patent WO 02/46241, 2002.
22. Palmacci, S., Josephson, L. US Patent 5,262,176, 1993.
23. Buchwald, V.F., Post, J.E., *Am. Mineral.* **1991**, *76*, 272.
24. Sugimoto, T., Matijevic, E. *J. Colloid Inter. Sci.* **1980**, *74*, 227.
25. Park, J., An, K., Hwang, Y., Park, J.-G., Noh, H.-J., Kim, J.-Y., Park, J.-H., Hwang, N.-M., Hyeon, T. *Nature Mater.* **2004**, *3*, 891.
26. Thaburet, J.F., Merbouh, N., Ibert, M., Marsais, F., Quenguiner, G. *Carbohydr. Res.* **2001**, *330*, 21.
27. Bragd, P.L., Besemer, A.C., van Bekkum, H. *Carbohydr. Res.* **2000**, *328*, 355.

Chapter 15

Polysaccharide–Drug Conjugates as Controlled Drug Delivery Systems

M. Nichifor and G. Mocanu

"Petru Poni" Institute of Macromolecular Chemistry, Aleea Gr. Ghica Voda nr. 41A, 700 487 Iasi, Romania

This contribution deals with the role of the polymeric carrier and the chemical bond between the drug and the carrier in the design of polymer-drug conjugates as efficient controlled drug delivery systems. Polysaccharide-drug conjugates prepared by different methods, together with their *in vitro* and *in vivo* behavior, are presented according to the reported results including author's work in the field.

Introduction

All controlled drug delivery systems aim to improve the therapeutic drug efficacy and reduce side effects by supplying the right drug concentration when necessary (control of time) and at the right place (control of distribution). The control of the release time is realized either by keeping a constant efficient drug concentration in circulation for long period of time (sustained release), or by providing the release under some physiological conditions (pulsatile or modulated release). Control of distribution can be realized by selective targeting of the drug to the ill tissues or organs (*1 - 3*).

The concept of the polymeric prodrugs or polymer-drug conjugates was first introduced by Ringsdorf in 1975 (*4*) to describe the systems where the therapeutic agent is chemically bound to a polymer. The system is inactive as such but can release the active drug after chemical or enzymatic hydrolysis of the chemical bond between polymer and drug (*5*). Ringsdorf have also highlighted the opportunities offered by these systems, which allow the attachment of the drug, targeting ligand and modulating ligand to the same polymeric carrier.

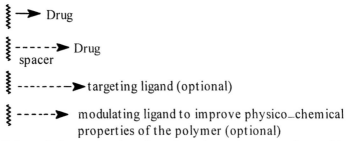

Scheme 1. Representation of a polymeric prodrug according to Ringsdorf

During the years, polymeric conjugates have received an increasing interest, due to their very exciting pharmaceutical and medical applications. The numerous advantages of polymer carriers were exploited. The binding of a drug to a polymer can result in an improvement of its physico-chemical (solubility, stability to metabolic degradation) and biological (bioavailability, toxicity) properties. A site-specific delivery of the conjugate can be achieved passively either at a tumor site, due to polymer enhanced permeation and retention (EPR effect) (6 - 9), or on the surface of the liver and kidney, by introducing cationic charges (10, 11, 12). Further improved selectivity is possible by active targeting, through attached targeting moieties which can be recognized by specific receptors found at the target site (13 - 16). The design of the chemical bond as a specific substrate to the enzymes specific to the target site can result in site-specific drug release (17, 18). The immunoprotective ability of polymeric prodrugs was also reported (19).

Chemical developments in polymer-drug conjugates and their therapeutic applications have been recently reviewed (6, 14, 20). Several polymer-drug conjugates based on poly(styrene-co-maleic anhydride), poly(N-hydroxypropyl methacylamide) and poly(ethylene glycol) are in different stages of clinical trials (6, 21 - 24).

In the following we will discuss two of the decisive aspects in the design of an efficient polymer-drug conjugate: the nature of the polymer and the type of chemical bond between the polymer and the drug.

Polymer Carrier

The main requirements for a polymer to be appropriate as drug carrier are the presence of a large number of reactive groups available for drug fixation, non-toxicity, biocompatibility, biodegradability or at least a low accumulation in the body (after its task as carrier has been fulfilled), lack of immunogenicity. In the case of systemic administration, the water-soluble (or dispersible) polymer should have the adequate molar mass, effective charge and hydrophobicity. These properties influence the conjugate body distribution, its uptake by the reticulo-endothelial system or urinary excretion (*17, 25*). Polysaccharides are often used in the synthesis of conjugates with drugs, proteins or enzymes, via the OH groups of the polysaccharides. Polymeric prodrugs based on dextran, carboxymethyl dextran, starch, cellulose, carboxymethyl cellulose, alginates, pullulan, chitosan, hyaluronic acid, and carragenan have been designed for systemic or oral administration (*15, 26, 27*). The chemical structure, M_w, electric charge, functional groups, polydispersity and branching of polysaccharides influence their *in vivo* behavior, such as the body distribution and elimination. Most of these polysaccharides have a long plasma half-life and an increased EPR effect, leading to the accumulation of conjugates in tumor tissues. Thus, studies on mitomycin-dextran conjugate have shown that the macromolecular prodrug presents higher tumor localization compared with normal tissue (*28*). On the other hand, a longer circulation time is realized by using dextran bearing appropriate anionic charges (carboxymethyl), because of lower hepatic uptake or urinary excretion. After *i.v.* administration in rats, camptothecin-carboxymethyl dextran conjugates had a longer plasma life-time, an enhanced accumulation in tumor tissues, where a sustained release of the drug occurred (*29*).

Polysaccharides like dextran are slowly hydrolyzed by specific enzymes, mainly in the liver (*27*). In the case of oral administration, polysaccharides are degraded only in colon, by specific enzymes (glucosidase, galactosidase, amylase, pectinase, xylanase, dextranase) (*30*). Water-insoluble crosslinked dextran can be also used for oral administration, and its degradation in the presence of α-1,6- glucosidases has been reported (*31*).

Chemical Bond Drug-Polymer

The nature and the stability of polymer-drug bond control the rate of the drug release, therefore the effectiveness of conjugate biological activity (*32*).

Polysaccharides have a high number of reactive groups (OH, NH_2, COOH), which can be involved in polymer analogous reactions with various reagents. The drugs have been bound mainly via polysaccharide OH groups, therefore the reactions involving these groups will be taken under consideration. The coupling procedures are limited by the types of drug reactive groups and should proceed under mild conditions, without alteration of the chemical structure and biological activity of the drugs (*33*). The coupling can take place by a straight reaction between polysaccharide and drug (direct binding), or after an appropriate activation of the functional groups involved in the binding (Scheme 2).

Direct Binding

The procedure can be applied to carboxyl group containing drugs, with formation of an ester group. It can be performed with acid chlorides of drugs

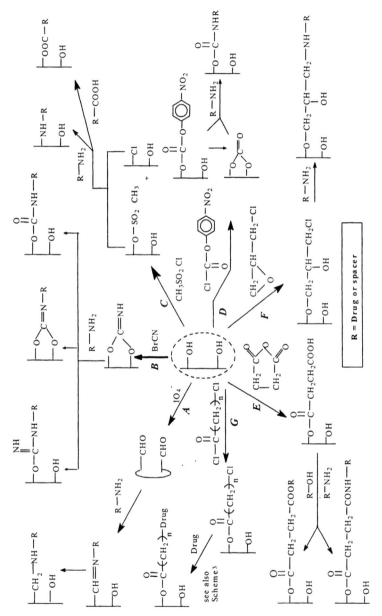

Scheme 2. Activation methods of OH groups of polysaccharides without or with introduction of spacers

in the presence of HCl acceptors, or by using carbonyl imidazol or carbodiimides as coupling reagents. Many studies have been devoted to the synthesis of polymer-drug conjugates by this procedure, and to their *in vitro* and *in vivo* characterization. So, nicotinoyl chloride have been reacted with dextran (*34*), cellulose (*35*) and chitin (*36*) to obtain prodrugs with potential antilipolytic activity. Acetylsalicylic (*37*) and suprofen (*38*) derivatives of the dextran showed lower acute toxicity than the parent drugs, but comparable good anti-inflammatory activity *in vivo*. Polysaccharide prodrugs of 5-aminosalicylic acid obtained using its imidazolyl derivative as intermediate can provide a sustained and specific release in colon (*39*). Binding of many anti-inflammatory drugs like naproxen, ketoprofen, fenoprofen, diclofenac, ibuprofen, or indomethacin to dextran was also realized in the presence of carbodiimide (*40, 41 - 43*). Generally, the ester bonds in these conjugates undergo a slow chemical hydrolysis, under mild conditions (*32*). As a result of the sterical hindrance, the carrier protects these bonds towards both chemical and enzymatic degradation (*25*). For instance, dextran-nalidixic acid conjugates with various drug contents do not release the drug neither at pH 1,2 or 6.8, corresponding to those of the stomach and small intestine, nor in tissue homogenates and small intestine contents. However, they release the free drug in the presence of cecal contents (*44*). The finding suggests that the polysaccharide was hydrolyzed by colonic enzymes into smaller oligozaharides, resulting in a higher accessibility of the ester bond to esterases (*45*). This behavior prompted to the design of polysaccharide conjugates as colon- specific drug delivery systems for naproxen (*46, 47*), ketoprofen (*48*) glucocorticoids (*30, 49, 50*), 5-amino salicylic acid (*51*). The colon-specific targeting mediated by degradable polysaccharides may be a more efficient treatment of many colon diseases (Crohn's disease, ulcerative colitis, amebiasis, carcinomas and infections) due to reduced influence of systemic side effects, lower drug doses of drugs, supply of the drug to the biophase only when is required and maintenance of the drug in the intact form till it reaches the target site (*52*).

Functional Group Activation

In order to increase the coupling yield under non-harmful conditions, activation of polysaccharide hydroxyl groups has to be performed. The new bonds type between polysaccharide and drug are formed, for instance carbonate, carbamate or amide bonds. These bonds are more hydrolytically stable than the ester ones, but they may be cleaved under certain *in vivo* conditions (*32*).

Periodat oxidation (*A*) of polysaccharides is a controllable, fast and quantitative method amenable to large-scale application (*53, 54*). The resulting acyclic dialdehyde groups react with amino groups, to form imino-conjugates

(Schiff bases, relatively unstable compound) or amino-conjugates (in the presence of reducing agents). In the latter case hydrolytically stable C-N bonds are formed, however many studies concerning the *in vivo* behavior of these conjugates have been reported. A nystatin-dextran conjugate showed a minimum inhibitory concentration value against *Candida Albicans* similar to that of the free nystatin, and was 25 times less toxic than it after a single injection given to mice (*55*). Periodate oxidized Gum Arabic gave a crosslinked conjugate with primaquine, due to the presence of two amino groups in the drug. The conjugate released the drug with a slow rate (*56*). The arabinogalactan-amphotericin B conjugates, obtained both as Schiff bases and as an amino derivatives (*57 - 60*), were 60 times less hemolytic for sheep erythrocytes than the parent drug, at the same inhibitory concentration (*57*) and can be used in treatment of invasive fungal infections (*58, 59*), due to its accumulation in the lungs (*60*). Conjugates of the polysaccharides with daunomycin (*61*) and procainamide (*53*) were also prepared. Dounomycin conjugate showed higher plasma concentration, higher anticancer activity and lower acute toxicity than parent drugs. The method has been also used to bind proteins or enzymes, including chimotripsin (*62*) and insulin (*63*), to polysaccharides, without a significant loss in the pharmacological activity of the proteins (*64*).

Cyanogen halide activation (*B*) gives high coupling yields, but the high dilution required to avoid crosslinks (*54*) does not recommend this method for large-scale applications. It has been used for covalent attachment of amino groups containing drugs to polysaccharides, despite of the many side products formed. The resulting polysaccharide-drug conjugates have been evaluated for their biological activity. Thus, after intravenous administration to mice, the dextran-glutathione conjugate was taken up by hepatic cells and intracellularly hydrolyzed to free gluthathione, which provided protection against hepatotoxicity induced by acetamidophen (*65*). Conjugate of dextran sulfate with 3'-azido-2 – didesoxythymidine (AZT) was synthesized in the expectation for a synergistic effect of dextran sulfate and AZT towards AIDS (*66*).

Sulfonyl chloride activation (*C*) results in the formation of sulfonyl esters able to react further with carboxylic or amino group containing drugs. Tosyl and mesyl chloride activated polysaccharides have been used for drug binding *in situ* (*67, 68*) or after the purification of the intermediate products (*69*). A more careful examination of the activation products obtained by reaction of pullulan with mesyl chloride showed us the formation of both chlorodesoxy- and mesyl substituted polysaccharide. The preponderance of one of these forms depends strongly on activation conditions (*70*). Several conjugates have been prepared by this activation method. Conjugates of pullulan with nicotinic acid or sulphatiazole released slowly the active compounds at physiological pH (*71*). This behavior can be attributed both to the hydrophobicity of the chlorodesoxy-substituted polysaccharide and to the close position of the cleavable bond towards the polymeric chain. Arabinogalactan-amphotericin-B conjugates with up to 20 % w/w drug had similar inhibitory concentration values against

Candida Albicans, but were less toxic than the free drug injected *i.v.* to mice (*69*).

Phosgene activation of the polysaccharides results in the formation of chlorocarbonate groups, which can react with hydroxyl groups (forming carbonate links) or amino groups (forming carbamate links). Application of this method is very limited due to the side reactions which can occur, but synthesis of carbonate esters of testosterone with hydroxypropyl cellulose (*72*) and carbamate ester of dextran with insulin (*73*) have been reported.

Chloroformate activation (*D*) of dextran with p-nitrophenyl chloroformate leads to aromatic carbonate ester and less reactive carbonate (*74*). Intramolecular carbonate formation is predominant, but intermolecular reaction can also occur. This activation method can be used to link amino-containing compounds (peptide spacers, peptide-drug conjugates) to polysaccharides via stable carbamate bonds (*53*). Non-reacted carbonate groups can lead to cross-linking.

Activation with Introduction of Spacers

The procedure allows to introduce both a new functional group, more reactive than OH group, and a spacer group, which reduces the sterical hindrance around the bond connecting the drug to polymer. This method makes use of bifunctional compounds which can react first with the polymer functional groups, then the drug is coupled to the new created functional group as such or after an appropriate activation. In another approach, the bifunctional compound reacts first with the drug, then the resulted drug derivative is linked to an appropriate activated polysaccharide. The activation methods are the same presented above.

Succinic or glutaric anhydride (*E*) have been used to activate either the polysaccharide chain, or to activate the small molecular drug containing OH groups, prior to coupling to the polymer. The method has been applied to the activation of dextran (*75, 76*), inulin (*75*), pullulan (*77*) or chitosan (*78*) followed in some cases by the coupling of amino or hydroxyl group containing compound to activated polysaccharide. The conjugates of dextran with glucocorticoids like prednisolone, methylprednisolone (*79*) or dexamethazone (*30, 80*) were prepared by this methods, and their potential as colon specific delivery systems was studied. Water soluble and water insoluble succinoyl chitosan-mitomycin C conjugates showed pH dependent *in vitro* release characteristics and good antitumor activity against various tumors in mice. They displayed a long systemic retention, low toxicity and specific accumulation in the tumor tissues (*78*).

Epichlorohydrine (*F*) reacts with polysaccharides in the presence of bases or $Zn(BF_4)_2$ as catalysts, with the formation of 3-chloro-2-hydroxypropyl derivatives. The method was used to prepare conjugates of inulin with procainamide (*53*) and those of dextran with 6-purine thiol or 5- fluorouracil derivatives (*81*).

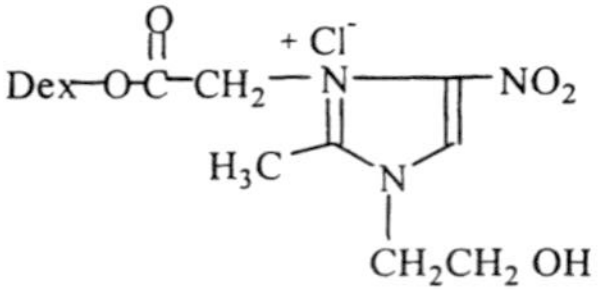

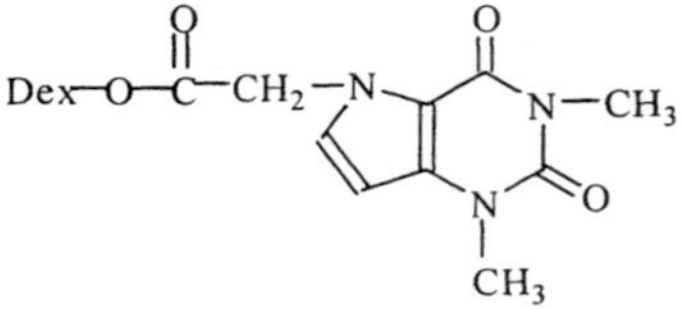

theophylline conjugate **3**

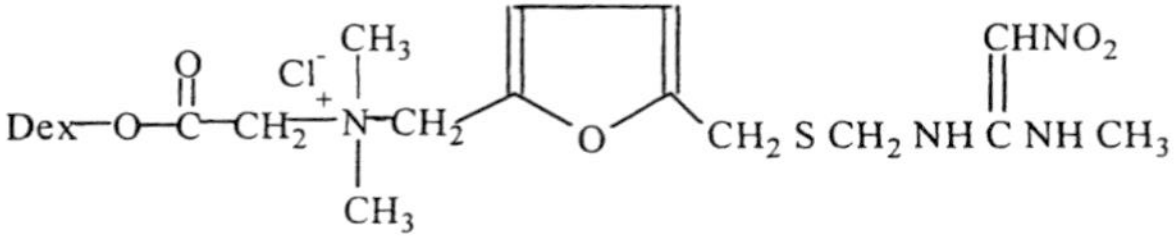

oxacillin conjugate **5**

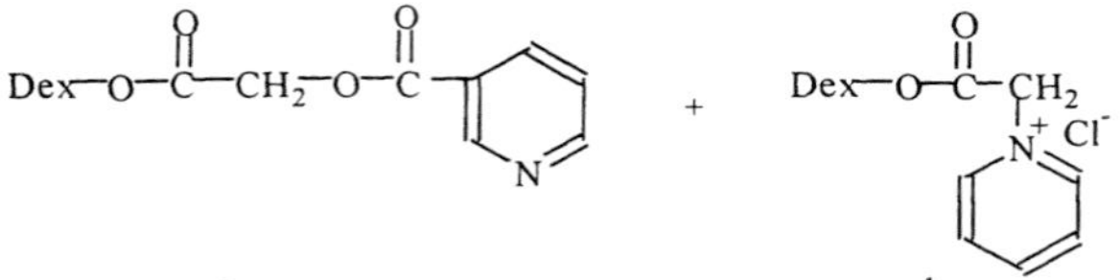

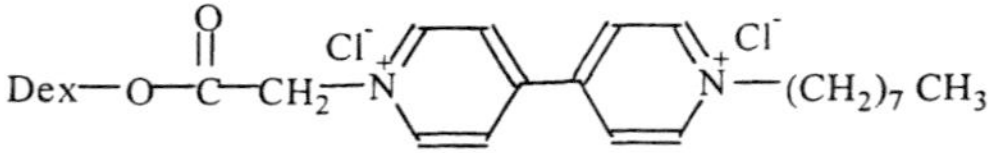

dipyridyl compound conjugate **7**

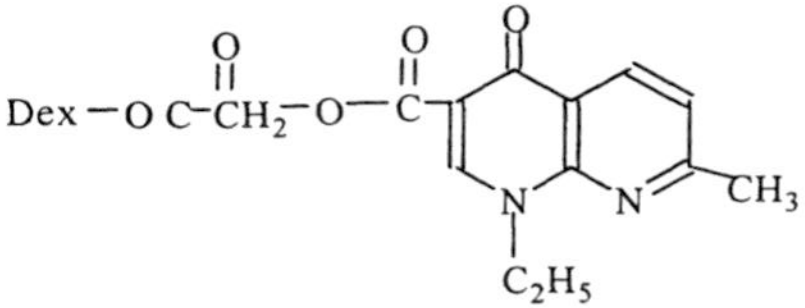

nalidixic acid conjugate **2**

ranitidine conjugate **4**

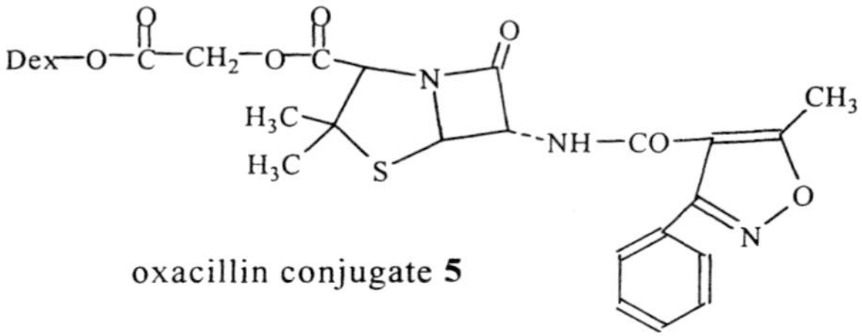

nicotinic acid conjugate **6**

metronidazole conjugate **1**

Scheme 3. *Chemical structures of the conjugates obtained from chloroacetylated dextran*

Chloroacyl chloride (*G*) modification of polysaccharides leads to chloroacyl derivatives of dextran (*82, 83*) and pullulan (*84*) able to react further with carboxyl and amino group containing compounds. For drugs containing both COOH and N –heterocylces (nicotinic acid, nalidixic acid or oxacillin), the binding takes place preferentially to more nucleophilic COOH groups (*85, 86*). According to the studies on the binding of bioactive carboxylic acids, the esterification reaction proceeds better with COOK than with COOH (*83*).

Chloroacetylated crosslinked polysaccharide microparticles have been used as supports for synthesis of conjugates with metronidazole (*87*), theophylline (*88*), nicotinic acid, nicotinamide (85) nalidixic acid, oxacillin (86), or ranitidine (*89*) (Scheme 3). Generally, the release of the drugs from their conjugates follows the mechanism proposed by Larsen for dextran-metronidazole succinate conjugates, according to the Scheme 4 (*25*), and is faster in basic (pH 7.4) than in (pH 1.6) acidic solutions.

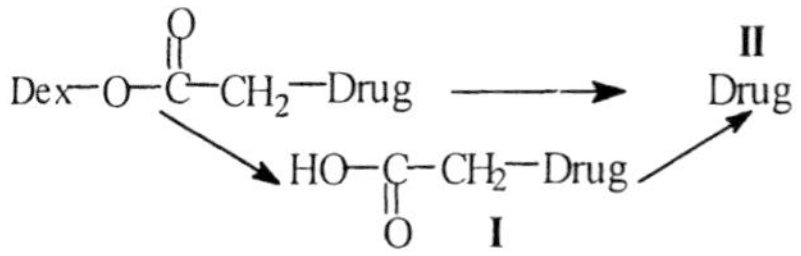

$$\text{Dex-O-}\overset{\overset{O}{\|}}{C}\text{-CH}_2\text{-Drug} \longrightarrow \begin{array}{c} \mathbf{II} \\ \text{Drug} \end{array}$$
$$\searrow \text{HO-}\underset{\underset{O}{\|}}{C}\text{-CH}_2\text{-Drug} \nearrow$$
$$\mathbf{I}$$

Scheme 4. Pathways for the chemical hydrolysis of the conjugates 1-7

The chemical hydrolysis of metronidazole-dextran conjugate 1 containing up to 26 wt% drug linked as a quaternary ammonium salt and of theophylline-dextran conjugate 3 containing 34 wt% drug bound as a tertiary amine compound, was followed by TLC, HPLC and H^1-NMR (*87, 88*). Analyses have shown that the conjugates release both the carboxymethyl derivative I (about 80% in the case of theophylline), through hydrolysis of the ester bond, and the free drug II. Pharmacokinetic studies performed after oral administration of metronidazole conjugate to rats showed a longer lasting constant plasma concentration of the drug compared with the controls treated with the free drug (*90*).

Nicotinic acid conjugate 6 contained about 43 wt% drug, from which 33% as a diester derivative 6a and 10% as a quaternary ammonium salt 6b. The polymeric conjugate administered with an endogastric catheter to rats released gradually the active substance in body, and induced a lowering of triglycerides level in plasma. The effect lasted a long period of time after each administration, proving a good hypolipidemic activity (*91*).

Nalidixic acid conjugate 2 contained up to 33 wt% drug linked by a diester bridge. The release of the drug occurred gradually, as a function of pH and time (Table I).

Comparatively, the oxacillin-dextran conjugate 5, which was obtained with a 60 wt% drug content, released the drug in lower amount, perhaps because of the higher hydrophobicity of the polymer matrix, which reduces the diffusion of the

Table I. Nalidixic acid release from conjugate 2, as function of pH and time

pH	% nalidixic acid released, at various time intervals[*]					
	1h	2h	3h	4h	8h	24h
1.6	1.5	3	8	12	18	25
7.4	20	24	28	36	47	54

[*] Data form reference (86)

external fluid to the hydrolysable bonds. The conjugate **4** of ranitidine (antiulcerous drug), containing up to 52 wt% active molecule bound to the support as a quaternary ammonium salt, was given orally to rats and the extent of gastric lesions inhibition was followed. Conjugate administration was more efficient than that of the free drug, both for the extent of lesion inhibition (61.4% comparing with 33.9%) and the period of action (more than 48 h) (92).

Antimicrobial dipyridyl compounds were linked up to 50 wt% as quaternary ammonium salts to chloroacetylated dextran (93). Conjugates **7** released easily the small molecule, especially in slightly basic pH. Preliminary studies have shown that viologen-containing polysaccharides have better antimicrobial activity against *S. lutea ATCC and E. Coli* than the small molecular products.

Amino acids and peptides are the most used spacers, as they allow a wider variation of the spacer length and chemical composition. The length of the amino acid spacer was decisive for the regeneration of mitomycin C from its conjugates with dextran, and for subsequent antitumor activity (94). Incorporation of enzymatic sensitive spacers like peptides between the drug and the carrier enables a more site specific release of the drug, for instance under the catalysis of lysosomal enzymes inside the cells (17, 18, 22, 24). Degradation of the drug-peptide linkage is influenced by the substrate-enzyme specificity, hydrophilicity and sterical crowding, as it was demonstrated by various release studies performed with conjugates having the drug bound via a peptidyl spacer (17, 32). Doxorubicin-peptide-carboxymethylpullulan conjugate showed improved antitumor effect compared with the drug itself (95). The peptidyl spacer length influenced the lysosome-mediated drug release from camptotecin-carboxymethyl or norfloxacin-dextran conjugates (96, 97).

The multi-prodrug approach might achieve a better controlled and a more specific drug release. According to this approach, the release of the drug by a step-wise (cascade) process should ensure a more specific drug release at the target site, comparing with the release implying the disruption of a single chemical bond (32, 98 - 101). For instance, an oligopeptide having 5-FU as an α substituent at its C terminus can be bound through its N-terminus to dextran and the polymer-drug conjugate so obtained can release the active drug (free 5-FU) by several degradation steps presented in Scheme 4. The first and rate determining process is an enzymatic hydrolysis catalyzed by enzymes specific to the tumor site (tumor-associated enzymes or lysosomal enzymes). *In vitro* release studies performed with conjugates **8** in the presence of tumor-associated enzymes like collagenase IV (CIV) or cathepsin B (CB) have shown that the degradation can occur at several peptide bonds (1, 2 or 3 in Scheme 5) with formation of free 5-FU (the amino acid **9** is very unstable and decomposes instantaneously after its formation), dipeptide or tripeptide derivatives of 5-FU. The place of the cleavage depended on the enzyme used as catalyst, the oligopeptide chemical composition and configuration (especially those of the amino acid X). Thus, CIV specifically cleaved the position 3, with the formation of tripeptides, and CB cleaved mainly the peptide bond in position 1, with formation of free 5-FU. The extent of peptidyl bond cleavage depended on the peptide composition and its distance from the dextran backbone (Fig. 1).

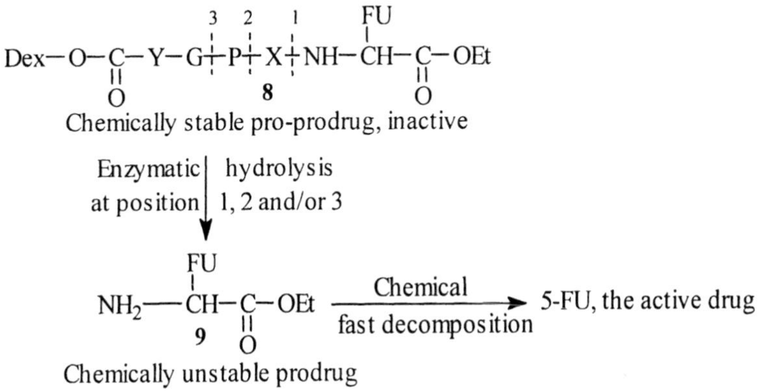

$$\begin{array}{ccccccccc} & & 3 & 2 & 1 & & FU & & \\ Dex-O-\underset{\underset{O}{\|}}{C}-Y-G\dashv P\dashv X\dashv NH-CH-\underset{\underset{O}{\|}}{C}-OEt \end{array}$$

8

Chemically stable pro-prodrug, inactive

Enzymatic hydrolysis
at position 1, 2 and/or 3

↓

$$NH_2\!-\!\!\!\underset{\underset{\textbf{9}}{}}{CH}\!-\!\!\underset{\underset{O}{\|}}{C}\!-\!OEt \xrightarrow[\text{fast decomposition}]{\text{Chemical}} \text{5-FU, the active drug}$$

with FU on CH

Chemically unstable prodrug

Scheme 5. Release of 5-FU from its conjugate s with dextran, according to the pro prodrug (cascade) approach. G is glycine and P phenylalanine.

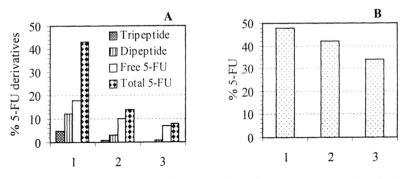

*Fig. 1. Release of 5-FU and its derivatives from dextran conjugates **8** with X = alanine (A) or glycine (B) and Y = aminocaproic acid (1), glycine (2), or a covalent bond (3), after 12 h incubation in the presence of collagenase IV, at pH 7.4 and 37 °C. (Data from reference 99)*

The cytotoxicity of the dextran conjugates **8** against the murine colorectal carcinoma cell line C26 was also influenced by the chemical structure of peptidyl spacer, the best results being obtained for the conjugates with X = glycine (*101*).

Conclusion

The presented data suggest that the polysaccharides-drug conjugates are promising systems for improving drug delivery. Most of the studies have been carried out with animals. Future studies should be focused on clinical tests, which can better support their pharmacological potential.

References

1. Ferruti, P.; Tanzi, M.C. *CRC Critical Reviews in Therapeutic Drug Carrier Systems* **1986**, *2*, 175.
2. Sinko, P.; Khon, J. in *Polymeric Delivery Systems: Properties and Applications*, El-Nokaly, M.A.; Piatt, D. M.; Charpentier, B. A. Eds. ACS Symposium Series 520, American Chemical Society, Washington, DC 1993, p 19.
3. Ulrich, K. E.; Canizzaro, S. M.; Labger, R. S.; Sgakesheff, K. M. *Chem. Rev.* **1999**, *99*, 3181.
4. Ringsdorf, H. *J. Polym. Sci. Symp.* **1975**, *51*, 135.
5. Grozdinski, J. J. *React. Funct. Polym.* **1999**, *39*, 99.
6. Duncan R. *Nature Reviews/ Drug Discovery* **2003**, *2*, 347.
7. Maeda, H. *Adv. Enzyme Regul.* **2001**, *41*, 189.
8. Maeda, M.; Takasuka, N.; Suga, T.; Uehara, N.; Hoshi, A. *Anti-Cancer Drugs* **1993**, *4*, 167.
9. Uchegbu, I. F. *Pharm. J*, **1999**, *263*, 355.
10. Harada, M.; Murata, J.; Sakamura, Y.; Sakakibara, H.; Okuno, S.; Suzuki, T. *J. Control. Rel.* **2001**, *71*, 71.
11. Nomura, T.; Saikawa, A.; Morita, S.; Sakaeda, T.; Yamashita, F.; Honda, K.; Takakura, Y.; Hashida, M. *J. Control. Rel.* **1998**, *52*, 239.
12. Matsumoto, S.; Yamamoto, A.; Takakura, Y.; Hashida, M.; Tanigawa, N.; Sezaki, H. *Cancer Res.* **1986**, *46*, 4463.
13. Luo, Y.; Prestwich, C. D. *Current Cancer Drug Targets* **2002**, *2*, 209.
14. Allen, M. *Nature Reviews,* **2002**, *2*, 750.
15. Sinha, V. R.; Kumria, R. *Int. J. Pharm.* **2001**, *224*, 19.
16. Hashida, M.; Nishikawa, M.; Takakura, Y. *J. Control. Rel.* **1995**, *36*, 99.
17. Soyez, H.; Schacht, E.; Vanderkerken, S. *Adv. Drug. Deliv. Rev.* **1996**, *21*, 81.
18. Ulbrich, K.; Subr, V. *Adv. Drug Deliv. Rev.* **2004**, *56*, 1023.
19. Hoste, K.; De Winne, K.; Schacht, E. *Int. J. Pharm.* **2004**, *277*, 119.

20. Veronese, F. M.; Morpurgo, M. *Il Farmaco* **1999**, *54,* 497.
21. Kopeček J. et al. U.P. Pat. 5 037 883, **1991.**
22. Kopeček, J.; Kopečkova, P.; Minko, T.; Lu Z. R. *Europ. J. Pharm. Biopharm.* **2000**, *50,* 61.
23. Modi, S.; Pratrash, J. *CRIPS* **2004**, *5,* 2.
24. Duncan, R.; Gac-Breton, S.; Keane, R.; Mushida, R.; Sat, Y.N.; Satchi, R.; Searle F. *J. Control. Rel.* **2001**, *74,* 135.
25. Larsen C. *Adv. Drug Deliv. Rev.* **1989**, *3,* 103.
26. Mehvar R. *Curr. Pharm. Biotechn.* **2003**, *4,* 283.
27. Mehvar, R. *J. Control. Rel.* **2000**, *69,* 1
28. Atsumi, R.; Endo, K.; Kakutani, T.; Takakura, Y.; M.; Hashida, M.; Sazaki, H. *Cancer. Res.* **1987**, *47,* 5546.
29. Okuno, S.; Harada, M.; Yano, T.; Kiuchi, S.; Tsuda, N.; Sakamura, Y.; Iami, J.; Kawaguchi, T.; Tsujihara, K. *Cancer Res.* **2000**, *60,* 2988.
30. Pang, Y.N.; Zhang Y.; Zhang, Z. *World J. Gastroenterol.* **2002**, *8,* 913.
31. Ceska, M. *Experientia* **1972**, *28,* 146.
32. D' Souza, A.J.M.; Topp, E. *J. Pharm. Sci.* **2004**, *93,* 1962.
33. Molteni, L. in *Drug Carrier in Biology and Medicine* G. Gregoriadis (Ed.), Academic Press: London, 1979, p. 107.
34. Ferruti, P.; Paoletti, R.; Puglisi, L. Fr. Pat. 2 346 016, **1979.**
35. McCormick, C. L.; Lichatowich, D. L. *ACS Symp. Ser.* **1980**, *121,* 37.
36. Sanchez-Chaves, M.; Arranz, F. *Macromol. Chem.* **1988**, *189,* 2269.
37. Papini, P.; Feroci, M.; Anzzi, G. *Ann. Chim.* **1969**, *59,* 890.
38. Sharivastava, S. K.; Jain, D. K.; Trivedi, P. *Pharmazie* **2003**, *58,* 804.
39. Zou, M.; Okamoto, H.; Cheng, G.; Hao, X.; Sun, J.; Cui, F.; Danjo, K. *Europ. J. Pharm. Biopharm.* **2005**, *59,* 155.
40. Bamford, C. H.; Middleton, I. P.; Al-Lamee, K. Polymer, **1986**, *27,* 1981.
41. Harboe, E.; Johansen, M.; Larsen, C. *Farm. Sci. Ed.* **1988**, *16,* 73.
42. Larsen, C.; Johansen, M. *Acta Pharm. Nordica* **1989**, *2,* 57.
43. Larsen, C. *Int. J. Pharm.* **1989**, *52,* 55.
44. Lee, J. S.; Jung, Y. I; Doh, M. J.; Kim, Y. M. *Drug Dev. Ind. Pharm.* **2001**, *27,* 331.
45. Larsen, C.; Harboe, E.; Johansen, M.; Olesen, H. P. *Pharm. Res.* **1989**, *6,* 995.
46. Vandame, Th. F.; Lenourry, A.; Charrueau, C.; Chaumeil, J. C. *Carboh. Polym.,* **2002**, *48,* 219.
47. Harboe, E.; Larsen, C.; Johansen, M.; Olensen H. P. *Pharm. Res.* **1989**, *6,* 919.
48. Larsen, C.; Jensen, B. H.; Olesen, H. P. *Acta Pharm. Nord.* **1991**, *3,* 41.
49. McLeod, A. D.; Friend, D.; Tozer, T. *J. Pharm. Sci.* **1994**, *83,* 1284.
50. Yano, H.; Hirayama, F.; Kamada, M.; Arima, H.; Uekama, K. *J. Control. Rel.* **2002**, *79,* 103.
51. Jung, Y. J.; Lee, J. S.; Kim, H. H.; Kim, Y. T.; Kim, Y. M. *Arch. Pharm. Res.* **1998**, *2,* 179.

52. Chourasia, M. K.; Jain, S. K. *J. Pharm. Pharmaceut. Sci.* **2003**, *6*, 33.
53. Schacht, E.; Ruys, L.; Vermeersch, J. *J. Control. Rel.* 1984, *1*, 33.
54. Yalpani, M. *CRC Critical Rev. Biotechn.* **1986** *3*, 375.
55. Domb, A.; Linden, G.; Polacheck, I.; Benita, S. *J. Polym. Sci. Part A; Polym. Chem.* **1996**, *34*, 1229.
56. Nishi, K. K.; Jayakrishnan, A. *Biomacromolecules* **2004**, *5*, 1489.
57. Ehrenfreund-Kleinman, T.; Azzam, T.; Falk, R.; Polacheck, I.; Golenser, J.; Domb, A. J. *Biomaterials*, **2002**, *23*, 1327.
58. Falk, R.; Domb, A. J.; Polacheck, I. *Antimicrob. Agents Chemother.* **1999**, 1975.
59. Golenser, J.; Frankenburg, S.; Ehrenfreud, T.; Domb, A.J. *Antimicrob. Agents Chemother.* **1999**, 2209.
60. Falk, R.; Grunwald, J.; Hoffman, A.; Domb, A. J.; Polacheck, I.; *Antimicrob. Agents Chemother.*, **2004**, 3606.
61. Bernstein, H.; Hurwitz, E.; Maron, R.; Arnon, R.; Sela, M.; Wilchek, M. *J. Natl. Cancer Inst.* **1978**, *60*, 379.
62. Torchilin, V. P.; Tischenko, E. G.; Smirnov, V. N.; Chazov, E. I. *J. Biomed. Mater. Res.* **1977**, *11*, 223
63. Torchilin, V. P.; Il'ina, E. V.; Mazaev, A. V.; Lebedev, B. S.; Smirnov, V. N.; Chazov, E. J. *J. Solid Phase Biochem*, **1977**, *2*, 187.
64. Takakura, Y.; Kaneko, Y.; Fujita, T.; Hashida, M.; Maeda, H.; Sezaki, H. *J. Pharm. Sci.* **1989**, *78*, 117.
65. Kaneo, Y.; Fujihara, Y.; Tanaka, T.; Kozawa, Y.; Mori, H.; Yguchi, S. *Pharm. Res.* **1989**, *6*, 1025.
66. Usher, T.; Patel, N.; Tele, C.; Wolk, L. WO 09500177, **1995.**
67. Sanchez-Chavez, M. ; Rodriguez, J. M. ; Arranz, F. *Macromol. Chem. Phys.* **1997**, *198*, 3465.
68. Sanchez-Chaves, M. ; Arranz, F. *Polymer,* **1997**, *38*, 2501.
69. Ehrenfreund-Kleinman, T.; Golenser, J.; Domb, A.J. *Biomaterials*, **2004**, *25*, 3049.
70. Mocanu, G.; Constantin, M.; Carpov, A. *Angew. Macromol. Chem.* **1996**, *241*, 1.
71. Mocanu, G.; Constantin, M.; Fundueanu, G.; Carpov, A. *S.T.P. Pharma Sci,* **2000**, *10*, 439.
72. Yolles, S.; Morton, J. F.; Sartori, M. F. *J. Polym. Sci.* **1979**, *17*, 4111.
73. Barker, S.A.; Disney, H.M.; Somers, P.J. *Carboh. Res.* **1972**, *25*, 237.
74. Schacht, E.; Vermeersch, J.; Vandorne, F.; Vercauteren, L. *J. Control. Rel.* **1985**, *2*, 245.
75. Vermeersch, J.; Vandoorne, F.; Permentier, D.; Schacht, E. *Bull. Soc. Chim. Belg* **1985**, *94*, 591.
76. Arranz, F.; Sanchez-Chaves, M.; Ramirez, J. C. *Angew. Macromol. Chem.* **1992**, *194*, 87.
77. Bruneel, D.; Schacht, E. *Polymer,* **1994**, *35*, 2656.
78. Kato, Y.; Onishi, H.; Machida, Y. *Biomaterials* **2004**, *25*, 907.

79. Zhang, X.; Mehvar, R. *Int. J. Pharm.* **2001**, *229*, 173.
80. McLeod, A. J.; Friend, D.; Tozer, T. N. *Int. J. Pharm.* **1993**, *92*, 105.
81. Mora, M.; Pato, J. *Macromol. Chem.* **1990**, *191*, 1051.
82. Mocanu, G.; Carpov, A. *Cell. Chem Technol.* **1992**, *26*, 675.
83. Ramirez, J.; Sanchez-Chaves, M. ; Arranz, F. *Polymer* **1994**, *35*, 2651.
84. Mocanu, G.; Vizitiu, D.; Mihai, D.; Carpov, A. *Carboh. Polym.* **1995**, *28*, 131.
85. Mocanu, G.; Airinei, A.; Carpov, A. *STP Pharma Sci.* **1994**, *4*, 287.
86. Mihai, D.; Mocanu, G.; Carpov A. *J. Bioact. Compat. Polym.* **2000**, *15*, 245.
87. Mocanu, G.; Carpov, A. *J. Bioact. Compat. Polym.* **1993**, *8*, 383.
88. Mocanu, G.; Airinei, A.; Carpov, A. *J. Control. Rel.* **1996**, *40*, 1.
89. Constantin, M.; Mocanu, G.; Mihai, D.; Carpov, A. *Pharmacol. Pharmaceut. Letters,* **1996**, *6*, 107.
90. Teslariu, E.; Nechifor, M.; Spac A.: Mocanu, G. unpublished results.
91. Filip, C.; Ungureanu, D.; Gheorghita, N.; Ghitler, N.; Mocanu, G.; Nechifor, M. *Rev. Med. Chir. Med. Nat. Iasi* **2003**, *107*, 179.
92. Hriscu, A.; Dorneanu, V.; Spac, A.F.; Mocanu, G.; Constantin, M. *Rev. Med.-Chir. Soc. Med. Nat. Iasi* **1997**, *101*, 174.
93. Avram, E.; Lacatus, C.; Mocanu, G. *Eur. Polym. J.* **2001**, 1901.
94. Takakura, Y.; Matsumoto, S.; Hashida, M.; Sezaki, H. *J. Controll. Rel.* **1989**, *10*, 97.
95. Nogusa, H.; Yano, T. T.; Kashima, N.; Yamamoto, K.; Okuno, S.; Hamana, H. *Bioorg. Med. Chem. Lett.* **2000**, *10*, 227.
96. Roseeuw, B. E.; Coessens, V.; Schacht, E.; Vrooman, B.; Domurano, D.; Marchal, G. *J. Mater. Sci.: Materials Med.* **1999**, *10*, 743.
97. Harada, M.; Sakakibara, H.; Yano, T.; Suzuki, T.; Okuno, S. *J. Control. Rel.* **2000**, *69*, 399.
98. Nichifor, M.; Coessens, V.; Schacht, E. *J. Bioact. Compat. Polym.* **1995**, *10*, 199.
99. M. Nichifor, E. Schacht, L. W. Seymour, *J. Control. Rel.* **1996**, *39*, 79.
100. Nichifor, M.; Schacht, E.; Seymour, L.W. *J. Control. Rel.* **1997**, *48*, 165.
101. Nichifor, M.; Seymour, L.W.; Anderson, D.; Shaoibi, M.; Schacht, E. *J. Bioact. Compat. Polym.* **1997**, *12*, 265.

Chapter 16

Polysaccharide Hydrogels for the Preparation of Immunoisolated Cell Delivery Systems

Jean-Michel Rabanel, Nicolas Bertrand, Shilpa Sant, Salma Louati, and Patrice Hildgen[*]

Faculté de Pharmacie, Université de Montréal, 2900, Edouard Montpetit, Montréal, Québec H3T 2M4, Canada
[*]Corresponding author: patrice.hildgen@umontreal.ca

In the last decades, the rapid advances in the understanding of the role of proteins and peptides in several diseases have propelled them as therapeutic entities. Protein delivery is a challenge and one of the new approaches to their administration is to implant cells producing the therapeutic protein of interest. However, the grafted cells or tissues have to overcome immunological defences of the organism. Cell microencapsulation is one of the most promising avenues to protect cells from immune rejection, while permitting their survival alongwith the protein secretion activity without deleterious effects of immunosuppressant drugs. Polysaccharides (PS) have been retained as biomaterial to prepare such devices, in a variety of applications because of their ability to form non-toxic and biocompatible hydrogels which can show desired properties for cell immunoisolation such as selective permeability and soft surfaces identical to, or mimicking natural biological extracellular matrices. This chapter will address the main classes of natural and modified PS used in this field and will discuss different issues in regard to particular applications. The performances of the encapsulation systems are largely dictated by the physico-chemical properties of the PS, such as sugar monomer composition, primary and secondary structure, degree and nature of substitutions, type of gel crosslinking, as well as microbead preparation techniques. However, all the current microencapsulation technologies display limitations particularly on long-term performances. A new approach is thus portrayed based on amylopectin hydrogel microencapsulation. With this new technique, the gel forming and encapsulation steps are dissociated, allowing use of a wider range of gel forming methods and biomaterials as well as complete capsule characterization before cell encapsulation.

Introduction

Cell therapy and cell encapsulation

The growing interest in cell therapy and cell encapsulation is fuelled by three main considerations. Since biomedical science is able to identify the mechanisms of illness and dysfunctions at the molecular level, new treatments based on replacement of natural peptides and proteins have gained importance over more conventional drugs. At the same time, advances in genetic engineering generate new cell lines able to deliver overexpressed proteins, not only to replace defective proteins but also to act as therapeutic agents. And finally, there is the acute problem of shortage of tissue and organ donors to replace diseased or malfunctioning organ or tissue.

The challenge in peptide and protein delivery

Production and administration of therapeutic proteins is a challenge to the pharmaceutical scientists because of their physicochemical and biological characteristics[1]. The delivery is particulary challenging as the usual parenteral route and sustained drug delivery systems are not completely satisfactory. Administration in some compartments is difficult, as in case of Central Nervous System (CNS), because of the blood-brain barrier (BBB)[2]. Furthermore, short half-life of protein also leads to over-dosage, increases costs and hences possible adverse effects. It also means repeated injections or use of infusion pump to maintain the therapeutic level. Finally, conventional delivery does not take into account fine *in vivo* regulation of protein concentration by complex biological feedbacks.

In situ production is attractive and several cell therapy approaches have been considered in this perspective. The first means was to graft the exogenous tissue producing the protein of interest. However, low human tissue availability, immune reaction against the non-autologous graft, the necessity for high dosage of immunosuppressant drugs are all limiting factors. The second means is the gene therapy, where genes coding for the proteins are inserted *in vivo*. A variation of this method is to genetically alter the cells lines *ex vivo* to overexpress the protein of interest and subsequently, inoculate these cells near the site of action. One of the major problems with this approach is histocompatibility and immunological reactions against the grafted cells if the cells are not originating from the patient (non-autologous graft).

Cell microencapsulation

To address all these concerns, cell therapy has to be associated with the creation of a immunoprivileged site for the grafted cells. Cell microencapsulation to deliver specific therapeutic biomolecules has received much attention for the continuous delivery of biomolecules to a well defined site[3]. Encapsulation of cells in a semi-permeable device (figure 1) allows the entry of molecules essential for cell metabolism (nutrients, oxygen, growth factors, etc...) and outward diffusion of cell metabolites such as waste products and active molecules. At the same time, cells and larger molecules of the immune system are kept away from the transplanted cells, avoiding use of highly toxic immuno-suppressant drugs[4, 5]. Immunoisolation also allows the use of animal cell lines (xenotransplantation) and immortalized cell lines, eventually genetically modified, instead of just primary human cell lines. The active molecule can be secreted continuously or in response to appropriate host stimulus, such as insulin secretion in response to glucose concentration by encapsulated pancreatic islets. Other advantage of microencapsulation is the possibility to target specific organ or body compartment achieving a high sustained local concentration, and thus, decreasing potential side-effects. Although the capsules are a highly artificial device not directly connected to host body and organs, it has been shown to support entrapped cell metabolism, growth and differentiation[3]. The capsule can be considered as a nest, a cell friendly microenvironment with the presence of an extracellular matrix, which can serve as cell scaffold. It can also prevent excessive cell growth and eventually may allow the removal of the cells if problems arise during the treatment course[3, 6].

Microcapsules parameters

Microencapsulation as potential means to isolate transplanted cells was first introduced by Lim and Sun in 1980[7]. The advantage of microencapsulation is that distance of diffusion into the device is decreased and the spherical shape of microcapsules is thought to be the ideal geometry for diffusive transport due to their high surface to volume ratio[6]. Presently, microcapsules are probably the preferable system for cell transplantation and represent significant promise for biotechnology and medicine. There are some important considerations for the success of cell microencapsulation:

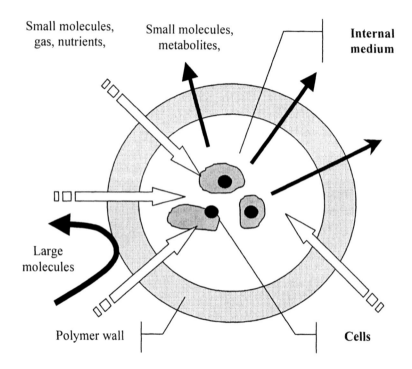

Small molecules, gas, nutrients,

Small molecules, metabolites,

Internal medium

Large molecules

Polymer wall

Cells

Figure 1: Schematic representation of a microcapsule
(Size: 100 microns to 2 mm)

- Mechanical stability (membrane strength): The microcapsules must withstand a certain amount of physical and osmotic stress, as breakage of the capsules will lead to immune rejection of the encapsulated cell and damaging reaction to the host tissues.

- Biocompatibility: A primary consideration is the totally biocompatible material which neither interferes with cell homeostasis (encapsulated cells and surrounding host tissue) nor triggers an immune response in the host.

- Permeability: The capsule must have an adjusted permeability, in term of the entry and exit of molecules. For well-controlled transport properties, it is imperative that the membrane wall thickness be uniform. Immunoprotection is also an issue as not only immune cells but also antibodies and cytokines should be excluded[4, 5].

- Size: The encapsulation of cells into a thin, spherical membrane results in microcapsules, which are small beads (0.2 to 0.8 mm in diameter). The thinner the hydrogel barrier, the faster is the nutrient diffusion. Indeed, it has been suggested that the biocompability of capsules can depend on their size[8]. In fact, smaller capsules are more prone to osmotic stress and subsequent swelling than larger capsules, due to their increased surface to volume ratio. Microencapsulation allows optimization of mass transfer (high surface/internal volume ratio), which is critical for cell viability and a fast secretory response to external signal. However, smaller size also limits cell loading capacity[9].

- Need for a gentle encapsulation technique: mild conditions (aqueous medium, no reactive species) are required to protect cell viability.

- Durable microcapsules: PSs are degradable polymers, therefore if the aim of the cell therapy is to provide sustained release of therapeutics over a long period of time, degradation rate should be assessed.

Rationale for use of polysaccharide hydrogel

The choice of membrane material is crucial for the preparation of microcapsules. The inner part is in contact with encapsulated cells while the outer surface morphology plays an important role in determination of the permeability and greatly influences the host reaction to the implant. For these reasons, microcapsules have been formulated typically using hydrogel-based materials[3, 6]. Hydrogels are networks of water-soluble polymers, and can be produced by different crosslinking methods compatible with cell viability. Polyanions can be gelled in the presence of divalent cations (ionotropic gel) or in the presence of PE of opposite charges to form a polyelectrolyte complex gel (PEC)[11]. Upon heating and cooling thermoreversible gels form hydrogen bonding leading to gelation. Use of interpenetrating networks (IPN) is another method to produce a hydrogel[11] but such gels are usually too weak to be used in cell immunoisolation techniques. However, they could intervene as second gelling mechanisms, stabilizing a pre-existing gel. Covalently bonded hydrogels should be avoided since the generally toxic chemical agents used in their preparation affect the viability of entrapped cells[10]. Morever, novel crosslinking methods are emerging such as photocrosslinkage that may be promising for cell entrapment[14]. Hydrogels may absorb water up to thousand times their weight, in turn, swelling without dissolving[10, 11].

Hydrogels provide an environment compatible to cell viability and functions, similar in many aspects to the normal tissues. This compatibility is due in part to the hydrophilic nature of the matrix and because hydrogel minimizes mechanical friction between cells and capsules. Moreover, as there is no or little interfacial tension between hydrogel and surrounding tissue, cell adhesion and protein precipitation at the interface is minimized[10]. Hydrogel matrix entrapment has been extensively described and used to stabilize cell culture by physically holding cells and protecting them from shear forces in bioreactors[12]. They are also used in tissue engineering applications[13] and if the hydrogels have capacity to provide immunoisolation, they can be used for cell therapy applications, particularly in microencapsulation techniques[3].

Permeability of hydrogels

Permeability of the hydrogel should provide immunoisolation and, at the same time, allow sufficient mass transfer for nutrients, gases and secreted therapeutic proteins. These conflicting requirements necessitates a certain level of size discrimination properties, i.e a pore size and pores size distribution permitting exclusion of immune cells, IgG, cytokines and complement proteins while small sized nutrients and secreted proteins can diffuse freely through the gel[4, 9]. Hydrogels have good diffusional properties for hydrophilic molecules (most of the nutrients and excreta are hydrophilic). These diffusional and mass transfer properties can be influenced by the degree of hydrophobicity of the hydrogel, degree of cross-linking, pH and temperature[12]. Usually, a molecular cut-off around 50 kDa to 100 kDa is sought to exclude immunoglobulins (IgG) and direct cellular contact between grafted cells and the host cells[9]. Further decrease in cut-off, can be beneficial to exclude cytokines and some complement proteins, particularly when implanting xenogenic cells. However, it may dramatically decrease the mass transport coefficient of essentials nutrients and excreta[15]. Although permeability of hydrogel, have been studied using size exclusion chromatography[16, 17] or in diffusional models[18], the physical characterization of hydrogel porosity is not completely elucidated[19, 20].

PS hydrogels

Hydrogel can be formed by natural or synthetic polymers[6, 10, 11]. Nonetheless, the use of PS as material for encapsulation has been considered for decades because they constitute an important class of physiological materials[21]. Besides, PS are involved in cell surface properties including tissue adhesion, signal transduction and transport mechanisms. Moreover, they can be easily functionalized with adhesion molecules, receptors, ligands in order to simulate the normal environment of the cell, stimulate adhesion, differentiation spreading and growth even production of therapeutic proteins[22]. Natural and modified PS alone or in combination with other natural polymers have been proposed for cell encapsulation[3, 6, 10].

In spite of their attractiveness, PS hydrogels have intrinsic limitations (Table 1). In some applications, where long lasting capsules are required, the PS degradation rate *in vivo* should be assessed and eventually adjusted[23]. The low mechanical strength, instability of ionic interactions in PE hydrogels and difficulty to handle them (e.g. viscosity, sterilization, etc.) are amongst other problems that could be circumvented by the proper choice of hydrogel and method of crosslinking. Natural products can show batch to batch variability leading to differences in permeability and compatibility properties. A substantial challenge related to the biomaterials used in cell encapsulation has been the non-availability of clinical-grade polymers[24]. Cell entrapment in hydrogel, although usually conducted in very mild conditions, does not mean that there is no stress on cells during encapsulation. Finally, sterilization is an important issue for implantable device. Usually, the PS solution is filtration sterilized before use implying that the viscosity should be kept low, limiting the choice of PS (particularly using lower M_w) which in turn, has a great influence on the final performance.

As we will see later, single component PS hydrogel cannot account alone for all the properties desired for a microencapsulation device. In most (if not all) of the cases, the hydrogel will be associated with other materials to obtain a proper microencapsulation system. In summary, microcapsules can be described as an inner spherical core of hydrogel, (composed of one or more polymers), entrapping and interacting with encapsulated cells, surrounded by one of more external layers of different hydrogel polymers to improve permeability and stability properties, to prevent cell release from the bead and to ensure biocompatibility (figure 2). In brief, the relevant properties of the microencapsulation device can be achieved by:

- The proper choice of the PS
- Adequate purity of the PS
- The control of gelling conditions (ionic concentration, temperature, pH)
- Combination with other natural or modified PS or synthetic molecules
- Chemical or enzymatic modification of the PS

312

Table 1: Advantages and disadvantages of natural PS hydrogel matrix for cell encapsulation

Advantages
    Biocompatibility
    Biodegradability
    Mimics natural surfaces (interactions with host tissue)
    Hydrophilicity
    Easy functionalization (as with adhesion ligands)
    Easy chemical modification (for modified permeability or gelling properties)
    Mild conditions of cell encapsulation
    Environment favourable to cell survival
    Control of permeability and porosity

Disadvantages
    Hydrophilicity : difficulties to cross membrane barriers, etc
    Biodegradable (for e.g. not desired in long lasting experiments)
    Instability of non-covalent gel (as ionotropic hydrogels)
    Mechanically weak
    Nonhomogeneity of gels (gelling domains of various porosity)
    Crosslinking methods limited to non-toxic methods
    Difficulties to sterilize
    Batch to batch variation (differences in permeability properties)
    Variable biocompatibility of materials
    Possible stress on cells during encapsulation procedures (depending on the preparation technique allowed by the PS of choice)

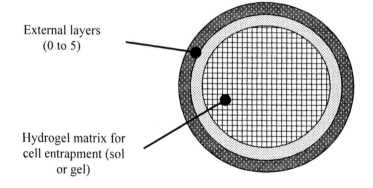

External layers
(0 to 5)

Hydrogel matrix for
cell entrapment (sol
or gel)

Figure 2: General structure of hydrogel-based cell encapsulation microbead
providing immunoisolation

Polysaccharides for Microencapsulation

PS of different origins have been investigated for their ability to form hydrogels and immunoisolation devices: from seaweeds: the alginates, carrageenans and agarose; from microbial origin: gellan gum and xanthan; from animal origin: hyaluronic acid, chondroitin and chitin; and finally from higher plant: cellulose, starch and pectin. PS can be used as hydrogel forming material, or as adjuvants to confer particular properties to the bulk of the hydrogel. They can also be adsorbed at the surface of the microcapsule to confer permeability or biocompatibility properties, or serve as cell scaffolding or cell adhesion materials. Amongst the various PS, alginates are the most widely used, therefore in this chapter we will discuss it in more detail.

Alginates

Alginates are products of marine brown algae. They constitute a large family of negatively charged linear PS and are copolymers of 1,4 linked β-D-mannuronic acid (M) and α-L-guluronic acid (G) differentiated by compositions and sequences. The two monomers are arranged in mannuronic (MM block) or guluronic (GG block) sequences or in alternate monomer sequences (MG block).

Gelling properties

Alginates, in the presence of divalent cations such as calcium, barium and strontium form ionotropic gel[21, 25]. Divalent cations interact with block of guluronic monomers to form bridges between polymer chains in a highly cooperative fashion (figure 3). Alginates can also be reticulated by ionic interactions with positively charged polymers such as poly-L-Lysine (PLL) to form a complex coacervate or polyion complex hydrogel[7, 26]. The block composition controls the gelling properties and, in turn, the permeability properties. The crosslinking density depends directly on the M to G ratio and M_w of the polymer. The M/G ratio also dictates the mechanical stability and pore size of the gel. The region of alternate monomer sequence is more flexible and contributes to decrease the viscosity, whereas high content G alginates are stiffer, more viscous in solution but form more stable gels, mechanically stronger and are more resistant to swelling caused by osmotic pressure[28]. Indeed, even if swelling does not cause the capsules to burst, it should be avoided since it increases the gel porosity and the possibility for cell protusion[28]. Nonetheless, high mannuronic acid (high-M) alginates could be of interest, because they are less viscous. Consequently, this allows for rounder particles with a higher alginate concentration that could be beneficial for both biocompatibility and stability. It has also been shown that PLL binds better to high-M alginate,

314

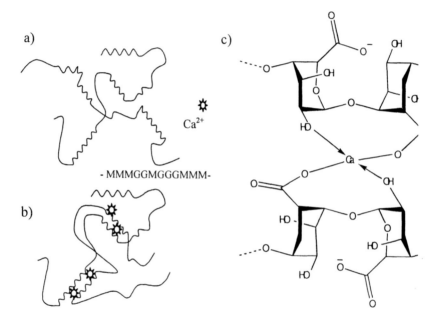

Figure 3: Formation of alginate gel
Hypothetical alginate gel formation: (a) free chain of alginate composed of
random sequence of manuronnic (M) and guluronic (G) acids; (b) hydrogel
with inter and intra coordination bonds between Ca^{2+} and G (formation of
junction zone "egg-box structure"); (c) details of the bridge between G regions.

allowing formation of a thicker and less permeable membrane. Therefore it can protect immobilized cells more efficiently against the host immune system[46]. It is noteworthy that the rate of diffusion of serum albumin in alginate capsules decreases with G monomer content as high-G alginates tend to form bigger pores[18]. On the other hand, high-M alginate chains are more flexible, less crosslinked by divalent cations, and therefore, they tend to form weaker gel[39]. Nonetheless, a recent report shows that high-M seems to be slightly more compatible than high-G alginates, although no difference in stability is found[27]. Hence, in the view of the two types of alginates, each having its own useful properties, optimum stability is often reached with intermediate composition[28]. In an effort to better control the characteristics of the alginates sources, it has been recently proposed to enzymatically tailor them[55, 56]. For example, enzymes known as epimerases can convert M to G in the polymer chain. One of these enzymes, AlgE4, converts MM blocks into blocks of strictly alternating M and G. These epimerized alginates showed increased resistance to osmotic swelling and decreased permeability, measured as reduced diffusion of IgG into Ca/Ba-alginate gel beads. Calcium is widely used to crosslink alginates. However in some applications, barium replaces calcium as it interacts more strongly with G rich region and forms more stable gel[25, 27, 28]. In vivo, the ionotropic gel degrades by slow exchange of divalent cations with monovalent cations leading to dissociation of the ionic bridges. One can control rigidity and degradation of alginate hydrogels via M_w distribution[29]. High molecular weight tends to increase the gel strength but its use is limited due to its high viscosity which makes it difficult to handle, producing more irregular surface and increasing the average bead size.

Process of encapsulation

The classical encapsulation process is to first mix the cells in an alginate solution (1-3%)[7, 21]. The mixture is transferred drop-wise in a calcium chloride bath. Several technologies have been developed to produce droplets of desired size (extrusion, coaxial air flow, electrostatic field generators) and some are commercially available[30, 31]. Then, the particles are precipitated in a bath containing a PLL solution to form an external layer of the capsule whose properties would define the permeability of the capsule[7] and prevent the release of capsule content upon reliquefaction of the core in medium containing calcium chelating agents. In some applications, the core of the capsule is liquefied before implantation by treatment with sodium citrate (substitution of Ca^{2+} by Na^+). This increases the diffusion of nutrients towards encapsulated cells. Typically, size of the capsules is between 500 and 800 microns but smaller particles can be obtained[8]. An alternative technique using an emulsification/internal gelation system to increase productivity has been recently described[32]. Gelation of alginate droplet dispersed in oil emulsion is induced by pH change. Insoluble Ca^{2+} salts dispersed in the alginate solution are solubilized upon pH change and trigger homogenous ionotropic gelation[32]. Similarly, homogenous gelation of alginates has been achieved by injection of $BaCl_2$ crystals into alginate droplets falling into a $BaCl_2$ bath[33].

Permeability of alginate capsules

The permeability of the capsule depends on alginate characteristics (as discussed above), bead preparation method and coatings materials. The permeability can be modulated by the concentration and molecular weight of PLL as well as its contact time with particles[34]. As mentioned earlier, a M_w cut-off of 100 kDa or below seems desirable to exclude IgG and had been claimed to be obtained with alginate/PLL capsules[34]. The alginate/PLL membranes were reported not permeable to IgG when PLL of M_w ranging from 21 kDa to 390 kDa were used[35]. Later studies showed that this type of capsule has higher permeability threshold, around 300 kDa[36] and that diffusion coefficients of molecules under 20 kDa in the gel are identical compared to their diffusion in water[18]. This result is supported by size exclusion chromatography studies[17]. Pore size and pore size distribution are highly dependent on the physico-chemical characteristics of the alginates, gelling conditions as well as coating characteristics. Alginate gel crosslinking is a kinetic process controlled by the relative rate of diffusion of Ca^{2+} and alginates towards an inward moving gelling front. In presence of low Ca^{2+} concentration, the result of this process is non-homogenous gel beads, showing an increase alginate gradient concentration from the centre towards the surface[18]. Addition of non-gelling cations such as Na^+ prevents this phenomenon, leading to more homogenous gel beads. This observation allows for the optimization of permeability in the absence of outer layer. Permeability could be decreased by increasing the gradient step, using lower M_w alginate, lowering concentration of gelling cations or lowering bead size[18, 28, 37]. The high surface alginate concentration also increases the charge density and as a result the binding of polycations on the outer membrane. Finally, this gradient also has an influence on the distribution and physiology of encapsulated cells within the beads[38].

Biocompatibility of alginates

One of the problems associated with implantation of encapsulation devices, is the foreign body reaction taking place at the site of implantation. Cellular overgrowth and fibrosis around the implant diminish the nutrient, gas and metabolites diffusing in and out of the capsules. This reaction depends on the nature of the material, the type of surface (smooth or irregular), and the presence of contaminants in the alginates having mitogenic activity[5]. Alginates should be free of pyrogens, of immunogenic molecules and of possible polyphenols[39]. Improved purification of alginate has been achieved by different groups[40-42]. A combination of filtration, extraction and precipitation has been shown to increase the capsule biocompatibility but with little effect on the survival of encapsulated islets[43]. An electrophoresis method has also been used and resulted in amitogenic alginates over lymphocytes[44]. Empty capsules made of purified alginates showed a long term (2 years) biocompatibility whereas capsules with allograft islets showed prolonged survival up to 200 days[45]. Commercial grade alginates are now available but the purity of the final product remains a concern[24]. Alginates of varying M/G ratio present variable biocompatibility[46]. Antibodies against high-M alginate capsules were detected in the sera of mice[47]. It has been shown that the cellular overgrowth in rat intraperitoneal cavity was more severe with high-G alginate capsules than with intermediate-G alginate capsules[48]. Later, it appeared that it was due to inadequate binding of high-G alginates to PLL. Indeed, PLL binds better to M residues[37] and in the absence of PLL, no fibrotic reaction was observed[37, 46]. PLL decreases biocompatibility[49, 50]. As high-G alginates have beneficial effects on gel stability, the use of polycations, which interact with high-G alginates more effectively than PLL could be recommended[46]. Another option widely adopted is to add an outer coating of alginate to shield the PLL from the host tissue improving biocompatibility[34] and decreasing permeability[36]. However, because this last alginate layer is exposed, questions have been raised concerning its stability. Indeed, results show that the outer layer of alginate deteriorates *in vitro* after two weeks, possibly by ion exchange mechanism[36].

Optimizing beads surface biocompatibility

To further optimize the biocompatibility of capsules, it has been proposed to coat them externally with a poly(ethylene glycol) (PEG). A copolymer of PLL and monomethoxy-poly(ethylene glycol) (MPEG) has been used to form microcapsule outer layer[51]. In a mouse intraperitoneal implant model, PLL-g-MPEG capsules were seen to be more biocompatible than the classical alginate/PLL/alginate microcapsules. At the same time, these grafted copolymers demonstrated lower affinity for alginate and increased microcapsule permeability. To rectify this, penta-layered alginate/PLL/alginate/PLL-g-MPEG/alginate microcapsules were fabricated, and these demonstrated both appropriate selectivity and enhanced biocompatibility[51]. In a different method, PEG hydrogel, formed by an interfacial photopolymerization technique, was immobilized on the microcapsule surface. *In vivo*, these microcapsules did not elicit a fibrotic response, as was the case with alginate-PLL microcapsules[52]. Finally, similar *in vitro* results were obtained when PEG (or poly(vinyl-alcohol), PVA) was chemically grafted to PLL adsorbed on the alginate capsules[53] or when PEG was adsorbed on PLL or chitosan tetra-layered alginate particles[54]. Finally, in the matter of biocompatibility, it has been observed that smaller beads tend to be more compatible[8, 27]. This could be explained as follows: since nutrient diffusion pathway through the gel is shorter in smaller beads, fewer cells should die and fewer antigens should be released in the implant environment. Also, because of the higher mechanical resistance of smaller particles, mechanical stability should increase with smaller size[8].

Stability of alginate beads

Blended capsule core has been proposed to increase stability of the alginate gel. Poly(ethyleneimine) (or PEI) was used to stabilize the alginate gel in the bulk as interactions between the two polymers became significant following gelation. Thus, it was shown that the mechanical strength increased with the M_w of PEI. The stable interactions between the alginate and PEI prevented variations of the pore structure of the gel, and slowed the deterioration of its properties[57, 58]. PEG was also tested as an adjuvant to increase gel strength via a dual crosslinking method.. The major advantage of this dually crosslinkable system was the chemical stability of the resultant gels due to the presence of covalent bonds that maintain the integrity of the gel as opposed to reversible ionic linkages. PEG could also change the pore structure of blended gel, favouring cell growth[12]. Indeed, islets encapsulated in these systems were healthy and retained both viability and insulin secretory function, *in vitro*[59].

Stability of the alginate capsule membrane

Since PE gel is not stable over a long period of time in a biological environment, alginate hydrogels are frequently surface-reinforced with secondary polymers. This consolidation often occurs via ionic interactions to enhance their mechanical rigidity and stability as well as to reduce the permeability of the capsules. As presented earlier, although the first to be used was PLL[7, 34], its implication in inflammation and tissue fibrosis[49,50] led to, either a shielding strategy (addition of biocompatible layer) or the use of other polycations like PEI[60, 61], poly-L-ornithine[62, 63] and poly(vinylamine)[16]. However, further studies are needed to confirm the real efficacy and biocompatibility of those polymers.

Chitosan, a natural polycation, has been evaluated as capsule membrane material. Indeed, compared to PLL, chitosan binds alginate in an almost irreversible way. Thus, its use in cell encapsulation could mean decreased inflammation and fibrotic response since the amount of polycation released from the capsules would be negligible[64]. Permeability of capsules to protein could be decreased when chitosan M_w and degree of acetylation were increased. Finally, the addition of several layers of alginate and chitosan resulted in capsules impermeable to IgG[64, 65]. However, because the presence of a positively charged polymer at the surface may favour foreign body reaction as well as cell and protein attachment, chitosan and PEG were used together to create a membrane around alginate beads, improving stability and biocompatibility[66].

To ensure further stability and in spite of potential toxicity, covalent reticulations of the external layer of the capsules have also been proposed :

- with a mild glutaraldehyde or carbodiimide crosslinking reaction[66]

- with the formation of stable protein membrane around the alginate beads using a transacylation reaction between propylene glycol alginate (PGA) or pectin, and various proteins[67, 68].
- with the crosslinking between adjacent layers of alginate-PLL-alginate using a heterobifunctional photoactivable cross-linker, N-5-azido-2-nitrobenzoyloxysuccinimide (ANB-NOS) linked to PLL[69].

In vivo concerns for alginate capsules applicationss

Alginate microencapsulation has been used with a variety of cell types including PC12 cells for the treatment of Parkinson's disease[70], hepatocytes for the liver failure[71], parathyroid tissue for hypoparathyroidism[72], and the use of genetically engineered cells being investigated for gene therapy[1, 73]. But the most studied application remains the transplantation of Islets of Langerhans for the insulin delivery[74,75]. The only well-documented study of transplantation of encapsulated islets in one diabetic patient was reported in 1991[76, 77]. One of the major concern is the great distance between the encapsulated islets and the blood vessels, which limit the oxygen supply. Resulting hypoxia can lead to islet dysfunction and necrosis. The PS capsule walls prevent the neovascularization, which usually occurs within a few weeks after grafting of naked islets. To increase oxygen supply, intravascular or intraportal transplantation has been investigated in large animal model[78, 79] and using smaller capsules[80]. Oxygen supply is also dependant on occurrence of fibrotic overgrowth creating a physical and metabolic avascular barrier.

In the treatment of acute liver failure, different approaches to restore liver functions have been proposed including extracorporeal cell immobilization, or intracorporeal cell encapsulation[71]. *In vivo* results following intraperitoneal injection of alginates encapsulated hepatocytes showed an increase in the survival time of rats with fulminant hepatic failure. However, maintenance of hepatocyte function for long period is disappointing because alginate is a poor material for hepatocyte attachment (hepatocytes are anchorage-dependent cells) and PLL is toxic to hepatocytes[81]. However, capsules composed of PS carboxymethyl cellulose, chondroitin sulfate A, chitosan, and polygalacturonate, were found to be able to sustain long-term survival and growth of encapsulated hepatocytes[82]. Alginate encapsulated hepatocytes have also been proposed for extracorporeal hepatic replacement in a fluidized bed motion bioreactor[83].

The prospect to implant non-autologous genetically modified cells to produce therapeutic molecules in immunoisolated capsules, offer a new way to deliver recombinant proteins[84, 85]. Genetically engineered cells secreting antiangiogenic substances encapsulated in alginate microcapsules were described in mouse models to fight brain tumor[86-88]. These capsules were well-tolerated by CNS tissues. However, the use of genetically modified immortalized cell lines instead of slow growing cell lines, put more stress on the alginate capsules. Thus, cell growth control or more resistant capsules may be needed for these applications[88]. Production of angiostatin in the alginate beads showed that the optimal capsule should be small (360 μm) and non-homogenous[84].

As natural products, alginates have batch-to-batch variations that could lead to observed variable permeability or compatibility. Pore size is affected by alginate composition and gel preparation methodology. One can find impurities that could have detrimental effects, such as inflammatory response. To conclude, since it is difficult to address all permeability, stability and biocompatibility issues with a single membrane, it appears that successive deposition of polyelectrolytes on alginate capsule beads will become the norm[89-91].

Table 2: Some examples of adjuvant polymer for alginates capsules

Core	Coating	Property	Ref.
Alginate	PLL/Alginates	Biocompatibility	34
Alginate	PLL/PEI	Stability	7, 58
Alginate	Silica	Mechanical Stability	92
Alginate	CS/PMCG	Stability	93
Alginate	Chitosan	Biocompatibility	66, 94
Alginate/PEG		Gel Stability	59
Alginate/Chitosan		Gel Stability	95
Alginate/Gelatin		Cell attachment	96

Note: CS: Cellulose Sulphate; PMGC: poly(methylene-co-guanidine) hydrochloride; PEG: Poly(ethylene glycol); PEI: poly(ethylene imine); PLL: poly-L-Lysine

Chitosan

Chitosan is a linear amino-polysaccharide D-Glucan, the product of chitin N-deacetylation. For years, chitin and chitosan have been used as pharmaceutical excipients[97, 98]. It is a natural cationic polymer, soluble in diluted acid solution (generally pH below 6). Structurally, it mimics extracellular matrix components such as glycoaminoglycans. Biocompatibility and immunocompatibility of chitosan is not clear at this time[24]. The cationic polymer is mucoadhesive and no controlled studies regarding the safety and tolerability of chitosans are reported.

Gelling properties

Chitosan can form gels if associated with divalent anions or anionic polymers. It also gelifies with polyelectrolyte of opposite charge to form polyelectrolyte complex (PEC). Chitosan can form gel if associated with polyanions, alginates k-carrageenan, chondroitin sulphate, carboxymethylcellulose, dextran sulphate, gellan, heparin, pectin and xanthan[94 99]. Chitosan can also form covalent matrix with reactive crosslinking agent such as glyoxal and glutaraldehyde but these methods are generally not compatible with cell viability[100]. Although chitosan has been used to encapsulate microbial and plant cells[6], mammalian cell encapsulation was less successful as chitosan is water soluble only at non-physiological pH.

Encapsulation in chitosan matrices

In a process similar to alginate, cells can be entrapped in chitosan matrix[101]. Chitosan has been proposed for encapsulation of hepatocyte spheroids. The inner core of the capsule made of chitosan was coated with a layer of alginates. In spite of exposure to acidic pH, hepatocytes show viability and retain functions, however, no further information about immunoisolation and biocompatibility was given[101].

Encapsulation of the pancreatic islets employing chitosan-alginate matrix was also used for transplantation of microencapsulated Langerhans Islets to streptozotocin (STZ)-induced diabetic mice. This experiment resulted in disappearance of hyperglycemia and restoration of normoglycemia during a 30 days follow-up, suggesting graft viability and functionality. Neither graft failure, nor fibrous overgrowth was observed. This was attributed to the fibroblast-growth inhibitory properties of chitosan by the authors[102].

The chitosan derivative, glycol chitosan, a cationic polymer is soluble at neutral pH. By forming strong polyion complex with alginates, multi-layered microcapsules of alginates/glycol-chitosan show promises for immunoisolation[90]. *In vitro* testing of a new chitosan-polyvinyl pyrrolidone (PVP) hydrogels, crosslinked with glutaraldehyde showed low protein absorption, permeability well suited for immunoisolation and biocompatibility with Islets of Langerhans[103].

Chitosan as a cell scaffold in capsules

Glycosaminoglycans play a critical role in the cell attachment, differentiation, and morphogenesis. Because of its structural similarity to these molecules, chitosan was evaluated as a scaffold biomaterial. Freshly isolated foetal porcine hepatocytes were cultured on modified chitosan scaffolds and transplanted into rat groin fat pads with or without growth factor-induced neovascularization. Hepatocyte viability and liver organoid formation were examined and the chitosan-based biomaterial surfaces showed good hepatocyte attachment properties. In the future, chitosan-based biomaterials may be useful as scaffolds for creating liver tissue organoids[104]. Chitosan was also used as an attachment matrix for other mammalian cells such PC12 and fibroblasts, and precipitated chitosan seemed like a viable milieu for different cell lines[105].

Chitosan as an external layer in PE capsules

As mentioned earlier, its polycationic nature allows chitosan to be used as a substitute for PLL with alginate-based capsules. Chitosan have been reported to form an effective stable membrane with polyanionic inner core polysaccharides[60, 82]. Likewise, a new one-step procedure to prepare alginate or iota-Carrageenans/oligochitosan capsules has been proposed as an alternative to the alginate/PLL system[106, 107]. The optimal capsule strength is obtained with oligochitosan M_w of 2 to 4 kDa. Similar results were obtained by others groups[31]. This system might be beneficial because it involves only natural PS and allows working near neutral pH[106].

Agarose

Agarose is a neutral polysaccharide extracted from red algae, composed of alternate units of ß-D-galactopyranosyl and α-L-galactopyranosyl. Different modifications and substitutions such as that of hydroxyl groups by sulphate hemiesters and methyl esters lead to polymorphism. Moreover, because of the desulphonation of the natural agar polymer during its extraction, a large proportion of the α-L-galactopyranosyl is in the 3,6-anhydro form[108].

Gelling properties

Agar and agarose gel upon heating and cooling. The gelling properties are determined by M_w, M_w distribution, content of charged groups (the propensity to form gel increases with degree of desulphonation) and regularity of alternate monomers. Upon dissolution in aqueous solution (as low as 0.2 %), agarose forms gel. The gel structure is created upon cooling by formation of intramolecular hydrogen bonding (below 60°C) which stiffens the polymer chain. Below 40°C, non-covalent co-operative intermolecular hydrogen bonding in ordered conformation form double stranded helices that assemble into microcrystalline junction zones forming large pentagonal pores[108]. Chemical modifications with charged groups can interfere with intermolecular hydrogen bonding, affecting the degree of binding and, thus, modify the gel strength, permeability and gelling temperature. Gelling temperature could be lowered down to 30°C with methoxy substitutions, a temperature compatible with the cell survival. Gelling temperature is also dependent on gel concentration. Once the gel is formed, it cannot melt again unless the temperature is raised above 60°C. Because of this stability and durability, agarose can be advantageously used even if agarose gel is mechanically weaker than alginate or carrageenan gels[12].

Cell encapsulation

Pioneer work on agarose for cell immunoisolation were conducted by Hiroo Iwata and colleagues[109-112]. Agarose microparticles can be produced by a simple one step extrusion procedure[113] or by water-in-oil emulsion technique followed by cooling to allow capsule hardening[111]. Allografted islets of Langerhans encapsulated in 5% agarose gel restored normoglycemia for duration comparable to alginates capsules[111]. However, reported viability decreased markedly with transplantation of xenograft islets. *In vivo* survival of xenograft was improved by higher agarose concentration (7.5%), which may be due to the restricted diffusion of host antibodies into the capsules[114]. Islet xenotransplantation was enabled by agarose microbeads with agarose concentration higher than 7.5% even in recipients having a high level of preformed antibodies[114]. Complement activation system has been blamed for the reduced viability of xenografts[115]. A blend of agarose and polystyrene sulphonic acid (pSSA) shows promise to increase xenografts survival time, while maintaining diffusion of nutrients and excreta[116, 117]. pSSA seems to prevent xeno-antigen exposure, increases capsule strength[117], interacts with complement system and prevents its activation[115]. Similar results were obtained with agarose/polyvinyl sulphate microbeads, which show lesser cytotoxicity than pSSA[118]. Agarose does not interact with polyanion pSSA. Therefore, to prevent its leakage, synthetic polycations such as HexaDimethrine Bromide[119] or polybrene[120, 121] are added to form an insoluble polyion complex layer at the bead surface. An external layer of carboxymethylcellulose completes the capsule to give anionic properties to the surface and enhances biocompatibility[120, 121]. These more complex three-layered agarose capsules showed a similar response to glucose stimulation of encapsulated islets compared to simple agarose capsules[121]. Another potential problem due to the technique is the cell protrusion across the agarose, causing immunogenic reactions. The emulsion technique not only gives size heterogeneity, but can also cause shear stress on the cells and lead to heterogeneous dispersion of the cells within the capsules. This shortcoming can be eliminated by coating the agarose bead with an additional layer of agarose or by installing a polyionic layer similar to alginates/PLL capsules[120, 121]. Agarose has also been tested to prepare macrocapsules (6 to 8 mm) to encapsulate around 1000 pancreatic islets per bead[122]. Surprisingly, normal glucose levels have been reported in spite of the problematic cell survival in particles of this size. This report also showed the importance of the presence of a matrix compatible with cell survival and attachment. Indeed, the best results were obtained using a collagen/agarose blend to entrap the islets before adding a second agarose layer known for its immunoisolation properties[122]. On the other side, similar to the alginates, the smaller agarose capsules (less than 100 μm) showed enhanced mechanical stability and molecular diffusion[123], although smaller size increases the risk of cell protrusion.

Carrageenans

Carrageenans are extracted from marine red algae and there are different types. All of them share a common structure of alternating monomer of β (1,3)-D-galactose and α(1,4) 3,6-anhydro-D-galactose. In the "kappa" family, the β(1,3)-D-galactose units are sulphated in position 4, whereas in "iota" carrageenan both monomers are sulphated (other carrageenans exist but cannot form gels). Carrageenans contain sulphate group up to 24 to 55% of their monomer, influencing their gelling properties. As agarose, kappa (κ) and iota (ι) carrageenans can form thermo-reversible gels, but the resulting gel tends to be weaker. κ-carrageenans produce improved gel in presence of calcium, potassium and sodium cations. The strength of gel increases with molecular weight, carrageenan concentration, concentration of electrolyte and desulphation [108]. κ-carrageenans tend to form strong and rigid gel, while ι-carrageenans produce soft and elastic gel, less prone to defects. They can also form complex coacervation PE gel with polycations[107]. Immobilization of cells in κ-carrageenans has been described using K^+ as gel inducing agent, however, ion exchange in physiological conditions tends to dissolve the gel[124]. The formation of new microcapsules based on PE complexes between carrageenans and oligochitosan has been investigated. ι-carrageenans were found to be the most suitable for the formation of mechanically stable capsules. The capsules combined extremely high deformability and elasticity with permeability control. It has been shown that the length of the oligochitosan deposition time also influences the mechanical properties, whereas carrageenan concentration and the temperature during the capsule formation affect both, the mechanical and porosity characteristics of the membrane[107]. As encapsulating agents, carrageenans have found applications in yeast entrapment[125] and pro-biotic encapsulation[126]. Carrageenans have been reported as good candidates for inner hydrogel matrix for immunoisolation capsules[61].

Pectin

Pectin can be found in the cell wall of higher plants. It is mostly composed of 1-4 linked α-D galactosyluronic acid residues, but other sugars like L-arabinose and D-galactose can be found in varying amounts. Native pectins have around 70% of their carboxyl groups esterified with methyl groups. Pectates obtained by de-esterification are able to form gel. Degree of methylation (high methyl or low methyl pectin) controls gelling properties[127]. Pectin swells and dissolves rapidly in the aqueous medium unless it is somehow crosslinked. Pectates of low esterification degree can form gel in the presence of Ca^{2+} in a mechanism similar to alginates gelation (see figure 3). Pectate can also gel by interaction with alginates, forming aggregates between alginate GG blocks and pectin methoxy groups. As an anionic PE, it can also interact with polycations like chitosan and PLL to form gel. Calcium pectate gels have been tested for simple cell immobilization. Comparison of respiratory activity, substrate uptake and biosynthetic capacity of immobilized cells showed that pectate gel permitted faster substance transfer compared to alginate gel[128]. However, preparation of pectin microcapsules is difficult and particles are not spherical. Tails and coalescence could be observed with high methyl pectins while low methyl pectins gave irregular surfaces[127]. Stable in water, pectin/calcium chloride capsules dissolve in culture medium within few days due to polycation substitution by electrolytes[61]. Nevertheless, pectin has been used as an additive to alginates/PLL capsules for oral delivery of pro-biotics. A layer of pectate with low esterification degree added in a sandwich structure of alternate layers of polyanion and polycation (alginate-PLL-pectate-PLL-alginate), has been shown to impart superior acid stability and improved strength to the capsules[91].

Dextran & Dextran sulphate

Dextrans are D-glucose polymers of microbial origin, which mainly consist of alpha-D-(1-6) linked monomers, branched by alpha-(1-3) links. While low M_w dextrans have less branching and narrower range of M_w distribution, dextrans with M_w greater than 10 kDa are highly branched. Most dextrans are highly water-soluble. They can be used as viscosity binders in PE gel capsule for immunoisolation[60]. Dextran sulphate is an anionic polymer, mimicking natural mucopolysaccharides (as chondroitin sulphate or dermatan sulphate). It has been shown that all dextran sulphates inhibit complement activation[115]. In addition to their common inhibitory effects, dextran sulphate with M_w of 10 kDa and a degree of sulphonation of 1.99 on its pyranose ring, specifically degraded C3 without complement activation and, thus, there was no inflammatory reaction. These facts suggest that a membrane including dextran sulphate could effectively protect islet cells from humoral immunity while not triggering inflammatory reactions[129].

Cellulose & Cellulose derivatives

Cellulose is a 1-4 linked ß-D-glucopyranose linear polymer. Cellulose is a mechanically and chemically stable, inert polymer which is insoluble in water[130]. Cellulose membranes were used as immunoisolation encapsulation devices[131]. In particular, whilst studied as a support for cell culture, commercially regenerated cellulose (amorphous cellulose) membranes showed good permeability properties (M_w cut-off of 14 kDa) for islet transplantation and good biocompatibility although *in vitro* study suggested possibility of minor foreign body reaction[132]. Water insolubility is a major hurdle for the use of cellulose in cell microencapsulation, because cellulose can only be dispersed and

Table 3: Natural and modified Polysaccharides in cell encapsulation

	Origin	Gelation properties
Polyanionic polymers		
Alginates	Seaweeds	Ionotropic, complex coacervation
Carrageenans	Seaweeds	Thermal/Ionotropic
Pectin	Plants	Ionotropic, complex coacervation
Cellulose sulphate *	Plants	Complex coacervation
Carboxymethyl cellulose *	Plants	Complex coacervation
Sulphoethyl cellulose *	Plants	Complex coacervation
Dextran sulphate *	Microbial	Complex coacervation
Xanthan	Microbial	Thermal/Ionotropic
Gellan	Microbial	Thermal/Ionotropic
Hyaluronic acid	Animal	Complex coacervation
Chondroitin sulphate	Animal	Complex coacervation
Polycationic polymers		
Chitosan	Animal	Ionotropic, complex coacervation
Neutral polymers		
Agarose	Seaweeds	Thermal
Cellulose	Plant	Precipitation
Starch	Plants	Cross-link
Dextran	Microbial	Cross-link

Note: * modified natural PS

precipitated in organic solvent. Nonetheless, pre-existing cellulose particles have been used as a cell carrier for hepatocytes[133, 134]. Upon sulphonation, cellulose is converted to polyanionic cellulose sulphate (CS), a water-soluble polyelectrolyte which can form PEC gel in contact with a polymer of opposite charge. The couple, sodium CS/poly-(diallyl-dimethylammonium) chloride (pDADMAC) has been proposed and validated in different models[135]. CS/pDADMAC capsules delivering antibody producing cells, implanted under the skin in a mouse model proved to be a simple encapsulation procedure, displaying large pores, and excellent mechanical properties *in vivo* on long term basis[136]. In another approach, CS and pDADMAC polymerized on a transient Ca^{2+}/alginate scaffold, form homogenous capsules following dissolution of the alginate core by Ca^{2+} chelating agents. These so-called "CellMAC" capsules exhibited excellent mechanical properties and showed a molecular weight cut-off between 43 and 77 kDa[137]. Similarly, functional porcine Langerhans islets were recently successfully encapsulated in sodium CS/pDADMAC microcapsules[138]. The use of cellulose derivative, sulphoethyl cellulose (SEC) has also been proposed to prepare hollow beads on the basis of PEC gel using SEC as polyanion and chitosan as polycation[139]. The characteristics of the capsules were dependent on the M_w of SEC and chitosan, the degree of sulphonation of the cellulose and the polycation concentration[139].

Other polysaccharides

Glycoaminoglycans

While achieving proper permeability properties are crucial, another important factor is to provide cells with an environment that allows scaffolding, adhesion and promotes normal tissue functions. Glycosaminoglycans are negatively charged, highly viscous molecules playing a role in organization and stabilization of extracellular matrix (ECM), cell adhesion and proliferation. The physiologically most important glycosaminoglycans are hyaluronic acid, chondroitin sulphate, heparin, and heparan sulphate. They have the ability to form hydrogel and they emulate cell environment. Hyaluronic acid has the ability to form capsules in the presence of chitosan[60, 82]. Chondroitin A and C have been identified as potential biomaterials for immunoisolation[60] and have been used to generate PE capsules associated with CS and chitosan[60,82]. Chondroitin hydrogel embedded chondrocytes remained viable after photoencapsulation, suggesting their possible use for cartilage tissue engineering and possibly for cell encapsulation[140, 141].

Bacterial polysaccharides

Gellan is a natural anionic exopolysaccharide of microbial origin (*Pseudomonas elodea*). The structure is composed of tetrasaccharide repeating units. Native polymer bears acyl group on glucose residue as glycerate (one per repeated unit) and acetate (one per every two repeated units). Gellan gum is obtained by de-esterification of native polymers and contains only a low level of acyl groups. Gellan can form gels at low pH and in the presence of salts. The gelation is promoted by cations, divalent cations being more effective than monovalent cations. Bacteria were encapsulated in gellan gum microbeads. Encapsulated cells degraded comparable amounts of gasoline within the same period compared to equivalent levels of free cells. Encapsulation provided a protective barrier against toxic materials, eliminating the adaptation period[142]. Similar to gellan, xanthan is an exopolysaccharide of bacterial origin (*Xanthomas campestris*), an acidic polymer made of pentasaccharide subunits forming a cellulose backbone, grafted with trisaccaharide side chains[144]. Moreover, pyruvate and acetyl substitution are present in variable percentages. Solutions of xanthan have high viscosity at low concentration. Xanthan is insensitive to a broad range of temperature, pH and electrolyte concentration[144]. Although it has not been extensively studied, these properties could be of interest in immunoisolation as a gel forming PEC[61] or as an adjuvant to hydrogel for viscosity control[60].

Polymers combination in PEC gel

A systematic approach was taken recently to test combination of natural and synthetic PEs for cell immunoisolation[60, 61]. A wide spectrum of natural and modified polysaccharides was tested in several steps for cytotoxicity, gel strength, capsule resistance, permeability and physical characteristics. Mixed gel can exist as an interpenetrating network (IPN), a binding mix or a phase separation gel. The optimal inner polymer is a highly charged polyanion with M_w over 10 kDa. Alginates, carboxymethyl cellulose, λ-carrageenan, CS, pectin, gellan and xanthan are optimal PS for this purpose[61]. Blends of two PS gave mechanically stable capsule compared to individual PS. Regarding polycations, M_w[65, 106], concentration and charge densities dictated the performance. Presence of hydrogen bonding, adds to the stability of the PEC gel. Synthetic polymers were not suitable and were found to be generally toxic to insulin producing cells. A combination of low and high M_w PE often results in more stable capsules and permits permeability adjustment. One of the most promising polymer blends was described as a combination of the polyanionic sodium alginate and CS and the polycationic poly (methylene-co-guanidine) hydrochloride in the presence of calcium chloride and sodium chloride[93].

Amylopectin hydrogel microcapsules

Stable starch hydrogels can only be established by chemical crosslinking, which may be very damaging to living cells. To overcome these limitations, a team developed a few years ago, a new concept with the aim to develop CNS applications. To overcome the BBB, implantation of microencapsulated cells to deliver peptides with treatment for different neurological disorders in a localized and regulated manner is a promising strategy[2]. To alleviate neurochemichal deficits, adrenal medullary chromaffin cells were chosen since they produce high levels of both opioid peptides and catecholamines[145, 146]. Generally, the composition of the capsules used for the protection of chromaffin cells is alginate-PLL-alginate[70, 147, 148]. The intrathecal implantation of micro-encapsulated chromaffin cells might be a useful method to treat pain and Parkinson's disease[70, 147, 149].

Similar to macroencapsulation, the new approach was to encapsulate cells after the synthesis of the immunoisolation capsules by automatic microinjection[150]. This approach separates the capsule preparation from the encapsulation procedure, thus, avoiding stress on the cells during the entrapment step and non-homogenous repartition of cells in the beads. However, to reach this goal, the critical step was to produce hydrogel hollow particles of the desired size, wall thickness and permeability, compatible with the injection procedure.

Few methods have been described to produce hollow particles in the micrometer/millimeter range and compatible with cell viability. Preparation of hollow particles has been described essentially for nano-capsules for drug and peptide delivery system, capsules for aroma in food preparation, contrast agent for medical imaging, and floating gastro-intestinal drug delivery device. Double emulsion and spray drying using synthetic polymers are the techniques most commonly used. Other techniques have been developed, including self assembling capsules, vesicular polymerization, template approach using melamine formaldehyde particles coated with PE and emulsion/suspension polymerisation, but were not designed specifically for cell encapsulation and offer limited cell loading capacity[151]. Recently, PE deposition on a silica template followed by acidic degradation of silica has been proposed[143].

The first method investigated, was to introduce air micro-bubbles in a specially designed reactor during amylopectin gel crosslinking reaction[150]. Amylopectin (AP) is a highly non-randomly branched polysaccharides, with a native M_w of several millions and is used here as a partially hydrolyzed product (Glucidex 2 from Roquette, France). The hollow particles obtained by this approach have allowed the validation of the concept with dopamine secreting PC12 cell lines for CNS delivery[152]. However, this method was not satisfactory, as yield remained very low.

We have proposed a new method to produce hydrogel hollow capsules, relying on gel coating of a degradable core particle, which serves as a template for capsule wall formation[153, 154]. The synthesis scheme (figure 4), is a three-step original procedure: first, synthesis of a degradable microspheric core, followed by coating of the core by a layer of crosslinked AP gel by emulsion-polymerization, and finally, selective hydrolytic degradation of the core by pH change to form the hollow particle cavity[154]. The size of the cavity targeted is between 200 and 250 microns, for a total hollow particle diameter of 500 microns (size determined for host implantation by direct injection). Partially hydrolyzed AP has been selected because it is an abundant and inexpensive natural polymer, it is approved for human use, non-irritant when implanted *in vivo* (soft surface & contaminant-free) and biocompatible because of low free surface energy and low protein and cell adhesion. Permeability of the AP gel can be controlled by varying the parameter of the crosslinking reaction[155]. The subsequent steps are to microinject the cells into the cavity, along with PS matrix for cell attachment and growth as there is no extracellular matrix in the hollow particle. Once the cells are encapsulated, a supplementary coating step could take place before implantation for puncture point repair or add functions on the surface (figure 4).

We have selected non-polysaccharide degradable material for the core of particle with the aim to obtain selective degradation by pH change. A lot of degradable polymers have been developed for micro-particles preparation. Amongst them, aliphatic polyesters, such as polylactide (PLA) are available commercially in a large range of molecular weights and compositions. They are relatively inexpensive, easy to synthesize and approved by the Food and Drug Administration for human use. PLA is largely considered as biocompatible, and in spite of its relatively hydrophobic nature, it is completely degradable, generating the lactate eliminated by the organism.

PLA particles were prepared according to the solvent emulsion/evaporation technique to produce large spherical and smooth particles (around 220 microns as determined image analysis). Coating was carried out by emulsion-polymerization of a solution of AP and polyester core particles in paraffin oil under stirring[153]. The coating was stabilized by an crosslinking reaction at high molar ratio of cross-linking agent/glucose residues[156]. The template was eliminated by specific degradation of polyester in basic conditions. The end result showed hollow particles with residual polyesters, after 48 hours incubation time (figure 5).

Physical characterization of hydrogel microcapsules *in vitro* has been reviewed recently[157, 158]. The first factor to be considered is hydrogel and capsule permeability and has been discussed earlier in this chapter[12, 16, 17, 20]. The resistance to osmotic pressure[159] and mechanical stress[160, 161] are also important parameters. The hollow particles obtained with the new method described above were characterized for size and morphology (image analysis) and permeability. Permeability studies were conducted using FITC-Dextrans, using two approaches: ingress diffusion from the external medium or egress diffusion[153].

The method seems to produce a range of permeability needed for immunoisolation and shows several advantages over PEC methods. The separation of capsule preparation from the encapsulation step significantly reduces stress on cells. Cell protrusion and inhomogeneous repartition of cell within the capsule are avoided. Moreover, it allows use of chemical crosslinking to produce more stable hydrogel. Finally, the capsules can be completely

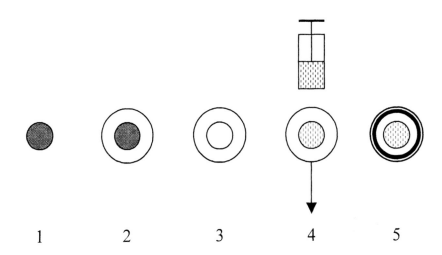

Figure 4: The five steps procedure to encapsulate cells
(1) Preparation of core polyester particle; (2) Coating of the particle with
hydrogel; (3) Selective degradation of the core; (4) Cell loading by
microinjection; (5) Optional modification of the surface (could take place after
or before cell injection).

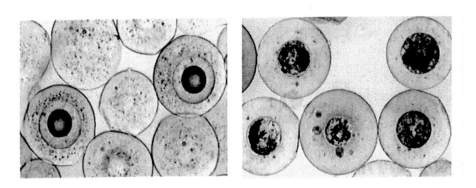

Figure 5: Core particle coating by emulsion/polymerization
Left : After coating step; Right : After degradation of the core particle.

characterized before their use. This approach could be extended to other materials for core particle and capsule walls.

Conclusion

Natural and modified PS having hydrogel forming ability are increasingly characterized and widely used in immunoisolation techniques. The chemical structure and the gelling mechanism of PS affect the properties of the microencapsulation devices. They are the materials of choice in many applications because of their biocompatibility and ability to entrap cells in mild conditions. In recent years, improvements in hydrogel and capsules have been made, such as use of polymer blends, development of new crosslinking methods, improved purification and methods to produce multi-layered capsules. In spite of a considerable literature on the use of encapsulated cells in animal models, there are still no significant clinical studies. More research has to be conducted to select the most suitable PS and preparation procedure for each application. At the same time, all the approaches suffer because of simultaneous cell entrapment and capsule formation steps, limiting the choice of crosslinking method. In an effort to overcome the intrinsic limitations of non-covalent PS gels, a new paradigm is proposed for the preparation of stable crosslinked starch hydrogel capsules that can be loaded by cells by automated microinjection after characterization. This new device should find its place particularly in the field of CNS delivery.

References

1. Aebischer, P.; Ridet, J. *Trends Neurosci* **2001**, 24, (9), 533-40.
2. Shoichet, M. S.; Winn, S. R. *Adv Drug Deliv Rev* **2000**, 42, (1-2), 81-102.
3. Uludag, H.; De Vos, P.; Tresco, P. A. *Adv Drug Deliv Rev* **2000**, 42, (1-2), 29-64.
4. Morris, P. J. *Trends Biotechnol* **1996**, 14, (5), 163-7.
5. Rihova, B. *Adv Drug Deliv Rev* **2000**, 42, (1-2), 65-80.
6. Li, R. H. *Adv Drug Deliv Rev* **1998**, 33, (1-2), 87-109.
7. Lim, F.; Sun, A. M. *Science* **1980**, 210, (4472), 908-10.
8. Robitaille, R.; Pariseau, J. F.; Leblond, F. A.; Lamoureux, M.; Lepage, Y.; Halle, J. P. *J Biomed Mater Res* **1999**, 44, (1), 116-20.
9. Cotton, C. K. *Trends Biotechnol* **1996**, 14, (5), 158-62.
10. Jen, A. C.; Conley-Wake, M.; Mikoas, A. G. *Biotech. and Bioeng.* **1996**, 50, 357-364.
11. Hoffman, A. S. *Adv Drug Deliv Rev* **2002**, 54, (1), 3-12.

12. Yi, Y.; Neufeld, R. J.; Poncelet, D. In *Polysaccharides, Structural diversity and functionnal versatility*, 2nd ed.; Dimitriu, S., Ed. Marcel Dekker: New York, 2005; pp 867-891.

13. Tsang, V. L.; Bhatia, S. N. *Adv Drug Deliv Rev* **2004**, 56, (11), 1635-47.

14. Hennink, W. E.; van Nostrum, C. F. *Adv Drug Deliv Rev* **2002**, 54, (1), 13-36.

15. de Haan, B. J.; Faas, M. M.; de Vos, P. *Cell Transplant* **2003**, 12, (6), 617-25.

16. Brissova, M.; Lacik, I.; Powers, A. C.; Anilkumar, A. V.; Wang, T. *J Biomed Mater Res* **1998**, 39, (1), 61-70.

17. Brissova, M.; Petro, M.; Lacik, I.; Powers, A. C.; Wang, T. *Anal Biochem* **1996**, 242, (1), 104-11.

18. Martisen, A. S., I.; Skjak-Braek, G. *Biotech. and Bioeng.* **1992**, 39, 186-194.

19. Gehrke, S. H.; Fisher, J. P.; Palasis, M.; Lund, M. E. *Ann N Y Acad Sci* **1997**, 831, 179-207.

20. Powers, A. C.; Brissova, M.; Lacik, I.; Anilkumar, A. V.; Shahrokhi, K.; Wang, T. G. *Ann N Y Acad Sci* **1997**, 831, 208-16.

21. Kierstan, M.; Bucke, C. *Biotechnol Bioeng* **1977**, 19, (3), 387-97.

22. Rowley, J. A.; Mooney, D. J. *J Biomed Mater Res* **2002**, 60, (2), 217-23.

23. Boontheekul, T.; Kong, H. J.; Mooney, D. J. *Biomaterials* **2005**, 26, (15), 2455-65.

24. Dornish, M.; Kaplan, D.; Skaugrud, O. *Ann N Y Acad Sci* **2001**, 944, 388-97.

25. Zekorn, T.; Siebers, U.; Horcher, A.; Schnettler, R.; Klock, G.; Bretzel, R. G.; Zimmermann, U.; Federlin, K. *Transplant Proc* **1992**, 24, (3), 937-9.

26. O'Shea, G. M.; Goosen, M. F.; Sun, A. M. *Biochim Biophys Acta* **1984**, 804, (1), 133-6.

27. Omer, A.; Duvivier-Kali, V.; Fernandes, J.; Tchipashvili, V.; Colton, C. K.; Weir, G. C. *Transplantation* **2005**, 79, (1), 52-8.

28. Thu, B.; Bruheim, P.; Espevik, T.; Smidsrod, O.; Soon-Shiong, P.; Skjak-Braek, G. *Biomaterials* **1996**, 17, (11), 1069-79.

29. Kong, H. J.; Kaigler, D.; KIm, K.; Mooney, D. J. *Biomacromolecules* **2004**, 5, 1720-1727.

30. Sugiura, S.; Oda, T.; Izumida, Y.; Aoyagi, Y.; Satake, M.; Ochiai, A.; Ohkohchi, N.; Nakajima, M. *Biomaterials* **2005**, 26, (16), 3327-31.

31. Serp, D.; Cantana, E.; Heinzen, C.; Von Stockar, U.; Marison, I. W. *Biotechnol Bioeng* **2000**, 70, (1), 41-53.

32. Poncelet, D. *Ann N Y Acad Sci* **2001**, 944, 74-82.

33. Zimmermann, H.; Hillgartner, M.; Manz, B.; Feilen, P.; Brunnenmeier, F.; Leinfelder, U.; Weber, M.; Cramer, H.; Schneider, S.; Hendrich, C.; Volke, F.; Zimmermann, U. *Biomaterials* **2003**, 24, (12), 2083-96.

34. Goosen, M. F.; O'shea, G.; Gharapetian, H. M.; Chou, S.; Sun, A. M. *Biotech. and Bioeng.* **1985**, 27, 146-150.

35. Halle, J. P.; Bourassa, S.; Leblond, F. A.; Chevalier, S.; Beaudry, M.; Chapdelaine, A.; Cousineau, S.; Saintonge, J.; Yale, J. F. *Transplantation* **1993**, 55, (2), 350-4.

36. Awrey, D. E. T., M. Hortelano, G.;Chang, P.L. *Biotech. and Bioeng.* **1996**, 52, 472-484.

37. Thu, B.; Bruheim, P.; Espevik, T.; Smidsrod, O.; Soon-Shiong, P.; Skjak-Braek, G. *Biomaterials* **1996**, 17, (10), 1031-40.

38. Bienaime, C.; Barbotin, J. N.; Nava-Saucedo, J. E. *J Biomed Mater Res A* **2003**, 67, (2), 376-88.

39. Strand, B.; Morch, Y. A.; Skjak-Braek, G. *Minerva Biotec.* **2000**, 12, 223-233.

40. Zimmermann, U.; Thurmer, F.; Jork, A.; Weber, M.; Mimietz, S.; Hillgartner, M.; Brunnenmeier, F.; Zimmermann, H.; Westphal, I.; Fuhr, G.; Noth, U.; Haase, A.; Steinert, A.; Hendrich, C. *Ann N Y Acad Sci* **2001**, 944, 199-215.

41. Wandrey, C.; Vidal, D. S. *Ann N Y Acad Sci* **2001**, 944, 187-98.

42. Klock, G. F., H.; Houben, R.; Zekorn, T.; Horcher, A.; Siebers, U.; Wöhrle, M.; Federlin, K.; Zimmermann, U. *Appl. Microbiol. Biotechnol.* **1994**, 40, (5), 638–643.

43. De Vos, P.; De Haan, B. J.; Wolters, G. H.; Strubbe, J. H.; Van Schilfgaarde, R. *Diabetologia* **1997**, 40, (3), 262-70.

44. Hasse, C.; Zielke, A.; Klock, G.; Schlosser, A.; Barth, P.; Zimmermann, U.; Sitter, H.; Lorenz, W.; Rothmund, M. *World J Surg* **1998**, 22, (7), 659-65.

45. de Vos, P.; van Hoogmoed, C. G.; van Zanten, J.; Netter, S.; Strubbe, J. H.; Busscher, H. J. *Biomaterials* **2003**, 24, (2), 305-12.

46. Klock, G.; Pfeffermann, A.; Ryser, C.; Grohn, P.; Kuttler, B.; Hahn, H. J.; Zimmermann, U. *Biomaterials* **1997**, 18, (10), 707-13.

47. Kulseng, B.; Skjak-Braek, G.; Ryan, L.; Andersson, A.; King, A.; Faxvaag, A.; Espevik, T. *Transplantation* **1999**, 67, (7), 978-84.

48. Soon-Shiong, P.; Otterlie, M.; Skjak-Braek, G.; Smidsrod, O.; Heintz, R.; Lanza, R. P.; Espevik, T. *Transplant Proc* **1991**, 23, (1 Pt 1), 758-9.

49. Strand, B. L.; Ryan, T. L.; In't Veld, P.; Kulseng, B.; Rokstad, A. M.; Skjak-Brek, G.; Espevik, T. *Cell Transplant* **2001**, 10, (3), 263-75.

50. Robitaille, R.; Dusseault, J.; Henley, N.; Desbiens, K.; Labrecque, N.; Halle, J. P. *Biomaterials* **2005**, 26, (19), 4119-27.

51. Sawhney, A. S.; Hubbell, J. A. *Biomaterials* **1992**, 13, (12), 863-70.

52. Sawhney, A. S.; Pathak, C. P.; Hubbell, J. A. *Biomaterials* **1993**, 14, (13), 1008-16.

53. Kung, I. M.; Wang, F. F.; Chang, Y. C.; Wang, Y. J. *Biomaterials* **1995**, 16, (8), 649-55.

54. Haque, T.; Chen, H.; Ouyang, W.; Martoni, C.; Lawuyi, B.; Urbanska, A. M.; Prakash, S. *Mol Pharm* **2005**, 2, (1), 29-36.

55. Strand, B. L.; Morch, Y. A.; Syvertsen, K. R.; Espevik, T.; Skjak-Braek, G. *J Biomed Mater Res A* **2003**, 64, (3), 540-50.

56. King, A.; Strand, B.; Rokstad, A. M.; Kulseng, B.; Andersson, A.; Skjak-Braek, G.; Sandler, S. *J Biomed Mater Res A* **2003**, 64, (3), 533-9.

57. Lee, C. S.; Chu, I. M. *Artif Organs* **1997**, 21, (9), 1002-6.

58. Kong, H. J.; Mooney, D. J. *Cell Transplantation* **2003**, 12, (7), 779-785.

59. Desai, N. P.; Sojomihardjo, A.; Yao, Z.; Ron, N.; Soon-Shiong, P. *J Microencapsul* **2000**, 17, (6), 677-90.

60. Prokop, A. H., D.; DimAri, S.; Haralson, M.A.; Wang, T.G. *Adv. Polym. Sci.* **1998**, 136, 1-51.

61. Prokop, A. H., D.; Powers, A.C.; Whitesell, R.R.; Wang, T.G. *Adv. Polym. Sci.* **1998**, 136, 52-73.

62. Calafiore, R.; Basta, G.; Sarchielli, P.; Luca, G.; Tortoioli, C.; Brunetti, P. *Acta Diabetol* **1996**, 33, (2), 150-3.

63. Heald, K. A.; Jay, T. R.; Downing, R. *Cell Transplant* **1994**, 3, (4), 333-7.

64. Gaserod, O.; Sannes, A.; Skjak-Braek, G. *Biomaterials* **1999**, 20, (8), 773-83.

65. Gaserod, O.; Smidsrod, O.; Skjak-Braek, G. *Biomaterials* **1998**, 19, (20), 1815-25.

66. Chandy, T.; Mooradian, D. L.; Rao, G. H. *Artif Organs* **1999**, 23, (10), 894-903.

67. Levy, M. C.; Edwards-Levy, F. *J Microencapsul* **1996**, 13, (2), 169-83.

68. Joly, A.; Desjardins, J. F.; Fremond, B.; Desille, M.; Campion, J. P.; Malledant, Y.; Lebreton, Y.; Semana, G.; Edwards-Levy, F.; Levy, M. C.; Clement, B. *Transplantation* **1997**, 63, (6), 795-803.

69. Dusseault, J.; Leblond, F. A.; Robitaille, R.; Jourdan, G.; Tessier, J.; Menard, M.; Henley, N.; Halle, J. P. *Biomaterials* **2005**, 26, (13), 1515-22.

70. Winn, S. R.; Tresco, P. A.; Zielinski, B.; Greene, L. A.; Jaeger, C. B.; Aebischer, P. *Exp Neurol* **1991**, 113, (3), 322-9.

71. Legallais, C.; David, B.; Dore, E. *J. Memb. Sc.* **2001**, 181, 81-95.

72. Hasse, C.; Bohrer, T.; Barth, P.; Stinner, B.; Cohen, R.; Cramer, H.; Zimmermann, U.; Rothmund, M. *World J Surg* **2000**, 24, (11), 1361-6.

73. Orive, G.; Hernandez, R. M.; Rodriguez Gascon, A.; Calafiore, R.; Chang, T. M.; de Vos, P.; Hortelano, G.; Hunkeler, D.; Lacik, I.; Pedraz, J. L. *Trends Biotechnol* **2004**, 22, (2), 87-92.

74. de Groot, M.; Schuurs, T. A.; van Schilfgaarde, R. *J Surg Res* **2004**, 121, (1), 141-50.

75. van Schilfgaarde, R.; de Vos, P. *J Mol Med* **1999**, 77, (1), 199-205.

76. Soon-Shiong, P. *Adv Drug Deliv Rev* **1999**, 35, (2-3), 259-270.

77. Soon-Shiong, P.; Heintz, R. E.; Merideth, N.; Yao, Q. X.; Yao, Z.; Zheng, T.; Murphy, M.; Moloney, M. K.; Schmehl, M.; Harris, M.; et al. *Lancet* **1994**, 343, (8903), 950-1.

78. Calafiore, R.; Basta, G.; Falorni, A., Jr.; Ciabattoni, P.; Brotzu, G.; Cortesini, R.; Brunetti, P. *Transplant Proc* **1992**, 24, (3), 935-6.

79. Toso, C.; Oberholzer, J.; Ceausoglu, I.; Ris, F.; Rochat, B.; Rehor, A.; Bucher, P.; Wandrey, C.; Schuldt, U.; Belenger, J.; Bosco, D.; Morel, P.; Hunkeler, D. *Transpl Int* **2003**, 16, (6), 405-10.

80. Leblond, F. A.; Simard, G.; Henley, N.; Rocheleau, B.; Huet, P. M.; Halle, J. P. *Cell Transplant* **1999**, 8, (3), 327-37.

81. Yin, C.; Mien Chia, S.; Hoon Quek, C.; Yu, H.; Zhuo, R. X.; Leong, K. W.; Mao, H. Q. *Biomaterials* **2003**, 24, (10), 1771-80.

82. Matthew, H. W.; Salley, S. O.; Peterson, W. D.; Klein, M. D. *Biotechnol Prog* **1993**, 9, (5), 510-9.

83. David, B.; Dufresne, M.; Nagel, M. D.; Legallais, C. *Biotechnol Prog* **2004**, 20, (4), 1204-12.

84. Visted, T.; Furmanek, T.; Sakariassen, P.; Foegler, W. B.; Sim, K.; Westphal, H.; Bjerkvig, R.; Lund-Johansen, M. *Hum Gene Ther* **2003**, 14, (15), 1429-40.

85. Cirone, P.; Bourgeois, J. M.; Austin, R. C.; Chang, P. L. *Hum Gene Ther* **2002**, 13, (10), 1157-66.

86. Read, T. A.; Sorensen, D. R.; Mahesparan, R.; Enger, P. O.; Timpl, R.; Olsen, B. R.; Hjelstuen, M. H.; Haraldseth, O.; Bjerkvig, R. *Nat Biotechnol* **2001**, 19, (1), 29-34.

87. Joki, T.; Machluf, M.; Atala, A.; Zhu, J.; Seyfried, N. T.; Dunn, I. F.; Abe, T.; Carroll, R. S.; Black, P. M., Continuous release of endostatin from microencapsulated engineered cells for tumor therapy. *Nat Biotechnol* **2001**, 19, (1), 35-9.

88. Rokstad, A. M.; Strand, B.; Rian, K.; Steinkjer, B.; Kulseng, B.; Skjak-Braek, G.; Espevik, T. *Cell Transplantation* **2003**, 12, (4), 351-364.

89. Schneider, S.; Feilen, P. J.; Slotty, V.; Kampfner, D.; Preuss, S.; Berger, S.; Beyer, J.; Pommersheim, R. *Biomaterials* **2001**, 22, (14), 1961-70.

90. Sakai, S.; Ono, T.; Ijima, H.; Kawakami, K. *J Microencapsul* **2000**, 17, (6), 691-9.

91. Ouyang, W.; Chen, H.; Jones, M. L.; Metz, T.; Haque, T.; Martoni, C.; Prakash, S. *J Pharm Pharm Sci* **2004**, 7, (3), 315-24.

92. Coradin, T.; Nassif, N.; Livage, J. *Appl Microbiol Biotechnol* **2003**, 61, (5-6), 429-34.

93. Wang, T.; Lacik, I.; Brissova, M.; Anilkumar, A. V.; Prokop, A.; Hunkeler, D.; Green, R.; Shahrokhi, K.; Powers, A. C. *Nat Biotechnol* **1997**, 15, (4), 358-62.

94. Polk, A.; Amsden, B.; De Yao, K.; Peng, T.; Goosen, M. F. *J Pharm Sci* **1994**, 83, (2), 178-85.

95. Orive, G.; Hernandez, R. M.; Gascon, A. R.; Igartua, M.; Pedraz, J. L. *Eur J Pharm Sci* **2003**, 18, (1), 23-30.

96. Kwon, Y. J.; Peng, C. A. *Biotechniques* **2002**, 33, (1), 212-4, 216, 218.

97. Illum, L. *Pharm Res* **1998**, 15, (9), 1326-31.

98. Kumar, M. N. V. R. *Reactive and Functional Polymers* **2000**, 46, 1-27.

99. Berger, J.; Reist, M.; Mayer, J. M.; Felt, O.; Gurny, R. *Eur J Pharm Biopharm* **2004**, 57, (1), 35-52.
100. Berger, J.; Reist, M.; Mayer, J. M.; Felt, O.; Peppas, N. A. *Eur J Pharm Biopharm* **2004**, 57, (1), 19-34.
101. Yu, S. B., R.; Kim, S. *Biotech. Tech.* **1999**, 13, 609-614.
102. Hardikar, A. A. R., M.V.; Bhonde, R.R. *J. of Biosc.* **1999**, 24, (3), 371-376.
103. Risbud, M.; Bhonde, M.; Bhonde, R. *Cell Transplantation* **2001**, 10, (2), 195-202.
104. Murat Elçin, Y.; Dixit, V.; Lewin, K.; Gitnick, G. *Artif. Org.* **1999**, 23, (2), 146-.
105. Zielinski, B. A.; Aebischer, P. *Biomaterials* **1994**, 15, (13), 1049-56.
106. Bartkowiak, A. *Ann N Y Acad Sci* **2001**, 944, 120-34.
107. Bartkowiak, A.; Hunkeler, D. *Colloids Surf B Biointerfaces* **2001**, 21, (4), 285-298.
108. Nijenhuis, K. T. *Adv. Polym. Sci.* **1987**, 136, 194.
109. Iwata, H.; Amemiya, H.; Matsuda, T.; Takano, H.; Hayashi, R.; Akutsu, T. *Diabetes* **1989**, 38 Suppl 1, 224-5.
110. Iwata, H.; Takagi, T.; Yamashita, K.; Kobayashi, K.; Amemiya, H. *Transplant Proc* **1992**, 24, (3), 997.
111. Iwata, H.; Takagi, T.; Amemiya, H.; Shimizu, H.; Yamashita, K.; Kobayashi, K.; Akutsu, T. *J Biomed Mater Res* **1992**, 26, (7), 967-77.
112. Iwata, H.; Takagi, T.; Amemiya, H. *Transplant Proc* **1992**, 24, (3), 952.
113. Gin, H.; Dupuy, B.; Baquey, C.; Ducassou, D.; Aubertin, J. *J Microencapsul* **1987**, 4, (3), 239-42.
114. Iwata, H.; Kobayashi, K.; Takagi, T.; Oka, T.; Yang, H.; Amemiya, H.; Tsuji, T.; Ito, F. *J Biomed Mater Res* **1994**, 28, (9), 1003-11.
115. Iwata, H.; Murakami, Y.; Ikada, Y. *Ann N Y Acad Sci* **1999**, 875, 7-23.
116. Iwata, H.; Takagi, T.; Kobayashi, K.; Oka, T.; Tsuji, T.; Ito, F. *J Biomed Mater Res* **1994**, 28, (10), 1201-7.
117. Miyoshi, Y.; Date, I.; Ohmoto, T.; Iwata, H. *Exp Neurol* **1996**, 138, (1), 169-75.
118. Suzuki, R.; Yoshioka, Y.; Kitano, E.; Yoshioka, T.; Oka, H.; Okamoto, T.; Okada, N.; Tsutsumi, Y.; Nakagawa, S.; Miyazaki, J.; Kitamura, H.; Mayumi, T. *Cell Transplant* **2002**, 11, (8), 787-97.
119. Kin, T.; Iwata, H.; Aomatsu, Y.; Ohyama, T.; Kanehiro, H.; Hisanaga, M.; Nakajima, Y. *Pancreas* **2002**, 25, (1), 94-100.
120. Tun, T.; Inoue, K.; Hayashi, H.; Aung, T.; Gu, Y. J.; Doi, R.; Kaji, H.; Echigo, Y.; Wang, W. J.; Setoyama, H.; Imamura, M.; Maetani, S.; Morikawa, N.; Iwata, H.; Ikada, Y. *Cell Transplant* **1996**, 5, (5 Suppl 1), S59-63.
121. Xu, B.; Iwata, H.; Miyamoto, M.; Balamurugan, A. N.; Murakami, Y.; Cui, W.; Imamura, M.; Inoue, K. *Cell Transplant* **2001**, 10, (4-5), 403-8.

338

122. Jain, K.; Yang, H.; Cai, B. R.; Haque, B.; Hurvitz, A. I.; Diehl, C.; Miyata, T.; Smith, B. H.; Stenzel, K.; Suthanthiran, M.; et al. *Transplantation* **1995**, 59, (3), 319-24.

123. Sakai, S.; Kawabata, K.; Ono, T.; Ijima, H.; Kawakami, K. *Biomaterials* **2005**, 26, (23), 4786-92.

124. Chibata, I.; Tosa, T.; Sato, T.; Takata, I. In *Methods in Enzymology*, Academic Press: New York, 1987; Vol. 135, pp 189-198.

125. Raymond, M. C.; Neufeld, R. J.; Poncelet, D. *Artif Cells Blood Substit Immobil Biotechnol* **2004**, 32, (2), 275-91.

126. Krasaekoopt, W.; Bhandari, B.; H., D. *Int. Dairy J.* **2003**, 13, 3-13.

127. Esposito, E.; Cortesi, R.; Luca, G.; Nastruzzi, C. *Ann N Y Acad Sci* **2001**, 944, 160-79.

128. Kurillova, L.; Gemeiner, P.; Vikartovska, A.; Mikova, H.; Rosenberg, M.; Ilavsky, M. *J Microencapsul* **2000**, 17, (3), 279-96.

129. Murakami, Y.; Iwata, H.; Kitano, E.; Kitamura, H.; Ikada, Y. *J Biomater Sci Polym Ed* **2003**, 14, (9), 875-85.

130. Miyamoto, T.; Takahashi, S.; Ito, H.; Inagaki, H.; Noishiki, Y. *J Biomed Mater Res* **1989**, 23, (1), 125-33.

131. Risbud, M. V.; Bhargava, S.; Bhonde, R. R. *J Biomed Mater Res A* **2003**, 66, (1), 86-92.

132. Risbud, M. V.; Bhonde, R. R.. *J Biomed Mater Res* **2001**, 54, (3), 436-44.

133. Kino, Y.; Sawa, M.; Kasai, S.; Mito, M. *J. of Surg. Res.* **1998**, 79, (1), 71-76.

134. Kobayashi, N.; Okitsu, T.; Maruyama, M.; Totsugawa, T.; Kosaka, Y.; Hayashi, N.; Nakaji, S.; Tanaka, N. *Transplant Proc* **2003**, 35, (1), 443-4.

135. Dautzenberg, H.; Schuldt, U.; Grasnick, G.; Karle, P.; Muller, P.; Lohr, M.; Pelegrin, M.; Piechaczyk, M.; Rombs, K. V.; Gunzburg, W. H.; Salmons, B.; Saller, R. M.*Ann N Y Acad Sci* **1999**, 875, 46-63.

136. Pelegrin, M.; Marin, M.; Noel, D.; Del Rio, M.; Saller, R.; Stange, J.; Mitzner, S.; Gunzburg, W. H.; Piechaczyk, M. *Gene Ther* **1998**, 5, (6), 828-34.

137. Weber, W.; Rinderknecht, M.; Daoud-El Baba, M.; de Glutz, F. N.; Aubel, D.; Fussenegger, M. *J Biotechnol* **2004**, 114, (3), 315-26.

138. Schaffellner, S.; Stadlbauer, V.; Stiegler, P.; Hauser, O.; Halwachs, G.; Lackner, C.; Iberer, F.; Tscheliessnigg, K. H. *Transplant Proc* **2005**, 37, (1), 248-52.

139. Rose, T. N., B.; Thielking, H.; Koch, W.; Vorlop, K.D. *Chem. Eng. Technol.* **2000**, 23, 769-772.

140. Bryant, S. J., Davis-Arehart, K.A., Luo, L.;Shoemaker, R.K.; Arthur, J.A.;Anseth, K.S. *Macromolecules* **2004**, 37, 6726-6733.

141. Li, Q.; Williams, C. G.; Sun, D. D.; Wang, J.; Leong, K.; Elisseeff, J. H. *J Biomed Mater Res A* **2004**, 68, (1), 28-33.

142. Moslemy, P.; Neufeld, R. J.; Guiot, S. R.*Biotechnol Bioeng* **2002**, 80, (2), 175-84.

143. Itoh, Y.; Matsusaki, M.; Kida, T.; Akashi, M. *Chem. Letters* **2004**, 33, (12), 1552-1553.
144. Becker, A.; Katzen, F.; Puhler, A.; Ielpi, L. *Appl Microbiol Biotechnol* **1998**, 50, (2), 145-52.
145. Pappas, G. D.; Lazorthes, Y.; Bes, J. C.; Tafani, M.; Winnie, A. P. *Neurol. Res.* **1997**, 19, 71-77.
146. Sagen, J.; Wang, H.; Tresco, P. A.; Aebischer, P. *J. Neurosci.* **1993**, 13, 2415-2423.
147. Kim, Y. M.; Jeon, Y. H.; Jin, G. C.; Lim, J. O.; Baek, W. Y. *Artif Organs* **2004**, 28, (12), 1059-66.
148. Xue, Y.; Gao, J.; Z., X.; Wang, Z.; Li, X.; Cui, X.; Luo, Y.; C., L.; Wang, L.; Zhou, D.; Sun, R.; Sun, A. M. *Art. Organs* **2001**, 25, (2), 131-135.
149. Lindner, M. D.; Emerich, D. F. *Cell Transplantation* **1998**, 7, (2), 165-174.
150. Mercier, P. Peyrot, M., Cauzac, F, Kalck, F.. Hauzaneau, F., Sallerin, B., Lazorthes, Y. International Patent Number WO 02/05943
151. Meier, W. *Chem. Soc. Rev.* **2000**, 29, 295-303.
152. Mercier, P.; Fernandez, F.; Tortosa, F.; Bagheri, H.; Duplan, H.; Tafani, M.; Bes, J. C.; Bastide, R.; Lazorthes, Y.; Sallerin, B. *J Microencapsul* **2001**, 18, (3), 323-34.
153. Rabanel, J. M.; Hildgen, P. *J Microencapsul* **2004**, 21, (4), 413-31.
154. Hildgen, P., Rabanel, J.M.; Mercier, P. Canada Patent Number 60/473,47, 05/28/03, 2003.
155. Hamdi, G.; Ponchel, G.; Duchene, D. *J Microencapsul* **2001**, 18, (3), 373-83.
156. Kartha, K. P. R.; Srivastava, H. C. *Starch* **1985**, 37, (9, S.), 297-306.
157. Hunkeler, D.; Rehor, A.; Ceausoglu, I.; Schuldt, U.; Canaple, L.; Bernard, P.; Renken, A.; Rindisbacher, L.; Angelova, N. *Ann N Y Acad Sci* **2001**, 944, 456-71.
158. Rosinski, S.; Grigorescu, G.; Lewinska, D.; Ritzen, L. G.; Viernstein, H.; Teunou, E.; Poncelet, D.; Zhang, Z.; Fan, X.; Serp, D. *J Microencapsul* **2002**, 19, (5), 641-59.
159. Van Raamsdonk, J. M.; Chang, P. L. *J Biomed Mater Res* **2001**, 54, (2), 264-71.
160. Leblond, F. A.; Tessier, J.; Halle, J. P. *Biomaterials* **1996**, 17, (21), 2097-102.
161. Rehor, A.; Canaple, L.; Zhang, Z.; Hunkeler, D. *J Biomater Sci Polym Ed* **2001**, 12, (2), 157-70.

Indexes

Author Index

Subject Index

A

Acetaminophen
  n exponent from swelling and release of tablets with, 131*t*
  permeability of acetate crosslinked high amylose starch (HASCL) and derivatives, 129, 131
  release from HASCL and derivatives, 126*f*
Acetate high amylose starch crosslinked 6
  acetaminophen release, 126*f*
  controlled release properties and dissolution kinetics, 125–128
  equilibrium swelling ratio depending on pH and swelling capacity, 130*f*
  synthesis, 125
  See also Crosslinked starch derivatives
Acetylation, calculating degree of, of chitosan, 233–234
Acetylsalicylic acid, release from crosslinked high amylose starch (HASCL) derivatives, 129*f*
Actinic keratoses, hyaluronan assisting treatment, 144–145
Activation methods
  introduction of spacers, 295–300
  OH groups of polysaccharides, 292
  See also Polysaccharide-drug conjugates
Adhesive strength, levan, 264–265
Administration
  depot systems, 202, 203*f*
  ideal depot system, 209
Agarose
  cell encapsulation, 323
  gelling properties, 322

See also Microencapsulation; Polysaccharides
Alginates
  biocompatibility, 316–317
  bridges between polymer chains, 314*f*
  examples of adjuvant polymer for, capsules, 320*t*
  gelling properties, 313, 315
  in vivo concerns for, capsules applications, 319–320
  optimizing beads surface biocompatibility, 317
  permeability of alginate capsules, 316
  process of encapsulation, 315
  stability of, beads, 318
  stability of, capsule membrane, 318–319
  See also Microencapsulation; Polysaccharides
Amino acids, polysaccharide activation by spacers, 298–300
Aminoethyl high amylose starch crosslinked 6
  acetaminophen release, 126*f*
  controlled release properties and dissolution kinetics, 125–128
  equilibrium swelling ratio depending on pH and swelling capacity, 130*f*
  synthesis, 124–125
  See also Crosslinked starch derivatives
Amylopectin
  hollow particles with residual polyesters, 331*f*
  hydrogel microcapsules, 328–330
  model of starch, 94, 96
  structure and architecture, 83*f*